Siliceous Deposits of the Tethys and Pacific Regions

James R. Hein Jelena Obradović
Editors

Siliceous Deposits of the Tethys and Pacific Regions

With 131 Figures

Springer-Verlag
New York Berlin Heidelberg
London Paris Tokyo

James R. Hein
US Geological Survey
Menlo Park, California, USA

Jelena Obradović
Faculty of Mining and Geology
University of Beograd
Beograd, Yugoslavia

The illustration on the front cover shows paleogeographic reconstruction for the boundary between the Middle and Late Jurassic (see page 124).

Library of Congress Cataloging-in-Publication Data
Hein, J.R. (James R.)
Siliceous deposits of the Tethys and Pacific Regions / James R. Hein, Jelena Obradovic.
p. cm.
"Selected presentations given at the Third International Conference on Siliceous Deposits, held in Yugoslavia, September, 1986"—Pref.
Includes index.
ISBN 0-387-96704-4 (alk. paper)
1. Rocks, Siliceous—Congresses. 2. Rocks, Siliceous—Pacific Area—Congresses. 3. Petrology—Congresses. 4. Petrology—Pacific Area—Congresses. 5. Tethys (Paleogeography)—Congresses.
I. Obradović, Jelena. II. International Conference on Siliceous Deposits in the Pacific Region (3rd : 1986 : Yugoslavia)
III. Title.
QE495.H45 1989
552′.5—dc19 88-31137
CIP

Typeset by Publishers Service, Bozeman, Montana.
Printed and bound by Arcata Graphics/Halliday, West Hanover, Massachusetts.
Printed in the United States of America.

9 8 7 6 5 4 3 2 1

ISBN 0-387-96704-4 Springer-Verlag New York Berlin Heidelberg
ISBN 3-540-96704-4 Springer-Verlag Berlin Heidelberg New York

Preface

This book is a product of selected presentations given at the Third International Conference on Siliceous Deposits, held in Yugoslavia, September 1986. This conference marked the culmination of the International Geological Correlations Project's (IGCP) research group 187, Siliceous Deposits of the Pacific and Tethys Regions. Project 187 is an outgrowth of IGCP Project 115, Siliceous Deposits of the Pacific Region, which together were active for 10 years, 1976 to 1986. Project 187 involved 142 scientists from 33 countries. The IGCP Program is sponsored by UNESCO and the International Union of Geological Sciences (IUGS).

This book can be considered as a companion volume to two other volumes produced by IGCP Projects 115 and 187. The first volume was edited by A. Iijima, J.R. Hein, and R. Siever (Elsevier, 1983), dealing with the geology, sedimentology, paleontology, and geochemistry of primarily circum-Pacific siliceous deposits. That volume, titled *Siliceous Deposits in the Pacific Region*, contains papers presented at the Second International Conference on Siliceous Deposits, which was held in Japan in 1981. That Conference was the culminating event of IGCP 115 and the beginning of the expanded IGCP 187. The second volume was a direct outcome of the new phase, IGCP 187, that placed emphasis on the ore and petroleum deposits associated with siliceous rocks. This second volume was edited by J.R. Hein, published by Van Nostrand Reinhold, 1987, and is called *Siliceous Sedimentary Rock-Hosted Ores and Petroleum*. This present volume deals with the geology, sedimentology, paleontology, and geochemistry of primarily Tethyan siliceous deposits but also includes papers on the Pacific region. A discussion and integration of each chapter in this book are presented in Chapter 1.

J.R. Hein, as International Group Leader of IGCP 115 and 187, and J. Obradović, as organizer of the Yugoslavian Conference, would like to thank all of our members for their contributions to the Program and to this book. We also thank Lisa Gein, U.S. Geological Survey, for technical and editorial help. All the papers in this volume were critically read by at least one reviewer as well as by both of us as the volume editors.

International and interdisciplinary cooperation is valuable for the understanding of complex and widespread geologic phenomena. Because of the biological origin of many siliceous deposits, their geochemical transformation in the marine environment, and their occurrence in many formations around the world, it is essential that oceanographers, paleontologists, geochemists, and geologists work together to reach a holistic understanding about the origin and evolution of these deposits. Not only is integration of disciplines needed, but it behooves workers from all nations whose territories contain siliceous deposits to cooperate in comparative studies. In this spirit of cooperation, IGCP 115 and 187 have been complete successes.

James R. Hein
Jelena Obradović

Contents

Part III: Pacific Siliceous Deposits

Contributors

Joyce R. Blueford
US Geological Survey, MS-144, 345 Middlefield Road, Menlo Park, CA 94025, USA

M.A. Bustillo
Museo Nacional de Ciencias Naturales, CSIC, José G. Abascal 2, 28006, Madrid, Spain

M. Dermitzakis
Department of Geology, University of Athens, Panepistimiopolis, Athens, 157 84, Greece

Patrick De Wever
CNRS-UA 319, University Pierre and Marie Curie, T15-16 E4, Laboratory of Stratigraphy, F-75-252 Paris, Cedex 05, France

A. Economou-Amilli
Department of Biology, University of Athens, Panepistimiopolis, Athens, 157 84, Greece

J.J. Elorza
Department of Geology, University of País-Vasco, Apart. 644, 48080 Bilbao, Spain

W. Frisch
Department of Geology and Paleontology, University of Tübingen, D-7400 Tübingen, Sigwartstrasse 10, FRG

Špela Goričan
Ivan Rakovec Paleontological Institute, ZRC SAZU, 61000 Ljubljana, Yugoslavia

Hans-Jürgen Gursky
Faculty de Ciencias de la Tierra, University Autónoma de Nuevo León, Apart. Postal 104, 67700 Linares, N.L., Mexico

Monika M. Gursky
Faculty de Ciencias de la Tierra, University Autónoma de Nuevo León, Apart. Postal 104, 67700 Linares, N.L., Mexico

Isamu Hattori
Geological Laboratory, Fukui University, 910 Fukui, Japan

James R. Hein
US Geological Survey, MS-999, 345 Middlefield Road, Menlo Park, CA 94025, USA

Azuma Iijima
Geological Institute, University of Tokyo, Hongo 7-3-1, Tokyo 113, Japan

Y. Kakuwa
Institute of Astronomy and Earth Sciences, University of Tokyo, Komaba 3-8-1, Tokyo 153, Japan

Susan M. Karl
US Geological Survey, 4200 University Drive, Anchorage, AK 99508-4667, USA

Andreas Magganas
Department of Geology, University of Athens, Panepistimiopolis, Athens 157 84, Greece

H. Matsuda
Geological Institute, University of Tokyo, Hongo 7-3-1, Tokyo 113, Japan

J.M. Molina
Department of Stratigraphy and Paleontology, Colegio Universitario, 23071, Jaen, Spain

Chavdar I. Nachev
Museum of Earth and Man, G. Traikov Str., No. 2, 1421, Sofia, Bulgaria

Ivan K. Nachev
Geological Institute, Bulgarian Academy of Sciences, Acad. G. Bonchev Str., block 24, 1113, Sofia, Bulgaria

Jelena Obradović
Faculty of Mining and Geology, University of Beograd, Djušina 7, 11000 Beograd, Yugoslavia

M. Pirzer
Department of Geology and Paleontology, University of Tübingen, D-7400 Tübingen, Sigwartstrasse 10, FRG

Pedro A. Ruiz-Ortiz
Department of Stratigraphy and Paleontology, Colegio Universitario, 23071, Jaen, Spain

Michael Stamatakis
Department of Geology, University of Athens, Panepistimiopolis, Athens, 157 84, Greece

A. Vecsei
Department of Geology, ETH Zentrum, CH-8092, Zürich, Switzerland

A. Wetzel
Department of Paleontology, University of Tübingen, D-7400 Tübingen, Sigwartstrasse 10, FRG

1
Siliceous Deposits of the Tethys and Pacific Regions

James R. Hein and Jelena Obradović

Overview

This introductory chapter briefly summarizes many aspects of siliceous deposits, from the introduction of silica into the marine environment to diagenesis and evolution of depositional basins. We also describe and contrast the Mesozoic Tethyan and Pacific depositional settings of siliceous deposits.

Silica enters the oceans primarily from rivers and is removed primarily in siliceous deposits via silica uptake by siliceous plankton. Less than one percent of the biogenic silica fixed in surface waters makes it into the geologic record. The solubility of the various silica polymorphs varies, and controls the dissolution or reprecipitation of silica in sea water, bottom sediments, and during late diagenetic stages. Recent studies in the geochemistry of siliceous deposits are shedding light on many long outstanding questions related to siliceous deposits. K-Ar and Rb-Sr age dating of cherts offer new and significant tools to better understand the timing of sedimentation and tectonics of chert-bearing sequences.

The plate tectonic regimes and depositional basins of the Mesozoic Tethyan seaway and the circum-Pacific region were fundamentally different. These differences resulted in deposition of siliceous sequences with different lithologic associations and lithologic successions.

Introduction

This book is concerned with fine-grained siliceous deposits predominantly of biogenic origin. Sediment and rock types of interest include radiolarian and diatom oozes and earths, diatomite, porcellanite, and chert including such varieties as nodular chert, bedded or ribbon chert, and radiolarite. All of these rocks and sediments occur with various degrees of purity, terrigenous and calcareous debris being the dominant accessory phases. The silica polymorphs that make up the deposits include biogenic silica (opal-A), opal-CT (disordered cristobalite and tridymite), quartz, and chalcedony (terminology of Jones and Segnit, 1971). Different silica phases appear during primary, diagenetic, and metamorphic stages in the evolution of the biogenic silica deposits. Diagenesis of primarily volcanogenic deposits will produce opal-C (cristobalite) as the silica phase rather than opal-CT.

Application of contemporary depositional mechanisms and environments of biogenic silica accumulation to the geologic record is difficult, because the nature and distribution of these deposits are closely tied to the evolution of organisms (calcareous and siliceous plankton), changes in chemistry of the atmosphere (oxygen and CO_2 contents), changes in sea water (oxygen, CO_2, temperature, upwelling, nutrients, current patterns), and global, regional, and local tectonics (distribution of continents determines dominant upwelling regions, basin geometries). More will be said in this chapter about these important modifying influences.

In this introductory chapter, we will briefly summarize salient points concerning the evolution of siliceous deposits. From the introduction of silica into the oceans, through burial, tectonism, and metamorphism, the cycle of silica is becoming better understood. We also describe and contrast the depositional and tectonic environments that hosted Tethyan and Pacific siliceous rocks.

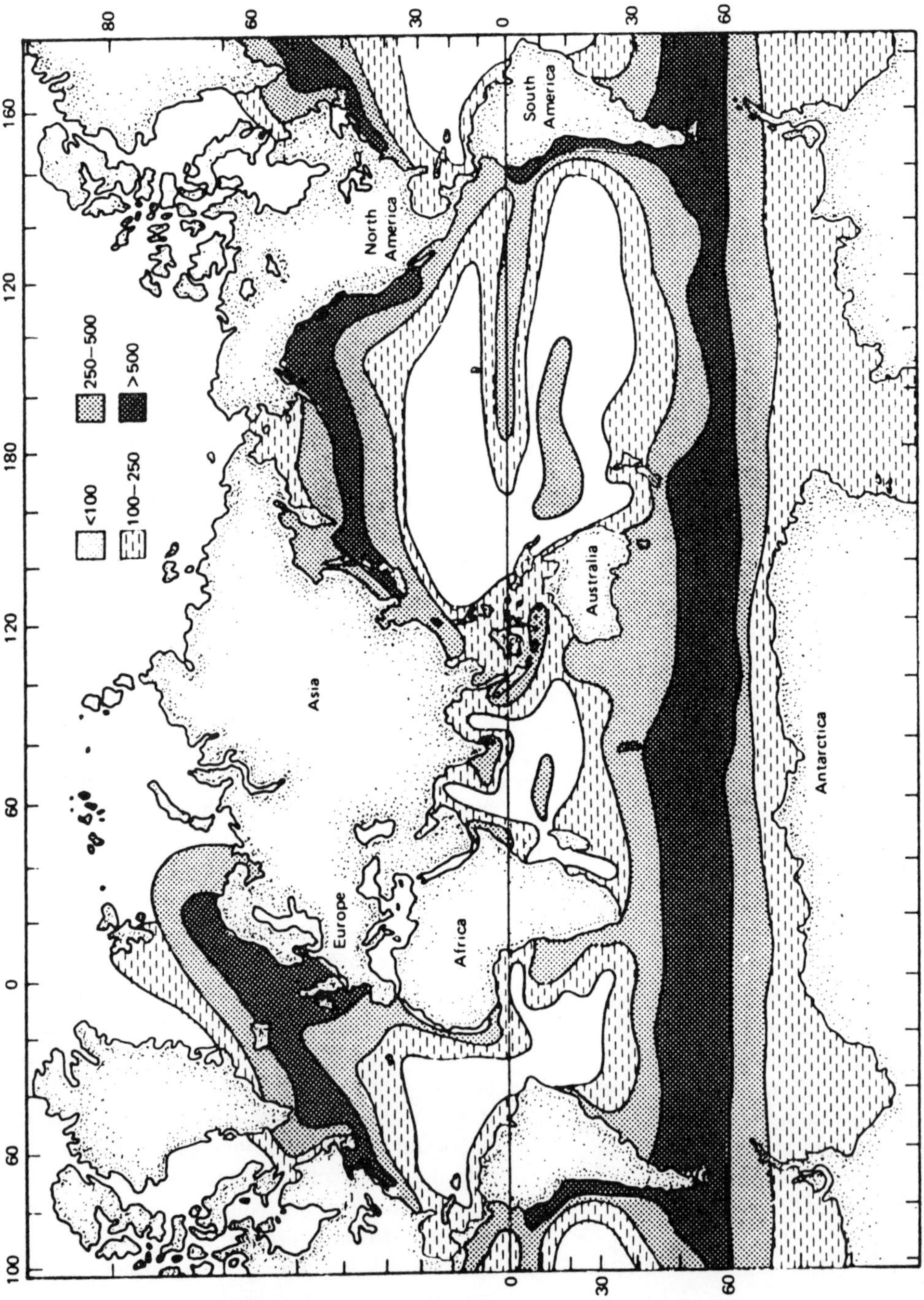

FIGURE 1.1. Annual production of silica by marine organisms ($SiO_2\ M^{-2}\ y^{-1}$) in the world ocean, constructed from primary production data. (From Lisitzin, 1972.)

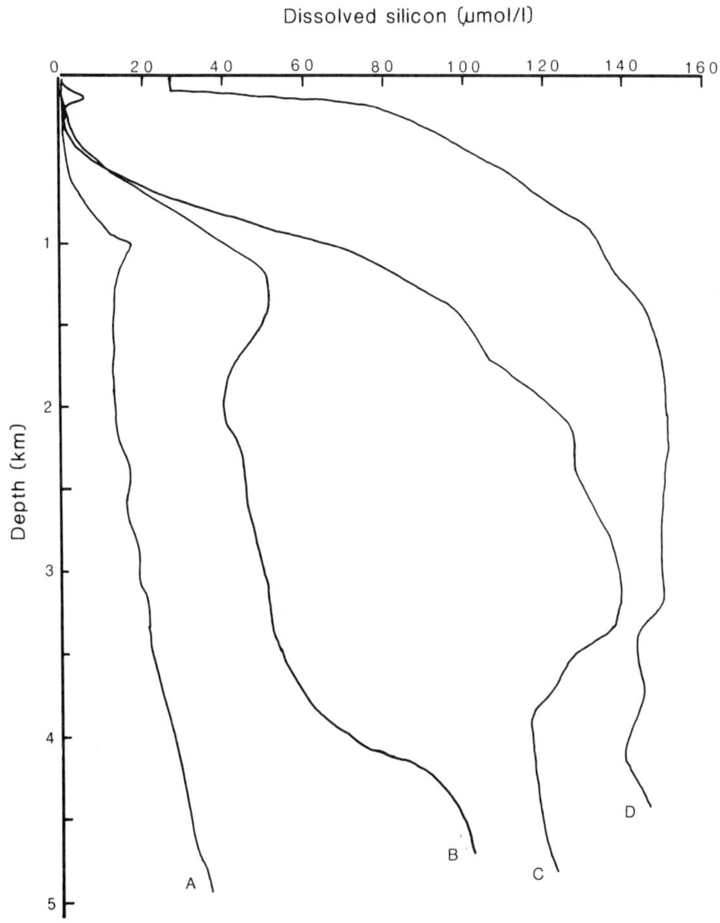

FIGURE 1.2. Dissolved silicon with depth in the oceans. (A) North Atlantic; (B) South Atlantic; (C) South Pacific; (D) North Pacific. (From Calvert, 1974.)

Silica in the Oceans

Dissolved silica occurs in seawater as silicic acid, $Si(OH)_4$. Silica is supplied to ocean water predominantly by rivers, with smaller contributions from submarine volcanism and submarine weathering. Silica is removed by biogenic siliceous deposits, uptake in estuaries, and inorganic uptake in the oceans (Siever, 1957; Mackenzie and Garrels, 1966; Harris, 1966; Calvert, 1968; Burton and Liss, 1973; Wollast, 1974; DeMaster, 1981). Siliceous plankton extract silica from surface seawater primarily in zones of oceanic and coastal upwelling, where nutrients are brought from midlevel waters to surface waters. Diatoms are the most abundant silica-secreting plankton today, followed by radiolarians and silicoflagellates. In pre-Cenozoic times, radiolarians were the primary siliceous plankton. In the present oceans, three latitudinal belts of high silica productivity occur, two high-latitude belts and a less well developed equatorial belt (Fig. 1.1). In addition, strong upwelling and high productivity occur along the west coast of continents (Fig. 1.1). In the past, this pattern of upwelling and plankton distribution changed as the position of the continents changed (Hein and Parrish, 1987).

Surface seawater is relatively depleted in dissolved silica because of silica extraction by photosynthetic plankton, but deeper waters are enriched in silica, because the plankton tests and frustules dissolve as the organisms die and sink through the water column (Fig. 1.2). The loss of diatom frustules with water depth can be as much as 90% within several hundred meters of the sea surface (Fig. 1.3), and commonly only 1% to 10% of the frustules make it to the sea floor, where most will dissolve within the first meter of burial (Bogoyavlenskiy, 1967; Hurd, 1973; Heath, 1974; Harper and Knoll, 1975). Thus, less than 1% of the biogenic silica produced in surface waters makes it into the geologic record.

Susceptibility to dissolution of biogenic silica in the oceans depends on particle size (Fig. 1.4), tem-

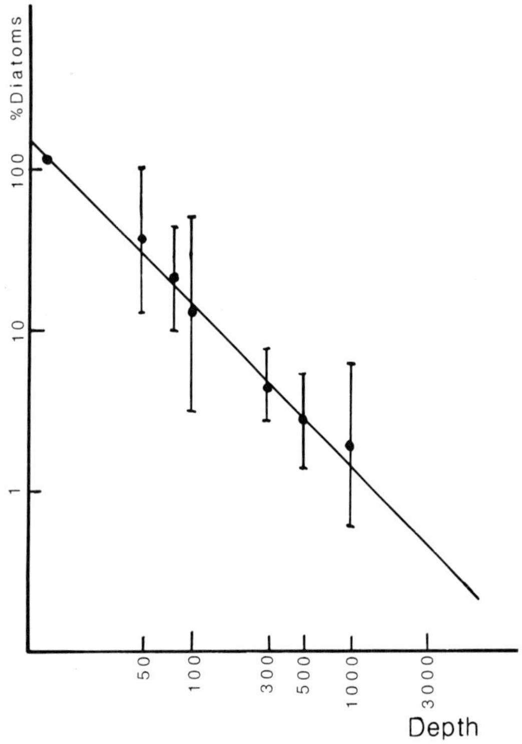

FIGURE 1.3. Frequency of occurrence of diatom skeletons at 30 stations of the Antarctic Sea as a function of water depth. (From Wollast, 1974; data from Kozlova, 1964.)

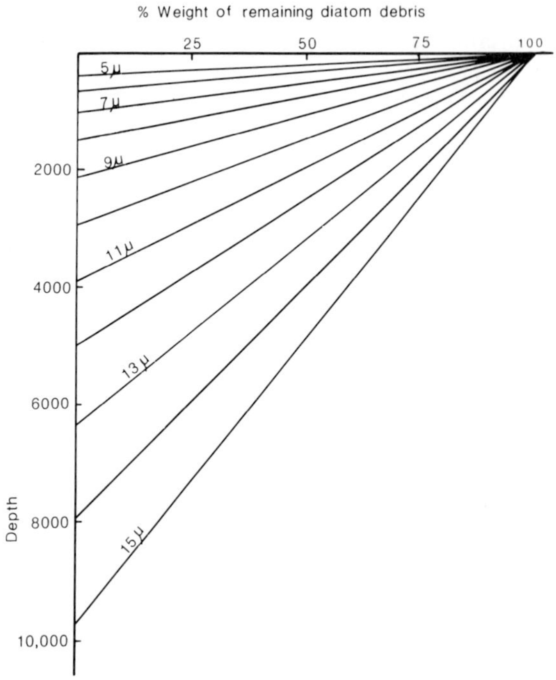

FIGURE 1.4. Weight fraction of diatom debris remaining at particular depths for various initial diatom radii. (From Wollast, 1974).

perature and pressure (Fig. 1.5), organic and inorganic protective coatings, absorbed ions, and degree of crystallographic order (Riedel, 1959; Lewin, 1961; Berger, 1968; Kamatani, 1971; Williams et al., 1985). The relative order of increasing susceptibility to dissolution of silica organisms is sponge spicules, radiolarians, silicoflagellates, and diatoms. Susceptibility to solution also varies between species within each major biologic group. Fossil specimens are less soluble than recently deposited plankton (Berger, 1968; Ernst and Calvert, 1969; Hurd, 1973).

The solubility of the various silica phases in seawater decreases from amorphous silica (opal-A) to quartz (Fig. 1.6; Table 1.1). On Figure 1.6, the opal-CT curve would fall between that of α-cristobalite and amorphous silica. The solubility of each phase increases with temperature and probably also with pressure as demonstrated for opal-A (Hurd, 1983) (Fig. 1.5) and for quartz (Jones and Pytkowicz, 1973). Apparently, quartz crystallizes directly from solution only when the concentration of silica in solution is below the equilibrium solubility of opal-CT (Murata and Larson, 1975; Kastner et al., 1977). When solutions become supersaturated with respect to opal-CT, this disordered phase will precipitate rather than the more stable quartz. Consequently, in deep-sea siliceous deposits, the order of silica phase development should be opal-A, opal-CT, and finally quartz (Kastner et al., 1977). In carbonate sediments, or during late-stage diagenesis, opal-A may transform directly to quartz.

Silica in Deep-Sea Sediments and Pore Waters

Opal-A in oceanic bottom sediments mirrors its distribution in surface waters, with the greatest amounts—over a fourth of the silica supplied by rivers and submarine volcanism (DeMaster, 1981)—in the circum-Antarctic region (Fig.1.7). The north Pacific and adjacent seas account for 10% of the dissolved silica input; estuaries, 10% to 15%;

TABLE 1.1. Solubility of silica in seawater at 25°C and 1 atm (mmole/L).

Form	Experimental data	Reference
Quartz	0.086	Mackenzie (1971)
Quartz	0.10	Morey et al. (1962)
Tridymite	0.63	Stöber (1967)
Cristobalite	1.00	Stöber (1967)
Amorphous silica	1.84	Stöber (1967)
Amorphous silica	1.75	Krauskopf (1956)
Finely divided silica gel	2.34	Siever (1962)

From Wollast, 1974.

and coastal upwelling areas, less than 5% (DeMaster, 1981). Regions of coastal upwelling have the highest accumulation rates of biogenic silica but are relatively restricted in area. These patterns of the distribution of opal-A have changed over the short term (Cenozoic) and the long term (Phanerozoic) as the result of changes in current patterns, plate motions, and the position of continents (Leinen, 1979; Hein and Parrish, 1987; Blueford, this volume).

The amount of silica in interstitial waters of deep-sea sediments varies widely depending on the sediment type. In the pore waters of near surface opal–A bearing sediment, the silica concentration increases markedly over that of bottom waters (Schink et al., 1975) and may become saturated with respect to opal-A within a meter of burial. In other deposits, pore water of near-surface sediment may remain greatly undersaturated in silica with respect to opal-A. The concentration of silica in pore waters of near surface deposits in the north Pacific is greater in oxidized deposits than it is in reduced sediments (Heath and Dymond, 1973).

Silica concentration profiles in pore waters of Deep Sea Drilling Project cores are variable but show supersaturation of silica with respect to opal-A within many different depositional settings (Fig. 1.8). These silica profiles reflect silica diagenesis within the sediment (Gieskes, 1983). Decreases in magnesium at some stratigraphic levels where porcelanite and chert formation occurred supports Kastner et al.'s (1977) and Kastner and Gieskes' (1983) idea of magnesium hydroxide nuclei as substrate for opal-CT precipitation. Areas of extensive silica precipitation show large decreases in porosity, which may result in large concentration gradients across the zones of silicification for many elements (Fig. 1.8E) (Gieskes, 1983).

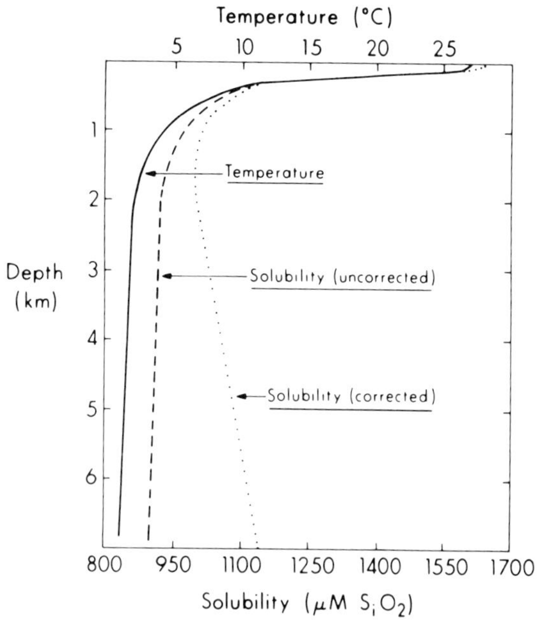

FIGURE 1.5. Estimated solubility of biogenic silica as a function of temperature and pressure (depth) in the central equatorial Pacific. Solid line, temperature vs. depth; dashed line, solubility estimated for change in temperature; dotted line, solubility estimated for change in temperature and pressure. (As presented by Hurd, 1983, from Griffin, 1980.)

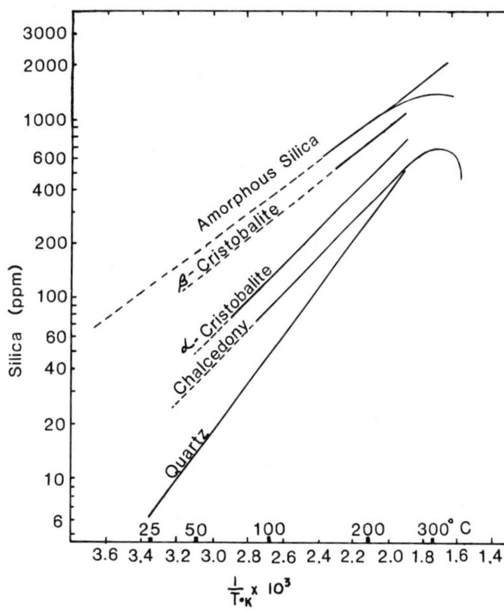

FIGURE 1.6. Solubility of silica phases along the two-phase curve, water plus vapor; plotted as log concentration vs. reciprocal absolute temperature. (From Fournier, 1970.)

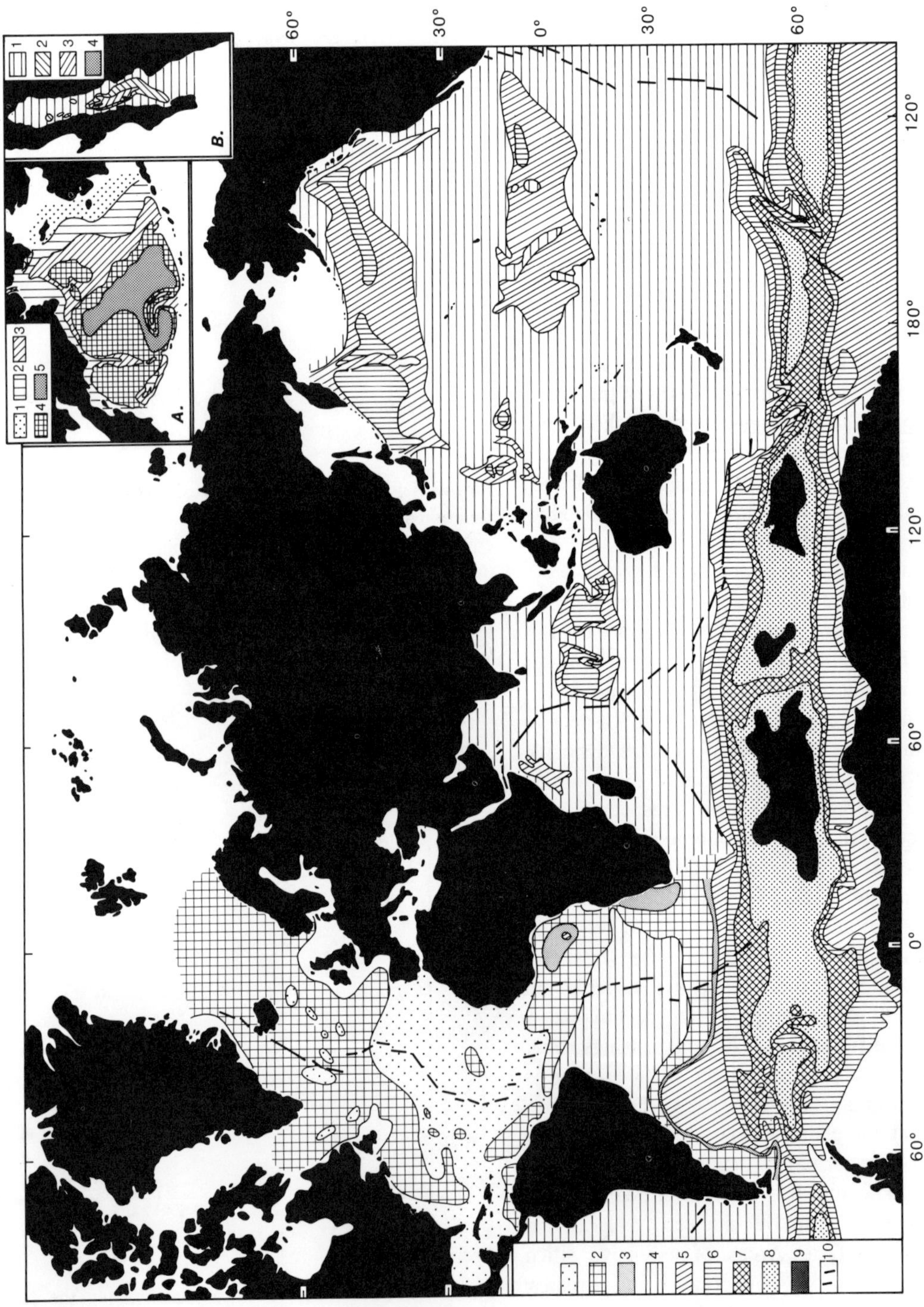

FIGURE 1.7. Distribution of amorphous SiO_2 in sediments from the ocean basins (in percent of dry weight). (1) <1; (2) 1–3; (3) 3–5; (4) <5 undifferentiated; (5) 5–10; (6) 10–30; (7) 30–50; (8) 50–70; (9) >70; (10) spreading axes. Inset A, Bering Sea; (1) <1; (2) 1–5; (3) 5–10; (4) 10–30; (5) >30. Inset B, Gulf of California; (1) <10; (2) 10–30; (3) 30–50; (4) >50. (Modified from Lisitzin, 1978.)

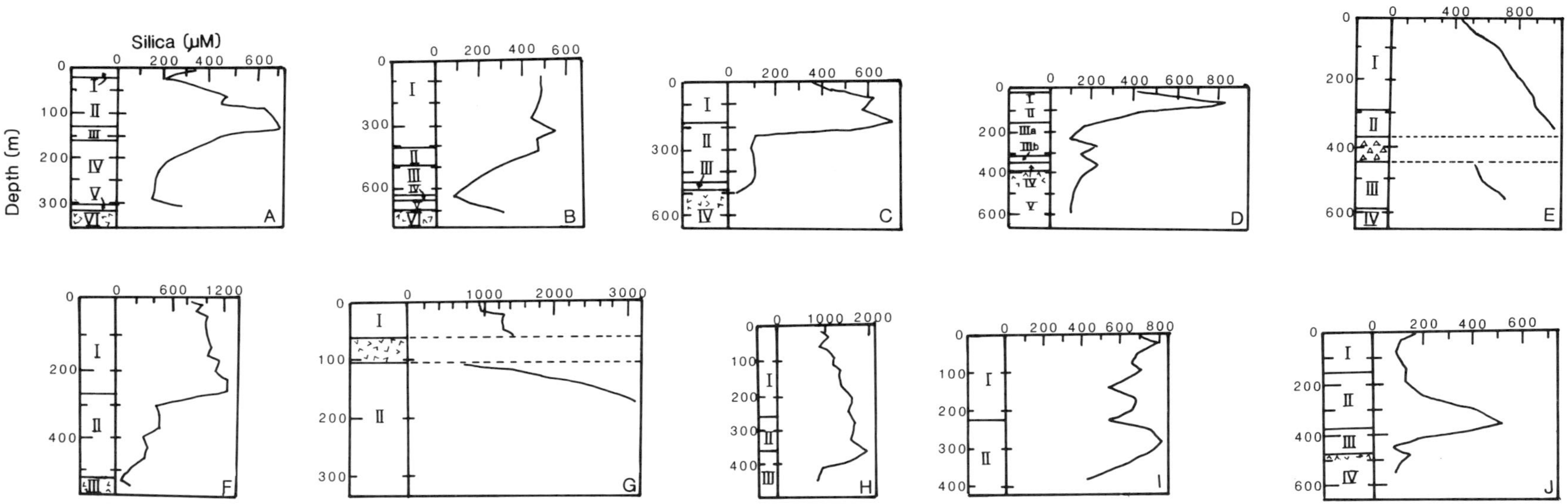

FIGURE 1.8. Concentration of dissolved silica in pore waters of a variety of sediments recovered by the Deep Sea Drilling Project. (A) Site 239, west Indian Ocean: I, silty clay; II, nannofossil ooze and clay; III, silty clay and nannofossil ooze; IV, silty clay and clayey silt; V, nannofossil clay; VI, basalt. (B) Site 323, east Pacific sector of Antarctic Ocean: I, clay, silt, diatom ooze; II, chert, claystone; III, claystone; IV nannofossil chalk; V, zeolitic clay; VI, basalt. (C) Site 336, NE Atlantic: I, gray mud, sandy mud, clay; II, mud, clay, mudstone; III, red and red-black clay, claystone, volcanic rubble; IV, basalt. (D) Site 446, Philippine Sea: I, terrigenous mud; II, siliceous pelagic clay, ash; IIIa, mudstone, siltstone, sandstone; IIIb, calcareous mudstone, turbidites; IV, calcareous mudstone, glauconite; V, basalt sills and intrusions. (E) Site 462, west-central equatorial Pacific: I, calcareous and radiolarian ooze, chalk, limestone; II, nannofossil chalk, radiolarian ooze, limestone, porcellanite, chert; III, volcaniclastic and zeolitic sandstone, radiolarian ooze, mudstone, limestone; IV, basalt flows. (F) Site 474, off tip of Baja California: I, diatomaceous ooze, mud, turbidites; II, silty claystone, clayey siltstone, mud turbidites; III, dolerite sills and claystone. (G) Site 477, Gulf of California: I, siliceous clay and mud; II, hydrothermally altered sediments, basaltic sill between I and II. (H) Site 479, Gulf of California: I, diatomaceous ooze and mud, dolomite beds start at about 80 m; II, limestone beds; III, claystone, limestone beds. (I) Site 496, slope off Guatemala: I, diatom and nannofossil-rich mud; II, sandy mud. (J) Site 528, SE Atlantic: I, foraminiferal nannofossil ooze; II, nannofossil chalk and ooze; III, chalk interbedded with turbidites; IV, interbedded sedimentary rocks in basalt basement complex. (Redrawn from Gieskes, 1983.)

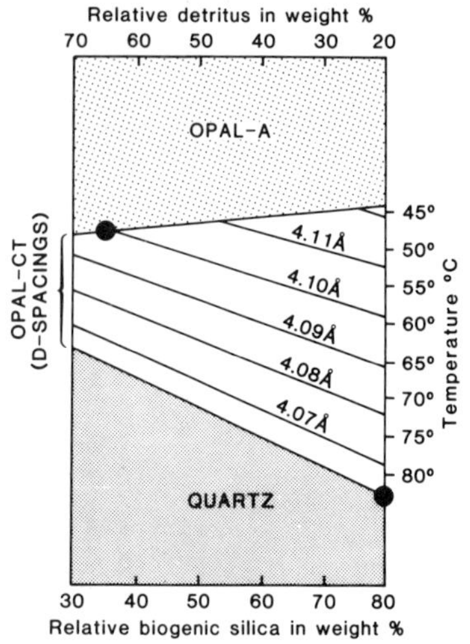

FIGURE 1.9. Preliminary temperature scale for silica diagenesis in the Monterey Formation, representing the combined effects of temperature and rock composition. Black dots are calibration points for this empirical silica phase zonation from the Point Conception Cost well. (From Isaacs, 1982; Keller and Isaacs, 1985.)

Silica Diagenesis

Many papers written during the past 15 years have addressed the subject of silica diagenesis in deep-sea deposits (e.g., Heath and Moberly, 1971; Von Rad and Rösch, 1974; Wise and Weaver, 1974; Garrison et al., 1975; Hein et al., 1978, 1983a), in marine deposits exposed on land (e.g., Murata and Nakata, 1974; Folk and McBride, 1976; Garrison et al., 1979; Iijima et al., 1978; Isaacs et al., 1983; Mizutani and Shibata, 1983; Baltuck, 1983; Stamatakis et al., this volume; Hattori, this volume), and under experimental conditions (e.g., Ernst and Calvert, 1969; Kastner et al., 1977; Stein and Kirkpatrick, 1976; Mizutani, 1977). Generally, with increasing temperature and time, opal-A → opal-CT → quartz or chalcedony. These are solution-reprecipitation reactions. This diagenetic mineral sequence is accompanied by textural changes in the rocks, which change from diatomite (or radiolarian earth) to porcellanite to chert. In places, depending on the sediment composition, this general diagenetic sequence of siliceous deposits varies, and opal-CT chert or quartz porcellanite is produced. Rates of diagenesis depend not only on temperature and time but also on the composition of the rock (Hein et al., 1978; Isaacs, 1982), interstitial water composition (Kastner et al., 1977; Williams and Crerar, 1985), and particle surface area (Hurd, 1983; Williams et al., 1985). With increasing time, lower temperatures are needed to initiate the opal-A to opal-CT transformation as shown by the T-t plot by Hein et al. (1978; reprinted by Stamatakis and Magganas, this volume).

Murata and Nakata (1974) and Murata et al. (1977) showed that the d(101)-spacing of opal-CT decreased (increased ordering) with increasing diagenesis. Isaacs (1982) showed that during diagenesis, rock composition strongly influenced the change in the opal-CT d-spacing, and Keller and Isaacs (1985) developed an empirical temperature scale for the changes in rock composition and opal-CT d-spacings (Fig. 1.9). At high initial diagenetic temperatures, highly disordered opal-CT can form, with d(101)-spacing as large as 4.131 Å (Hein et al., 1983a). The reported range in temperatures for the opal-A to opal-CT transformation is 18 to 58°C and for opal-CT to quartz, 31 to 110°C (summarized by Pisciotto, 1981). However, calculations of these temperatures associated with the mineral changes are based on a large number of assumptions (including nonequilibrium thermal gradients), and the actual range of temperatures for the silica transformations is probably much narrower, and they may not overlap (Keller and Isaacs, 1985). Reasonable ranges of temperatures for the silica transformations under common geologic conditions are 46 ± 2°C for opal-A to opal-CT and 77 ± 5°C for opal-CT to quartz. At higher temperatures and during metamorphism, the quartz attains a higher degree of crystallinity (Gursky and Gursky, this volume).

Because of these two silica transformations, porosity reduction in siliceous deposits occurs in two distinct steps, with little compaction occurring at other times (Fig. 1.10A). This type of compaction and porosity reduction is different from that of calcareous and terrigenous deposits (Fig. 1.10B) and has a significant effect on sedimentary basin evolution and petroleum occurrences in Cenozoic siliceous rocks (see Isaacs and Petersen, 1987, for a summary).

The chemical and isotopic compositions of siliceous deposits have been presented in many recent studies. These studies are shedding light on many long-outstanding questions, such as depositional mechanisms and environments (Shimizu and Masuda, 1977; Steinberg and Marin, 1978; Karl, 1982; Sugisaki et al., 1982; Rangin et al., 1981; Baltuck, 1982; Hein et al., 1983b; Steinberg et al.,

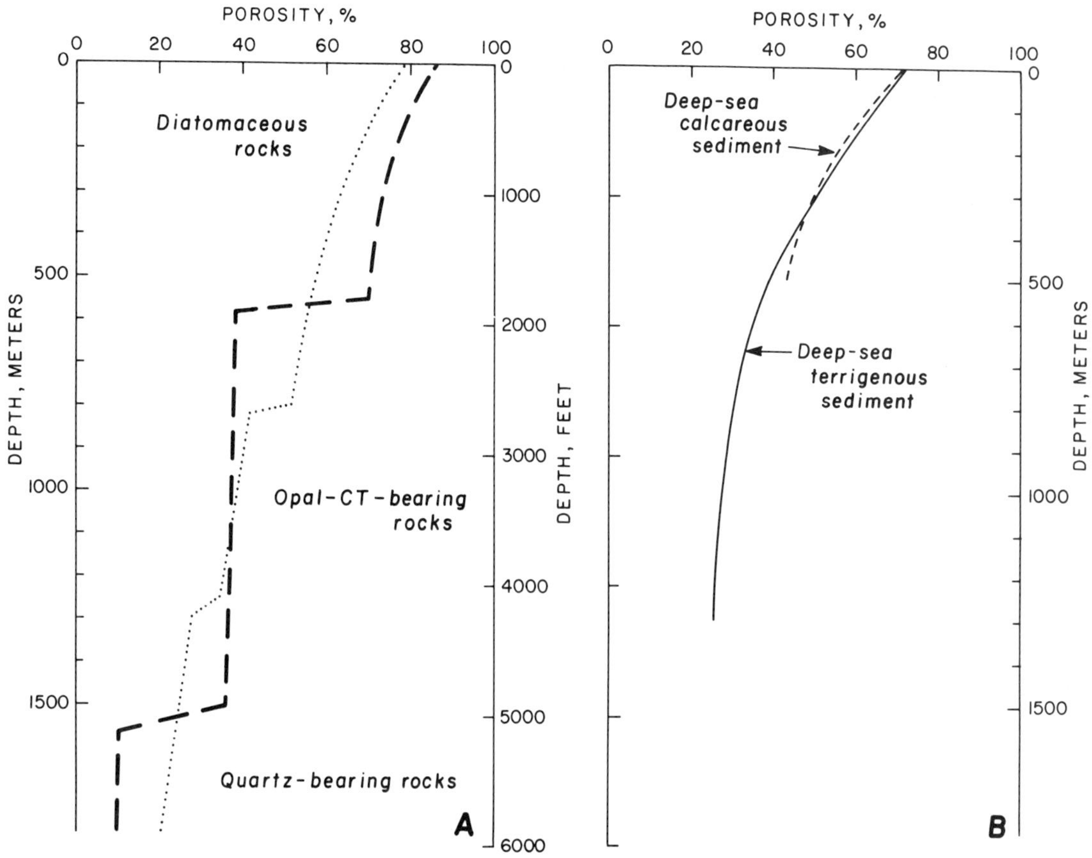

FIGURE 1.10. (A) Porosity reduction for highly diatomaceous rocks and their diagenetic equivalents (dashed line), showing two abrupt reduction steps corresponding to the two silica phase transformations. Dotted line for sparsely siliceous calcareous rocks of the Monterey Formation. (From Isaacs, 1981, as presented by Isaacs et al., 1983.) (B) Porosity reduction for terrigenous and calcareous deep-sea sediments. (From Hamilton, 1976, as presented by Isaacs et al., 1983.)

1983; Matsumoto and Iijima, 1983; Elorza and Bustillo, this volume; Ruiz-Ortiz et al., this volume), chemical evolution and diagenesis (Brueckner and Snyder, 1985; Hein et al., 1981, 1983a), and radiometric age dating (Shibata and Mizutani, 1980; Hart, 1981; Mizutani, 1983; Chyi et al., 1984; Weis and Wasserburg, 1987a,b; Hein and Pickthorn, unpublished data). If K-Ar and Rb-Sr age dating techniques for cherts can be refined, they will offer an ideal method to better understand the timing of sedimentation and the tectonics of chert-bearing sequences. This development could create a revolution in the study and use of cherts as applied to regional geologic and tectonic problems, similar to the revolution that occurred when techniques were developed to extract radiolarians from cherts and biostratigraphic intervals were defined (Pessagno and Newport, 1972; De Wever, this volume). From the few studies completed, Rb-Sr age dating appears to provide better results than K-Ar techniques.

Tethyan and Pacific Siliceous Deposits

The Tethys was an east-west oriented seaway situated between Gondwana and Laurasia. The Mesozoic Tethys extended from the Caribbean region to the Indonesian region. Before development of the Tethys in the Jurassic, the Paleotethys (Laubscher and Bernoulli, 1977) existed as an embayment of the Pacific Ocean during Triassic times (Fig. 1.11). The Tethys began in the Early Jurassic as the result of opening of the Atlantic Ocean. The seaway was restricted in some areas and open in others and con-

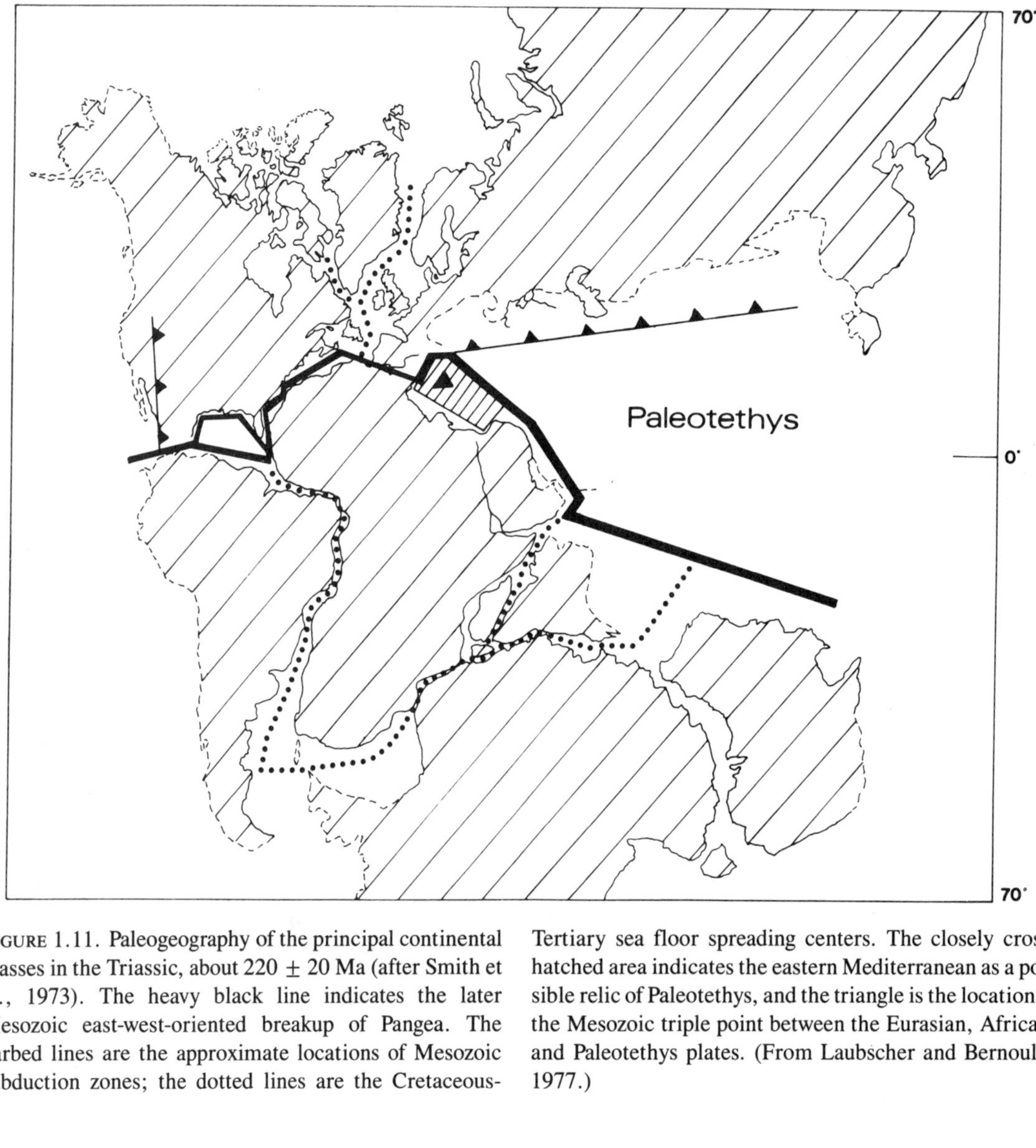

FIGURE 1.11. Paleogeography of the principal continental masses in the Triassic, about 220 ± 20 Ma (after Smith et al., 1973). The heavy black line indicates the later Mesozoic east-west-oriented breakup of Pangea. The barbed lines are the approximate locations of Mesozoic subduction zones; the dotted lines are the Cretaceous-Tertiary sea floor spreading centers. The closely cross-hatched area indicates the eastern Mediterranean as a possible relic of Paleotethys, and the triangle is the location of the Mesozoic triple point between the Eurasian, African, and Paleotethys plates. (From Laubscher and Bernoulli, 1977.)

sisted of several independently moving microplates located between the two major continental plates (Fig. 1.12) (Smith, 1971; Dewey et al., 1973; Biju-Duval et al., 1977; Laubscher and Bernoulli, 1977). Development of the Tethys involved a constantly evolving collage of subsiding continental margins, migrating spreading centers, transform faults, trenches, island arcs, and back-arc basins (Dewey et al., 1973). Consequently, extensional and compressional movements were closely juxtaposed in space and time (Laubscher and Bernoulli, 1977). The Tethyan seaway was eventually closed by plate tectonic motions in the Atlantic and Indian Oceans, northward drift of Gondwana fragments, and northward movement of Africa relative to Eurasia.

Some Tethyan depositional basins were small and short-lived; others were larger, persisting for much of the Mesozoic (Figs. 1.13, 1.14) (Bernoulli and Lemoine, 1980). In the external zones (landward of the arc-trench system, or that part of the orogenic belt nearest the craton or foreland), thick platform carbonates supported Mesozoic pelagic facies (Figs. 1.13, 1.14) (Bernoulli and Jenkyns, 1974). The pelagic facies include limestone, marl, radiolarite, and argillite that were deposited in both shallow-water and basinal areas. Deposits from these differ-

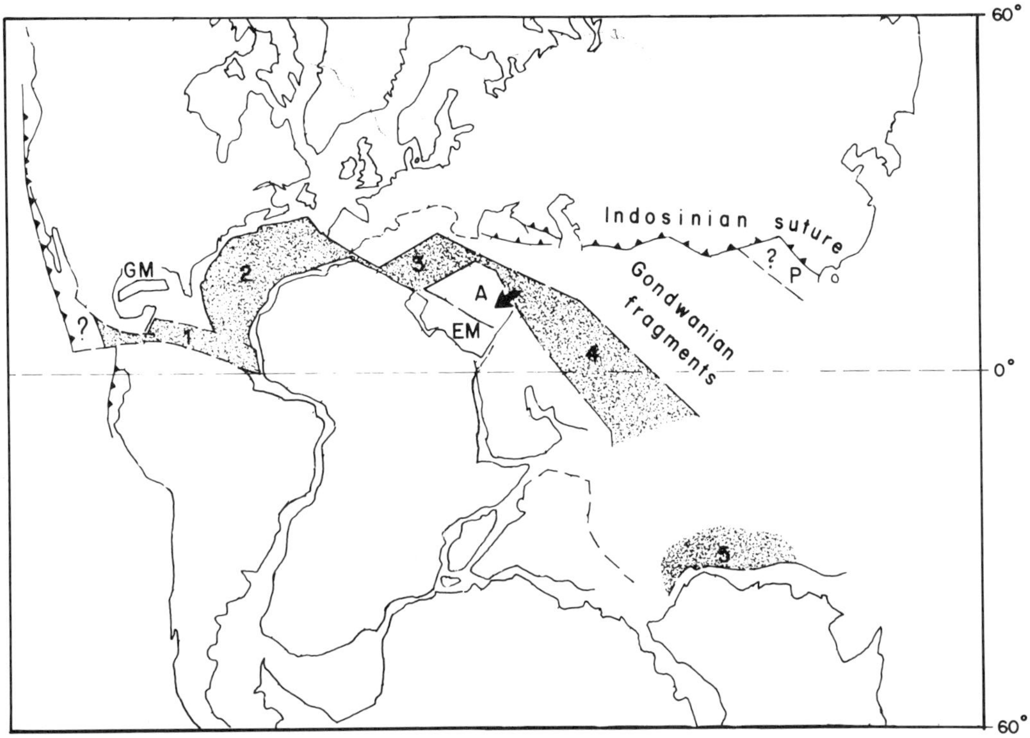

FIGURE 1.12. The Late Jurassic Tethys, about 150 ± 10 Ma. Stippled, Tethyan Jurassic oceanic crust. (1) Caribbean Tethys. (2) Central Atlantic. (3) Piemont-Ligurian Ocean. (4) Oceanic Tethys from the Vardar zone to the Indus-Tsang Po zone. (5) NW Australian margin. GM, Gulf of Mexico; A, Apulian block; EM. eastern Mediterranean; P, possible remnant of Paleotethys; arrow, local obduction of ophiolites onto the east Apulian margin. (From Bernoulli and Lemoine, 1980.)

ent depositional environments can be difficult to distinguish (Bernoulli and Jenkyns, 1974). Both nodular and bedded cherts (radiolarites) occur, but with few exceptions, they are limited in lateral and stratigraphic extent (Bernoulli and Jenkyns, 1974; Barrett, 1982; Nachev and Nachev, this volume; Obradović and Goriačan, this volume). Carbonate rocks associated with cherts are much more common in the Tethyan sequences (Figs. 1.13, 1.14) (Bernoulli and Jenkyns, 1974; Jenkyns and Winterer, 1982) than they are in circum-Pacific sequences (Hein and Karl, 1983; Murchey et al., 1983; see Karl, this volume, for typical sections from Alaska). Two general periods of radiolarite deposition occurred in the Tethys, Triassic and/or Liassic (Early Jurassic), and Dogger (Middle Jurassic) and/or Malm (Late Jurassic) (De Wever, this volume). Radiolarites from both periods are associated with sedimentary and ophiolite sequences. After the Late Jurassic, radiolarites became of minor importance, and nannofossil limestone became a common pelagic facies. Fluctuations in the calcite compensation depth, circulation patterns, competition with nannoplankton, and basin geometries controlled the distribution of the siliceous facies within the Tethyan seaway (Garrison and Fischer, 1969; Bernoulli and Jenkyns, 1974; Bosellini and Winterer, 1975; Jenkyns and Winterer, 1982; De Wever, this volume; Vecsei et al., this volume).

Circum-Pacific orogenic belts were formed by the docking of a complex mosaic of long- to short-traveled tectonostratigraphic terranes and by the development of complex ensimatic and ensialic volcanic arc systems. Tectonostratigraphic terranes may have originated from distant oceanic sources via plate motion (Nur and Ben-Avraham, 1977; Howell et al., 1985) or from continental margins via strike-slip motion or oceanic spreading (Vink et al., 1984). Bedded chert sequences are commonly part

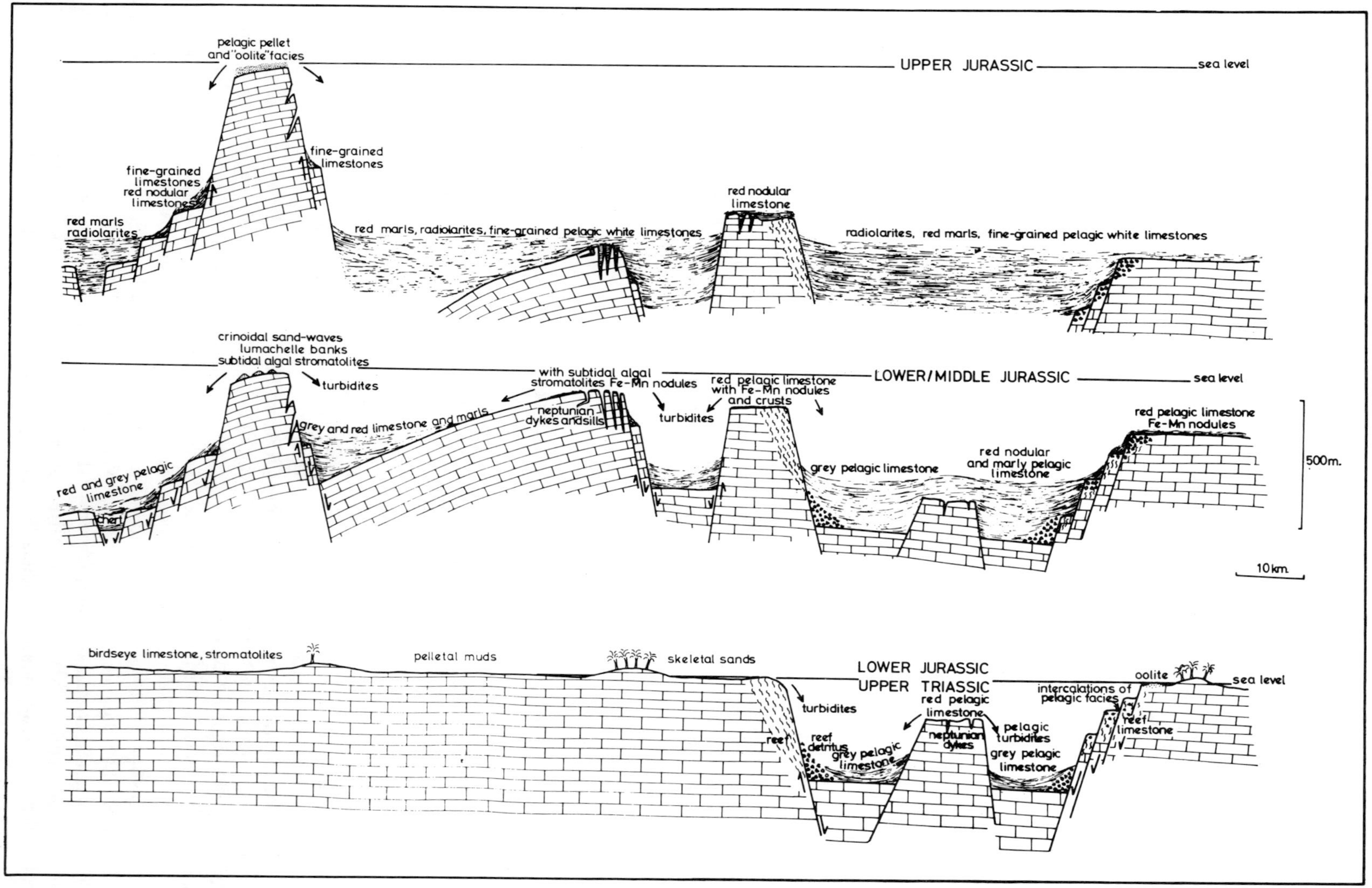

FIGURE 1.13. Generalized diagram of paleogeographic evolution of part of the southern continental margin of the Jurassic Tethys (southern margin of areas 3 and 4 in Fig. 1.12). Radiolarites formed in the deeper water areas, probably below the calcite-compensation depth, and were widespread until Tithonian time. (From Bernoulli and Jenkyns, 1974.)

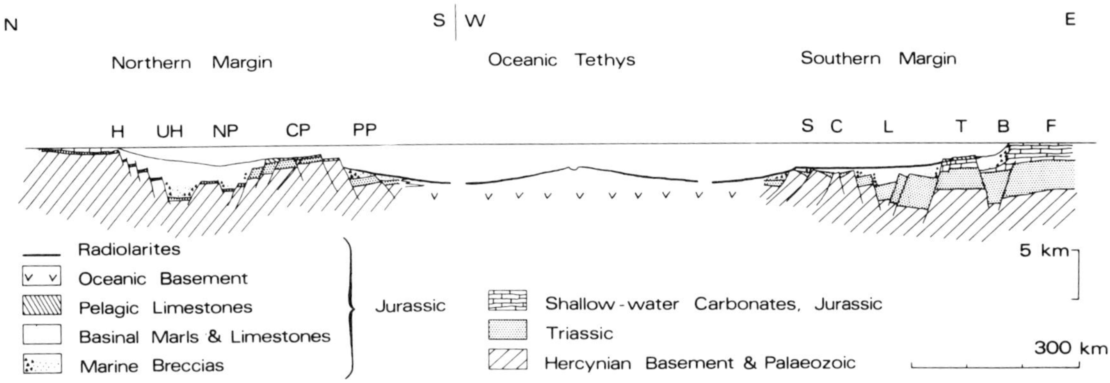

FIGURE 1.14. Palinspastic section through the Liguria-Piemont Ocean (section through area 3 of Fig. 1.12) for the Late Jurassic (Kimmeridgian). The main features: (1) Thick Alpine shallow-water sediments of Triassic age in the central Pennine area (CP) on the northern continental margin. (2) Early Jurassic block-faulting of the embryonic continental margins, well documented by marine breccia formations in the Ultrahelvetic–North Pennine (UH and NP) Trough and the southern scarp bordering the central Pennine platform (CP) in the Austro-Alpine nappes, and in the southern Alps. (3) Onlap of pelagic deep-sea deposits over the deeply submerged continental margins following the initial phase of block-faulting. These deposits are characterized by a sequence of solution facies indicating prolonged subsidence of the distal continental margins. (4) Development of Bahamian-type carbonate platforms along the southern continental margin (see Fig. 1.13) with adjacent troughs with pelagic and turbiditic carbonate sedimentation. (5) Development of the Tethys Ocean, documented by the oceanic ophiolite suite. Its width and shape are somewhat arbitrary, but 1,000 km as a rough figure is consistent with the history of seafloor spreading in the Atlantic. The width would be reduced by a transform fault into the eastern Mediterranean or increased by changes in the position of the poles of rotation or influences of Paleotethys plate movements. Jurassic magmatism at the continental margin of the southern Alps and in the North Pennine Trough has been omitted. H, Helvetic; UH, ultrahelvetic; NP, north Pennine; CP, central Pennine; PP, pre-Piemont zone; S, Sesia zone-Dent-Blanche nappe; C, Canavese zone; L, Lombardia zone; T, Trento zone; B, Belluno zone; F, Friuli zone. (From Laubscher and Bernoulli, 1977.)

of these Phanerozoic exotic terranes and are associated with a wide variety of lithologies, suggesting a host of primary depositional environments including open ocean, island arc, back arc, incipient ocean basin, rifted continental margin, and shelf and slope environments (Murchey et al., 1983; Tan, 1983; Jones and Murchey, 1986; Karl, this volume; Iijima et al., this volume).

It is clear that fundamental differences exist between the depositional and tectonic environments of Mesozoic Tethyan and Pacific orogenic belt cherts. The thermal structure of the crust and associated depositional basins varied, because the amount of crustal cooling associated with the large plates of the Pacific region was greater than it was for the smaller oceanic plates of the Tethys. The widths of volcanic arc complexes and depths of subduction for the Pacific were greater than they were for the Tethyan region (Szádeczky-Kardoss, 1977). Thrusting, nappe formation, and perhaps folding were greater in the Tethys region. Oceanographic conditions for the two regions differed—e.g., the level of the calcite compensation depth, circulation patterns, and upwelling zones. Intense monsoonal conditions may have characterized early Tethys which evolved into a zonal coastal upwelling system in the Cretaceous (Parrish et al., 1986; Hein and Parrish, 1987). These differences associated with the relatively restricted Tethys seaway and the open circum-Pacific region influenced the deposition and evolution of siliceous deposits, including the rates of deposition, rates of diagenesis, lithologic associations, lithologic successions, and deformational styles. The chapters in this book and in those in the companion book edited by Iijima, Hein, and Siever (1983) describe these features on a local and regional scale.

Acknowledgments

We thank Phyllis Swenson, Carla Chamberlain, and Marjorie Morrison for technical help with this paper. Caroline Isaacs and Margaret Keller provided reviews. We thank all of our colleagues in IGCP 187 for helpful discussions and advice.

References

Baltuck M (1982) Provenance and distribution of Tethyan pelagic and hemipelagic siliceous sediments, Pindos Mountains, Greece. Sedimentary Geology 31:63–88.

Baltuck M (1983) Some sedimentary and diagenetic signatures in the formation of bedded radiolarite. In: Iijima A, Hein JR, Siever R (eds) Siliceous Deposits in the Pacific Region. Elsevier, Amsterdam, pp 299–315.

Barrett TJ (1982) Stratigraphy and sedimentology of Jurassic bedded chert overlying ophiolites in the North Apennines, Italy. Sedimentology 29:353–373.

Berger WH (1968) Radiolarian skeletons: Solution at depths. Science 159:1237–1238.

Bernoulli D, Jenkyns HC (1974) Alpine, Mediterranean, and central Atlantic Mesozoic facies in relation to the early evolution of the Tethys. In: Dott RH, Shaver RH (eds) Modern and Ancient Geosynclinal Sedimentation. Society of Economic Paleontologists and Mineralogists Special Publication 19, pp 129–160.

Bernoulli D, Lemoine M (1980) Birth and early evolution of the Tethys: The overall situation. In: Aubouin J, Debelmas J, Latreille M (eds) Geology of the Alpine Chains Born of the Tethys. Colloque C5, 26th International Geological Congress, Mémoire BRGM 115, pp 168–179.

Biju-Duval B, Dercourt J, Le Pichon X (1977) From the Tethys Ocean to the Mediterranean Seas: A plate tectonic model of the evolution of the western Alpine system. In: Biju-Duval B, Montadert L (eds) International Symposium on Structural History of the Mediterranean Basins, Split, Yugoslavia, 1976. Editions Technip, Paris, pp 143–164.

Bogoyavlenskiy AN (1967) Distribution and migration of dissolved silica in oceans. International Geology Review 9:133–153.

Bosellini A, Winterer EL (1975) Pelagic limestone and radiolarite of the tethyan Mesozoic: A genetic model. Geology 3:279–282.

Burton JD, Liss PS (1973) Processes of supply and removal of dissolved silicon in the oceans. Geochimica et Cosmochimica Acta 37:1761–1773.

Brueckner HK, Snyder WS (1985) Chemical and Sr-isotopic variations during diagenesis of Miocene siliceous sediments of the Monterey Formation, California. Journal of Sedimentary Petrology 55:553–568.

Calvert SE (1968) Silica balance in the ocean and diagenesis. Nature 219:919–920.

Calvert SE (1974) Deposition and diagenesis of silica in marine sediments. In: Hsu KJ, Jenkyns HC (eds) Pelagic Sediments: On Land and Under the Sea. International Association of Sedimentologists Special Publication 1, pp 273–299.

Chyi MS, Crerar DA, Carlson RW, Stallard RF (1984) Hydrothermal Mn-deposits of the Franciscan Assemblage. II. Isotope and trace element geochemistry, and implications for hydrothermal convection at spreading centers. Earth and Planetary Science Letters 71:31–45.

DeMaster DJ (1981) The supply and accumulation of silica in the marine environment. Geochimica et Cosmochimica Acta 45:1715–1732.

Dewey JF, Pitman WC III, Ryan WBF, Bonnin J (1973) Plate tectonics and the evolution of the Alpine system. Geological Society of America Bulletin 84:3137–3180.

Ernst WG, Calvert SE (1969) An Experimental study of the recrystallization of porcelanite and its bearing on the origin of some bedded cherts. American Journal of Science 267A:114–133.

Folk RL, McBride EF (1976) The Caballos Novaculite revisited. I. Origin of Novaculite members. Journal of Sedimentary Petrology 46:659–669.

Fournier RO (1970) Silica in thermal waters: Laboratory and field investigations. Proceedings of the International Symposium on Hydrochemistry and Biochemistry, Tokyo, pp 122–139

Garrison RE, Fischer AG (1969) Deep-water limestones and radiolarites of the Alpine Jurassic. In: Friedman GM (ed) Depositional Environments in Carbonate Rocks. Society of Economic Paleontologists and Mineralogists Special Publication 14, pp 20–56.

Garrison RE, Rowland SM, Horan LJ, Moore JC (1975) Petrology of siliceous rocks recovered from marginal seas of the Western Pacific Leg 31, Deep Sea Drilling Project. In: Karig DE, Ingle JC, et al., Initial Reports of the Deep Sea Drilling Project 31. U.S. Government Printing Office, Washington, D.C. pp 519–529.

Garrison RE, Mack LE, Lee YG, Chun HY (1979) Petrology, sedimentology, and diagenesis of Miocene diatomaceous and opal-CT mudstones in the Pohang area, Korea. Geological Society of Korea Journal 15:229–251.

Gieskes JM (1983) The chemistry of interstitial waters of deep sea sediments: Interpretation of deep sea drilling data. In: Riley JP, Chester R (eds) Chemical Oceanography 8. Academic Press, New York, pp 221–269.

Griffin J (1980) The effect of pressure on the solubility of 11 silicates in seawater at 2°C, pH 8. Unpublished Master's Thesis, University of Hawaii, Honolulu.

Hamilton EL, (1976) Variations of density and porosity with depth in deep-sea sediments. Journal of Sedimentary Petrology 46:280–300.

Harper HE, Knoll AH (1975) Silica, diatoms, and Cenozoic radiolarian evolution. Geology 3:175–177.

Harris RC (1966) Biogenic buffering of oceanic silica. Nature 212:275–276.

Hart R (1981) K-Ar studies of cherts from Deep Sea Drilling Project site 464,Northern Hess Rise. In: Thiede J,

Vallier TL, et al., Initial Reports of the Deep Sea Drilling Project 62. U.S. Government Printing Office, Washington, pp 779–780.

Heath GR (1974) Dissolved silica and deep-sea sediments. In: Hay WW (ed) Studies in Paleo-Oceanography. Society of Economic Paleontologists and Mineralogists Special Publication 20, pp 77–93.

Heath GR, Dymond J (1973) Interstitial silica in deep-sea sediments from the north Pacific. Geology 1:181–184.

Heath GR. Moberly R (1971) Cherts from the western Pacific, leg 7, Deep Sea Drilling Project. In: Winterer EL et al., Initial Reports of the Deep Sea Drilling Project 7. U.S. Government Printing Office, Washington, pp 991–1007.

Hein JR, Karl SM (1983) Comparisons between open-ocean and continental margin chert sequences. In: Iijima A, Hein JR, Siever R (eds) Siliceous Deposits in the Pacific Region. Elsevier, Amsterdam, pp 25–43.

Hein JR, Parrish JT (1987) Distribution of siliceous deposits in space and time. In: Hein JR (ed) Siliceous Sedimentary Rock-Hosted Ores and Petroleum. Van Nostrand Reinhold, New York, pp 10–57.

Hein JR, Scholl DW, Barron JA, Jones MG, Miller J (1978) Diagenesis of late Cenozoic diatomaceous deposits and formation of the bottom simulating reflector in the southern Bering Sea. Sedimentology 25:155–181.

Hein JR, Vallier TL, Allen MA (1981) Chert petrology and geochemistry, mid-Pacific Mountains and Hess Rise, Deep Sea Drilling Project Leg 62. In: Thiede J, Vallier TL, et al., Initial Reports of the Deep Sea Drilling Project 62. U.S. Government Printing Office, Washington, pp 711–748.

Hein JR, Sàncetta C, Morgenson LA (1983a) Petrology and geochemistry of silicified upper Miocene chalk, Costa Rica rift, Deep Sea Drilling Project Leg 69. In: Cann JR, Langseth MG, et al., Initial Reports of the Deep Sea Drilling Project 69. U.S. Government Printing Office, Washington, pp 395–422.

Hein JR, Kuijpers EP, Denyer P, Sliney RE (1983b) Petrology and geochemistry of Cretaceous and Paleogene cherts from western Costa Rica. In: Iijima A, Hein JR, Siever R (eds) Siliceous Deposits in the Pacific Region. Elsevier, Amsterdam, pp 143–174

Howell DG, Jones DL, Schermer ER (1985) Tectonostratigraphic terranes of the circum-Pacific region. In: Howell DG (ed) Tectonostratigraphic Terranes of the Circum-Pacific Region. Circum-Pacific Council for Energy and Mineral Resources Earth Science Series 1, Houston, TX, pp 3–30.

Hurd DC (1973) Interaction of biogenic opal, sediment and seawater in the central equatorial Pacific. Geochimica et Cosmochimica Acta 37:2257–2282.

Hurd DC (1983) Physical and chemical properties of siliceous skeletons. In: Aston SR (ed) Silicon Geochemistry and Biogeochemistry. Academic Press, New York, pp 187–244.

Iijima A, Kakuwa Y, Yamazaki K, Yanagimoto Y (1978) Shallow-sea, organic origin of the Triassic bedded chert in central Japan. Journal of the Faculty of Science of the University of Tokyo, Section II, 19/5:369–400.

Iijima A, Hein JR, Siever R (eds) (1983) Siliceous Deposits in the Pacific Region. Elsevier, Amsterdam, 472 pp.

Isaacs CM (1981) Porosity reduction during diagenesis of the Monterey Formation, Santa Barbara coastal area, California. In: Garrison RE, Douglas RG (eds) The Monterey Formation and Related Siliceous Rocks of California. Pacific Section Society of Economic Paleontologists and Mineralogists Publication 15, pp 257–271.

Isaacs CM (1982) Influence of rock compositions on kinetics of silica phase changes in the Monterey Formation, Santa Barbara area, California. Geology 10:304–308.

Isaacs CM, Petersen NF (1987) Petroleum in the Miocene Monterey Formation, California. In: Hein JR (ed) Siliceous Sedimentary Rock-Hosted Ores and Petroleum. Van Nostrand Reinhold, New York, pp 83–116.

Isaacs CM, Pisciotto KA, Garrison RE (1983) Facies and diagenesis of the Miocene Monterey Formation, California: A Summary. In: Iijima A, Hein JR, Siever R (eds) Siliceous Deposits in the Pacific Region. Elsevier, Amsterdam, pp 247–282.

Jenkyns HC, Winterer EL (1982) Paleoceanography of Mesozoic ribbon radiolarites. Earth and Planetary Science Letters 60:351–375.

Jones DL, Murchey B (1986) Geologic significance of Paleozoic and Mesozoic radiolarian chert. Annual Reviews of Earth and Planetary Sciences 14:455–492.

Jones JB, Segnit ER (1971) The nature of opal I. Nomenclature and constituent phases. Journal Geological Society of Australia 18(1):57–68.

Jones MM, Pytkowicz RM (1973) Solubility of silica in seawater at high pressures. Bulletin of the Royal Society of Science of Liege 42:118–120.

Kamatani A (1971) Physical and chemical characteristics of biogenous silica. Marine Biology 8:89–95.

Karl SM (1982) Geochemical and depositional environments of upper Mesozoic radiolarian cherts from the northeastern Pacific rim and from Pacific DSDP cores. Unpublished Ph.D. Thesis, Stanford University, Stanford, CA 365 pp.

Kastner M, Gieskes JM (1983) Opal-A to Opal-CT transformation: A kinetic study. In: Iijima A, Hein JR, Siever R (eds) Siliceous Deposits in the Pacific Region. Elsevier, Amsterdam, pp 211–227.

Kastner M, Keene JB, Gieskes JM (1977) Diagenesis of siliceous oozes. I. Chemical controls of the rate of opal-A to opal-CT transformation—an experimental study. Geochimica et Cosmochimica Acta 41:1041–1059.

Keller MA, Isaacs CM (1985) An evaluation of temperature scales for silica diagenesis in diatomaceous sequences including a new approach based on the

Miocene Monterey Formation, California. Geo-Marine Letters 5:31–35.

Kozlova OG (1964) Diatoms of the Indian and Pacific sectors of the Antarctic. Publication of House of Akad., Nauka, Moscow.

Krauskopf KB (1956) Dissolution and precipitation of silica at low temperatures. Geochimica et Cosmochimica Acta 10:1–26.

Laubscher H, Bernoulli D (1977) Mediterranean and Tethys. In: Nairn EM, Kanes WH, Stehli FG (eds) The Ocean Basins and Margins 4A. Plenum, New York, pp 1–28.

Leinen M (1979) Biogenic silica accumulation in the central equatorial Pacific and its implications for Cenozoic paleoceanography. Geological Society of America Bulletin 90(II):1310–1376.

Lewin JC (1961) The dissolution of silica from diatom walls. Geochimica et Cosmochimica Acta 21: 182–198.

Lisitzin AP (1972) Sedimentation in the world ocean. Society of Economic Paleontologists and Mineralogists Special Publication 17.

Lisitzin AP (1978) Processes of Oceanic Sedimentation, Lithology, and Geochemistry. Publication of Nauka, Moscow.

Mackenzie FT (1971) Synthesis of quartz at earth-surface conditions. Science 173:533–535.

Mackenzie FT, Garrels RM (1966) Silica-biocarbonate balance in the ocean and early diagenesis. Journal of Sedimentary Petrology 36:1075–1084.

Matsumoto R, Iijima A (1983) Chemical sedimentology of some Permo-Jurassic and Tertiary bedded cherts in central Honshu, Japan. In: Iijima A, Hein JR, Siever R (eds) Siliceous Deposits in the Pacific Region. Elsevier, Amsterdam, pp 175–191.

Mizutani S (1977) Progressive ordering of cristobalitic silica in the early stage of diagenesis. Contributions to Mineralogy and Petrology 61:129–140.

Mizutani S (1983) Duration of chemical diagenesis. The Journal of Earth Sciences of Nagoya University 31:17–35.

Mizutani S, Shibata K (1983) Diagenesis of Jurassic siliceous shale in central Japan. In: Iijima A, Hein JR, Siever R (eds) Siliceous Deposits in the Pacific Region. Elsevier, Amsterdam, pp 283–297.

Morey GW, Fournier RO, Rowe JJ (1962) The solubility of quartz in water in the temperature interval from 25° to 300°C. Geochimica et Cosmochimica Acta 21:1029–1043.

Murata KJ, Larson RR (1975) Diagenesis of Miocene siliceous shales Temblor Range, California. U.S. Geological Survey Journal of Research 3:553–566.

Murata KJ, Nakata JK (1974) Cristobalite stage in the diagenesis of diatomaceous shale. Science 184:567–568.

Murata KJ, Friedman I, Gleason JD (1977) Oxygen isotope relations betwen diagenetic silica minerals in Monterey Shale, Temblor Range, California. American Journal of Science 277:259–272.

Murchey B, Jones DL, Holdsworth BK (1983) Distribution, age, and depositional environments of radiolarian chert in western North America. In: Iijima A, Hein JR, Siever R (eds) Siliceous Deposits in the Pacific Region. Elsevier, Amsterdam, pp 109–125.

Nur A, Ben-Avraham Z (1977) Lost Pacific continent. Nature 270:41–43.

Parrish JM, Parrish JT, Ziegler AM (1986) Permian-Triassic paleogeography and paleoclimatology and implications for therapsid distribution. In: Hotton N, MacLean PD, Roth JJ, Roth EC (eds) The Ecology and Biology of Mammal-Like Reptiles. Smithsonian Institution Press, Washington, pp 109–132.

Pessagno EA, Newport RL (1972) A technique for extracting radiolaria from radiolarian cherts. Micropaleontology 18:231–234.

Pisciotto KA (1981) Distribution, thermal histories, isotopic compositions, and reflection characteristics of siliceous rocks recovered by the Deep Sea Drilling Project. In: Warne JE, Douglas RG, Winterer EL (eds) The Deep Sea Drilling Project: A Decade of Progress. Society of Economic Paleontologists and Mineralogists Special Publication 32, pp 129–148.

Rangin C, Steinberg M, Bonnot-Courtois C (1981) Geochemistry of the Mesozoic bedded cherts of central Baja California (Vizcaino–Cedros–San Benito): Implications for paleogeographic reconstruction of an old oceanic basin. Earth and Planetary Science Letters 54:313–322.

Riedel WR (1959) Siliceous organic remains in pelagic sediments. In: Ireland HA (ed) Silica in Sediments. Society of Economic Paleontologists and Mineralogists Special Paper 7, pp 80–91.

Schink DR, Guinasso NL, Fanning KA (1975) Processes affecting the concentration of silica at the sediment-water interface of the Atlantic Ocean. Journal of Geophysical Research 80:3013–3031.

Siever R (1957) The silica budget in the sedimentary cycle. The American Mineralogist 42:821–841.

Siever R (1962)Silica solubility, 0°–200°C, and the diagenesis of siliceous sediments. Journal Geology 70:127–150.

Shibata K, Mizutani S (1980) Isotopic ages of siliceous shale from Hida-Kanayama, central Japan. Geochemical Journal 14:235–241.

Shimizu H, Masuda A (1977) Cerium in chert as an indication of marine environment of its deposition. Nature 266:346–348.

Smith AG (1971) Alpine deformation and the oceanic areas of the Tethys, Mediterranean, and Atlantic. Geological Society of America Bulletin 82:2039–2070.

Smith AG, Briden JC, Drewry GE (1973) Phanerozoic world maps. Paleontology Special Paper 12, pp 1–42.

Stein CL, Kirkpatrick RJ (1976) Experimental porcelanite recrystallization kinetics. A nucleation and growth model. Journal of Sedimentary Petrology 46:430–435.

Steinberg M, Marin CM (1978) Classification géochimique des radiolarites et des sédiments siliceux océaniques, signification paléo-oceanographique. Oceanologica Acta 1:359–367.

Steinberg M, Bonnot-Courtois C, Tlig S (1983) Geochemical contribution to the understanding of bedded chert. In: Iijima A, Hein JR, Siever R (eds) Siliceous Deposits in the Pacific Region. Elsevier, Amsterdam, pp 193–210.

Stöber W (1967) Formation of silicic acid in aqueous suspensions of different silica modification. In: Equilibrium Concepts in Natural Water Systems. American Chemical Society Series 67, pp 161–182.

Sugisaki R, Yamamoto K, Adachi M (1982) Triassic bedded chert in central Japan are not pelagic. Nature 298:644–647.

Szádeczky-Kardoss E (1977) Metallogeny in the northeastern Mediterranean as interpreted by plate tectonics. In: Janković S (ed) Metallogeny and Plate Tectonics in the Northeastern Mediterranean. Faculty of Mining and Geology, University of Belgrade, Publication by International Geological Correlations Project 3, pp 201–214.

Tan DNK (1983) Cherts of southeast Asia. In: Iijima A Hein JR, Siever R (eds) Siliceous Deposits in the Pacific Region. Elsevier, Amsterdam, pp 79–91.

Vink GE, Morgan WJ, Zhao W-L (1984) Preferential rifting of continents: A source of displaced terranes. Journal of Geophysical Research 89:10072–10076.

Von Rad U, Rösch H (1974) Petrography and diagenesis of deep-sea cherts from the central Atlantic. In: Hsu KJ, Jenkyns HC (eds) Pelagic Sediments on Land and Under the Sea. International Association of Sedimentologists Special Publication 1, pp 327–347.

Weis D, Wasserburg GJ (1987a) Rb-Sr and Sm-Nd systematics of cherts and other siliceous deposits. Geochimica et Cosmochimica Acta 51:959–972.

Weis D, Wasserburg GJ (1987b) Rb-Sr and Sm-Nd isotope geochemistry and chronology of cherts from the Onverwacht Group (3.5 AE), South Africa. Geochimica et Cosmochimica Acta 51:973–984.

Williams LA, Crerar DA (1985) Silica diagenesis II. General mechanisms. Journal of Sedimentary Petrology 55:312–321.

Williams LA, Parks GA, Crerar DA (1985) Silica diagenesis. I. Solubility controls. Journal of Sedimentary Petrology 55:301–311.

Wise SW, Weaver FM (1974) Chertification of oceanic sediments. In: Hsu KJ, Jenkyns HC (eds) Pelagic sediments on Land and Under the Sea. International Association of Sedimentologists Special Publication 1, pp 301–326

Wollast R (1974) The silica problem. In: Goldberg ED (ed) The Sea 5. John Wiley and Sons, New York, pp 359–392.

2
Radiolarian Evidence: Late Cretaceous Through Eocene Ocean Circulation Patterns

Joyce R. Blueford

Overview

Radiolarian assemblages, in Late Cretaceous through Eocene biosiliceous sediments, can help trace major paleocirculation patterns. Abundant radiolarians today indicate normal saline, oceanic conditions, so their presence in the fossil record documents seaways analogous to areas of present-day radiolarian deposition. The circulation patterns indicated by radiolarian distributions support and slightly modify previous paleoceanographic models. During the Late Cretaceous, circulation in the equatorial belt was continuous from the Pacific through the regions of the present-day Atlantic and into the Caspian Sea-Ural Mountains of the Soviet Union. Seaways connecting the Arctic and North Pacific Ocean enabled a wide distribution of radiolarian fauna. Paleocene radiolarian deposits are not abundant, and may indicate a circulation pattern that became restricted. Oceanographic conditions that prevent siliceous deposition are many, including regressions, weakened upwelling, lack of dissolved silica, temperature, and other biological or physical conditions. The distribution of Eocene radiolarians indicates renewed wide circulation in the Paleogene oceans. Radiolarian distribution needs to be considered with other fossil assemblages and with paleomagnetic data to accurately reconstruct the positions of the continents and the oceanic circulation patterns.

Introduction

In the fossil record, radiolarians are useful environmental indicators. Fossil radiolarians reflect normal salinity and oceanic conditions. Many times radiolarians are the only fossils found in chert and limestone in structurally complex areas. Radiolarian assemblages may provide information on reconstructing paleoceanographic conditions.

Siliceous sediments in the ocean basins accumulate in three major belts: a southern oceanic belt encompassing the globe in an almost uninterrupted band around Antarctica, a northern belt in the Pacific, and a near-equatorial belt that is better defined in the Pacific and Indian Oceans than in the Atlantic. The siliceous organisms that make up these deposits are diatoms, radiolarians, and silicoflagellates. The most abundant siliceous organisms in the water column are diatoms; they make up 31% to 75% of the total amount of silica-concentrating organisms in suspension in tropical regions (Lisitzin, 1972). Radiolarians make up 2% to 59%, and silicoflagellates 9% to 21% of the total number. The Antarctic region has 99% diatoms in suspension; of these, less than 1% are radiolarians and silicoflagellates.

Although diatoms are the most abundant organism by number, radiolarians are the most abundant by weight. Individual radiolarians weigh more than individual diatoms. On the average, one radiolarian skeleton is approximately 50 times as heavy as one diatom cell. One milligram of silica corresponds to 1,000 to 3,000 radiolarian skeletons or 3,300 to 260,000 diatom cells. The weight ratio of silicoflagellates:diatoms:radiolarians is 1:5:250. Using this relationship, the weight percentage of siliceous organisms in the tropics is 1% to 35% diatoms, 62% to 98% radiolarians, and 0.1% to 2% silicoflagellates. In the Antarctic, diatoms make up 99% to 99.9% of the total weight, and the radiolarians and silicoflagellates 0.1% to 1%.

Suspension studies also show a discrepancy between the ocean's surface assemblages and the sedi-

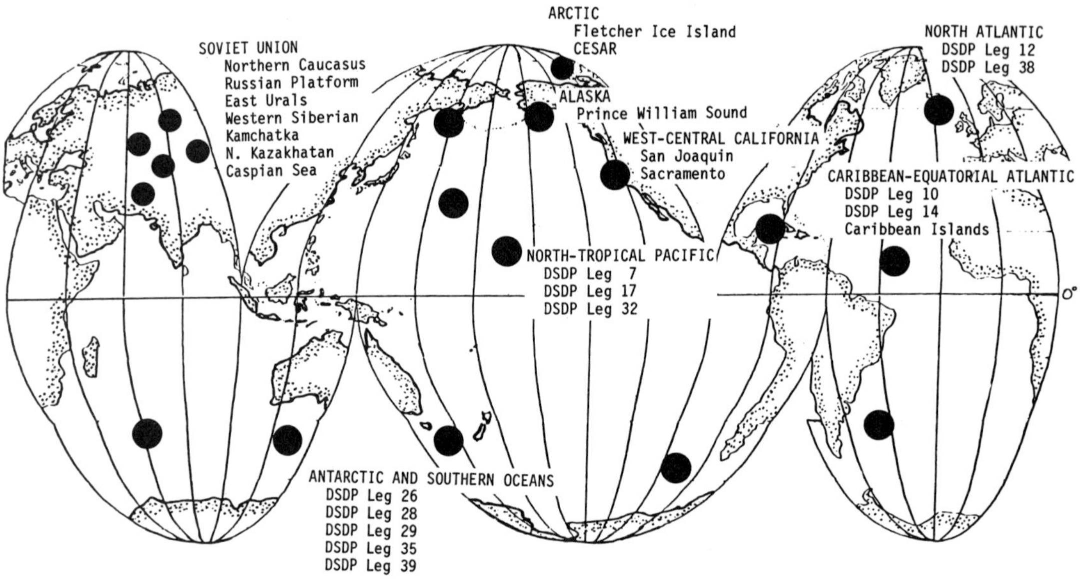

FIGURE 2.1. Map of locations referred to in this study. Black circles indicate the general areas referred to here.

ment assemblages. In a comparison of 1,000 individuals in the water column and in the sediment, Petrushevskaya (1971a,b) found that some species are abundant in plankton but rare in sediment, whereas other species are equally abundant in plankton and sediment. Skeletons of less frequently occurring species in sediment than in plankton are built of thin bars forming an open lattice. The skeletons of the species that are more frequent in sediment have thick walls or spongy skeletons, suggesting that the robustness of the skeletons is the controlling factor. Other factors that contribute to this discrepancy include the residence time of the organisms in suspension, ocean currents, dissolution rates, and dissolution after deposition. Although radiolarians can live throughout the water column, the majority live within the upper 200 meters and are therefore associated with the surface current patterns. Distinct species assemblages characterize different water masses in the surface water. Biogeographical provinces have been documented by Petrushevskaya (1967) and Renz (1973). Studies of radiolarian distributions in surface sediment in the Pacific (Nigrini, 1968, 1970; Casey 1971; Petrushevskaya, 1971a; Molnia-Cruz, 1977; Blueford, 1983), in the Atlantic (Goll and Bjørklund, 1974), and in the Indian Ocean (Nigrini, 1967; Petrushevskaya, 1971b) confirm that the surface sediment assemblages indicate major surface-circulation patterns. Therefore, radiolarian assemblages represent exclusively normal salinity marine oceanic conditions. Oceanic conditions are defined as ocean basins, seaways directly connecting oceanic basins, and coastal areas. Radiolarians are more abundant in areas of oceanic upwelling than in other places (Petrushevskaya, 1971a).

Radiolarian assemblages from 19 areas around the globe were plotted to deduce circulation patterns from the Late Cretaceous through Eocene oceans (Fig. 2.1). These 19 areas are all described in the literature, but other areas where Late Cretaceous and Paleogene radiolarians occur, such as southern England, northern France, Italy, Japan, Tunisia, Algeria, Morocco, Israel, and the Arctic, have not been fully documented (Hein and Parrish, 1987; Lipman, personal communication, 1985). These 19 sites offer clues to the overall oceanographic conditions during the Late Cretaceous through the Eocene.

Sites

West Central California

Biosiliceous sedimentary rocks of Cretaceous to Pliocene age are found in the San Joaquin and Sacramento Valleys of west-central California. The middle Eocene faunas have been well documented (Clark and Campbell, 1942, 1945; Blueford and White, 1984; Blueford and Brunner, 1984; Blueford, in press). Late Cretaceous radiolarian taxonomy has been determined (Clark and Campbell, 1944; Foreman, 1968; Pessagno, 1973), but overall, Cretaceous assemblage studies and species abundances have not been well documented. Foraminifers are present in and define the Paleogene sections like the Tumey Shale (Oligocene–Stewart, 1946), the Moreno Shale (Late Cretaceous and Paleocene–Payne, 1951), and the Capay Shale (early Eocene–Clark, 1935) rather than radiolarians. Thus, although radiolarian-bearing sections are present, the lack of documentation of radiolarian species and assemblage evolution through the Cretaceous and the Paleogene hampers comparisons of radiolarians from California with other areas.

Soviet Union

Localities in the Soviet Union that are of Paleogene and Late Cretaceous age are abundant and well documented (Lipman, 1950, 1952, 1957, 1959, 1960, 1971, 1972; Krasheninnikov, 1960; Borisenko, 1958; Mamedov, 1970; Runeva, 1975). Zonations have been established for the Russian Platform, Caspian Sea Region, Northern Caucasus, East Urals, Western Siberian Lowland, Kamchatka Peninsula, and North Kazakhatan. Although more than 300 sections have been studied, none are continuous, so the stratigraphy is a composite of these sections. Lipman (personal communication, 1985) stated that about 100 species, of the 400 species found in Paleogene deposits of Eurasia, are found in other worldwide Paleogene deposits. These Soviet zonations have been used for correlation of sections within large regions of the Soviet Union. Comparisons with assemblages in other parts of the world is problematic for the author because of incomplete English translations especially in reference to the taxonomy, and the lack of good photographs in the published monographs. Also, there is disagreement over the boundaries defined by the Soviet workers and boundaries defined by Berggren et al. (1985) for the middle and late Eocene.

Alaska

Continuous Late Cretaceous through Eocene sections are difficult to find in Alaska because of Cretaceous and Paleogene tectonics in this area. However, using the Russian stratigraphy, Alaskan and Russian assemblages can be compared (Plafker et al., 1985; Blueford, 1986).

The radiolarian assemblage of the Orca Group of Prince William Sound was found to be significant in unraveling the stratigraphy and oceanographic events that occurred from the Late Cretaceous through the Eocene (Plafker et al., 1985; Blueford, 1986). The assemblages in the Prince William Sound samples resemble the Late Cretaceous through Eocene faunas from west-central California, the southern oceans, the North Atlantic, and the Soviet Union. Paleogene assemblages from the equatorial–North Pacific and Caribbean–Equatorial Atlantic have some species in common with the Alaskan assemblages, but the lack of assemblage descriptions and abundance data makes definitive correlations impossible. The central California affinities of the species are more apparent in the younger sections, but they are not useful for precise dating of the Paleogene, because the stratigraphy of the middle Eocene is the only time period documented. Knowing the distribution of Paleogene radiolarians in the Prince William Sound area can explicate the evolution and movement of land masses and might also explain some aspects of the present-day distribution of radiolarians.

The faunas are basically dominated by spongodiscids, porodiscids, and actinommids. Their abundance implies an environment dominated by cool or cold water (Blueford and White, 1984). Warm water usually contains more nassellarians and armed spongodiscids and porodiscids that are not common in the Prince William Sound assemblage. However, this lack of warm-water species can mean either that the ocean was at a high latitude or that the waters were dominated by cool, upwelled water.

North-Tropical Pacific

The stratigraphy of the Late Cretaceous to the Holocene has been established in the tropical Pacific by using Deep Sea Drilling Project (DSDP) cores. Foreman (1971) found well-preserved Cretaceous radiolarians at sites 61 and 66 of DSDP leg 7. She arrived at this age by comparing radiolarian assemblages with those from Fresno County, Alameda County, Tesla Quadrangle, and Yolo County in California; east of Rousseau, South Dakota; Bavaria, Germany; South Atlantic; Cuba; Island of Roti, Timor; and Val Dorbias, Italy. However, she did not compare individual faunas, so paleoceanographic information cannot be interpreted from her published reports. Riedel and Sanfilippo (1971), using fauna from DSDP leg 7, found late Paleocene and early Eocene radiolarians at site 67 and middle to late Eocene radiolarians at sites 64 and 65. They found that the western Pacific assemblage has many components similar to the European Eocene (Italy and Israel) and to the Caribbean Eocene (Barbados, Trinidad, and Cuba) assemblages.

Moore (1973), using material from DSDP leg 17, found that the lower Paleogene was either missing or devoid of radiolarians. Late Cretaceous radiolarian assemblages, although spotty and poorly preserved, were present. Although Moore correlated DSDP leg 17 assemblages with other Pacific and California assemblages, he drew no conclusions as to which assemblages were most similar.

Foreman (1975), using assemblages from DSDP leg 32, site 313, found Late Cretaceous (site 310) and Early Eocene radiolarians. However, because Early Cretaceous assemblages were also found at sites 303, 304, 305, 306, and 307, she devoted most of the article to development of an Early Cretaceous radiolarian stratigraphy. She concluded that the Cretaceous assemblages have components of the zonation developed by Pessagno (1973) for California.

Riedel and Sanfilippo (1978) developed a Cenozoic low-latitude stratigraphy using various DSDP sections and land sections. As expected, their stratigraphy is not comparable to the entire radiolarian assemblage from high-latitude locations.

North Atlantic

Continuous sections throughout the Late Cretaceous and Paleogene are lacking from the North Atlantic. Benson (1972), in DSDP leg 12, sites 111, 112 (Labrador Sea), 117 (Bay of Biscay), and 118 (Rockall Basin), found Eocene and Paleocene assemblages that are moderately to poorly preserved. Zeolitized Mesozoic radiolarians were found in hole 118. Benson felt that the North Atlantic radiolarian assemblages in Paleocene and early Eocene sediments were reworked from nearby sea-floor deposits or from as far away as the French or Spanish continental margins.

Late Eocene and middle Eocene assemblages are well described from DSDP leg 38 in the Norwegian Sea. However, two paleontologists looking at the same material developed two slightly different stratigraphies. Bjørklund (1976) felt that most of the Eocene was represented in four zones: *Arachnocalpis tumulosa* (early middle Eocene), *Artostrobus quadriporus* (middle middle Eocene), *Calocyclas talwanii* (late middle Eocene), and *Lophocorys novegiensis* (late Eocene). He pointed out that his boundaries were not well defined and preliminary. Although his taxa are variable, they do show some similarities to the taxa from California, but he did not describe the entire assemblage, so correlation is not possible. Bjørklund (1976) felt that the Norwegian Sea fauna was deposited in an isolated basin.

Kozlova, in her supplementary contribution on Paleogene radiolarians (Dzinoridze et al., 1978), disagreed somewhat with Bjørklund's zonation. She used samples from leg 38, sites 336, 337, 338, 339, 340, and 343. Her zonations consist of five levels. Level I *(Tripodiscium tumulosa)*, middle Eocene, has similarities to the east side of the Ural Ridge and West Siberian lowland. Level II *(Phacodiscus testatus)* has some species common to the eastern part of the Caspian district and some parts of the south Russian platform. Level III *(Lophocory (?) norvegiensis)* has about 50% of the California Eocene species described by Clark and Campbell (1942, 1945). Level IV *(Heterosetrum tschujenkoi)* is different from any other described Eocene assemblage, but it has a few species common in the Soviet Union and California assemblages. Level V *(Theocyrites litos "Cr")* has species common in both California and Soviet Union assemblages. The *Ellipsoxiphus chabokovi* and *Heliodiscus lentis* zones of Lipman were not found in the Norwegian Sea, although Blueford (in press) found that much of the California Eocene is part of the *E. chabakovi* zone.

Kozlova's taxonomic descriptions are much more complete than Bjørklund's, and they strongly suggest affinity with the California sections. She stated that there is a great similarity of the Norwegian Sea assemblages with the California Eocene assemblages described by Clark and Campbell (1942, 1945) and that her *Lophocory norvegienesis* zone (early upper Eocene) might be equivalent to the Eocene radiolarians of California. The Norwegian basin radiolarian assemblages also show a close affinity with the assemblages from West Siberia, Aral Sea, and Ural Ridge that Lipman described and developed into a Paleogene stratigraphy (Lipman, 1957, 1960, 1972). Kozlova suggested that radiolarians appeared in the Norwegian basin during the late Eocene because of the east-west currents in the seas of central Europe and middle Asia that were connected as a consequent of tectonic movements. Some radiolarian assemblages may have been restricted, but many persisted throughout the Oligocene. Kozlova concluded that the Eocene radiolarian assemblages indicated influences by cold-water currents from or into the West Siberian Sea.

Caribbean and West Equatorial Atlantic

Foreman (1973) refined an Eocene radiolarian low-latitude stratigraphy using materials from DSDP leg 10. The early Campanian, late Paleocene, Eocene, and Oligocene (with the exception of late-middle Eocene), are recorded in continuous cores that contain radiolarians. Sanfilippo and Riedel (1973), using the same material as Foreman, extend the ranges of the radiolarian species within Foreman's zones. However, neither paper described some of the species, especially the spongodiscids and porodiscids; species of those families would have helped correlate the Caribbean assemblages with the California and Soviet assemblages.

Deep Sea Drilling Project leg 14 in the equatorial Atlantic, from Puerto Rico to Portugal, does not contain a continuous Eocene section. Petrushevskaya and Kozlova (1972) reported that the following radiolarian zones were present: *Podocyrtis chalara, Podocyrtis goetheana,* and the uppermost *Thyrsocyrtis bromia* at sites 140 and 144. Sites 136, 137, 138, 140, and 144 contain Late Cretaceous fauna.

Land-based radiolarian occurrences in the Caribbean region are summarized by Sanfilippo and Riedel (1974). The assemblages are from the low latitudes, and the stratigraphy is applicable to the equatorial Pacific. A closer affinity may exist between these fauna and the California and Soviet Union fauna than has previously been acknowledged, but the abundance of each species is not given, so it is not possible to make close comparisons.

Antarctic and Southern Oceans

Petrushevskaya (1974), using samples from DSDP leg 29, sites 278, 280A, and 281, in the Antarctic region of the Pacific (south of New Zealand), concluded that the late Eocene faunas have characteristic species that are similar to those of California and the Caribbean Basin. These species are also found in the boreal Eocene radiolarian faunas of the Soviet Union. However, she established Paleocene rather than late Eocene radiolarian zones. Pessagno (1974) documented a latest Campanian radiolarian fauna of low diversity that can be correlated with the late Campanian of California. Radiolarian deposition did not occur from the Late Cretaceous or Late Eocene.

In the southern Indian Ocean DSDP leg 26, Riedel and Sanfilippo (1974) found predominately Neogene and Cretaceous radiolarian assemblages at most sites. At sites 253 and 257 there occur poorly preserved Eocene radiolarian assemblages.

Chen (1975) reported Eocene radiolarians at DSDP leg 28, site 264, which is located near the Naturalistic Plateau off the west coast of Australia. The assemblage is similar but less diverse than the Eocene assemblages of the equatorial Pacific, but it has more in common with the Eocene assemblages of California. Chen did not develop a zonation for this area.

Weaver (1976) found well-preserved Neogene radiolarians in the area of the southeast Pacific Basin, except for poorly preserved but common occurrences of Cretaceous radiolarians at DSDP leg 35, site 323. No Paleogene radiolarians were found.

Eocene to Cretaceous radiolarians are present at DSDP leg 39, site 354, Cesera Rise (middle to late Eocene); site 355, Brazil Basin (middle Eocene–early Eocene); site 356, Sao Paula Plateau (middle Eocene-Late Cretaceous), and site 357, Rio Grande Rise (middle-late Eocene). The fauna is not described, and correlations are made by using low-latitude Paleogene zonations.

Arctic

The Arctic is a key area that could reveal how circulation patterns affected radiolarian distributions in the northern hemisphere. Fletcher Ice Island cores 422 and 437 of the U.S. Geological Survey and the Canadian Expedition to Study the Alpha Ridge (CESAR) 6 cores of the Canadian Geological Survey are the only cores available that contain Cretaceous to Paleogene siliceous sediments. Core 422 from the Alpha Ridge region is considered to be early Paleogene by Ling (1985) and middle or late Eocene by Bukry (1984), who based ages on silicoflagellates and ebridians. Core 437 is considered Cretaceous based on silicoflagellates (Bukry, 1981). CESAR Core 6, contained well preserved Late Cretaceous silicoflagellates (Bukry, 1985) and diatoms (Barron, 1985). Barron (1985) used published diatom studies on the Cretaceous of California (Long et al., 1946) to correlate the diatoms from the Arctic. Although the age assignment is not agreed on, these cores do document Arctic biosiliceous sediments. Radiolarians were not routinely looked for in these samples; I found broken radiolarians on smear slides of core 422.

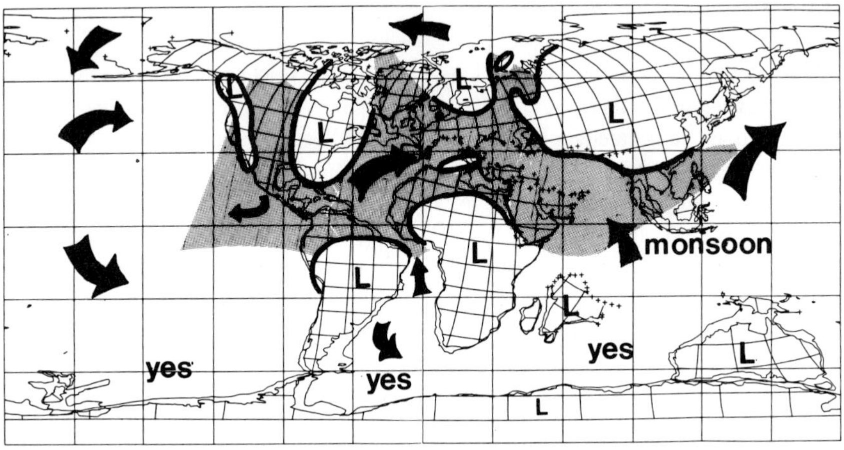

FIGURE 2.2. Santonian (Late Cretaceous) paleogeographic map. Arrows represent major surface circulation patterns. "L" represents land. Shaded areas represent oceans. Blank areas represent those areas where there are few data or open oceans. Heavy lines represent approximate areas of paleoshorelines. "Monsoon" represents areas of possible monsoon conditions. "Yes" means that there was siliceous deposition recorded in DSDP cores. Map is adapted from Smith et al. (1981), base map reprinted with the permission of Cambridge University Press.

Paleocirculation Reconstructions

Many paleocirculation reconstructions show the continents as emergent during the Cretaceous to late Eocene (Luyendyk et al., 1972; Kitchell and Clark, 1982; Seidov, 1986; Oberhansli and Hsü, 1986). These reconstructions conflict in part with the distribution of Cretaceous through Eocene siliceous deposits around the world. Circulation patterns of ocean currents are difficult to model for present-day oceans and even more difficult to model for past oceans. The causes of ocean currents have been debated among oceanographers for more than 50 years. The major debate centers around whether the thermal structure of the oceans (ocean/air theory) or wind intensity (air/ocean theory) is mostly responsible for inducing the surface movements that occur in the oceans. Other causal influences include rotation of the earth (which determines the basic flow pattern of the currents), topography (submarine and continental), climatic factors, geostropic balance, and turbulence (Tolmazin, 1985). The techniques of modeling global circulation for Cenozoic and Mesozoic oceans are summarized by Barron (1986) and Seidov (1986).

Many workers (Oberhansli and Hsü, 1986; Haq, 1981; Luyendyk et al., 1972; Kitchell and Clark, 1982) have developed circulation models that maintain the present continental configurations, but only Barron et al. (1981) and Barron and Washington (1982) show paleogeographic reconstructions with seaways similar to those determined by the distribution of radiolarians (Figs. 2.2–2.4). As an example, the laboratory model of Luyendyk et al. (1972), using planetary vorticity models developed by Von Arx (1952), constructed a rotating tank coupled with a wind system. In their model, the configurations of the continents were as they are today. Their experimental results showing direction of currents during the Cretaceous reveals the bias produced by continental positions. If you superimposed radiolarian distribution and other geologic information, the Luyendyk et al. (1972) reconstruction is not realistic. The continental configuration not only affects the direction of major circulation patterns but also affects the intensity of currents and areas of upwelling.

I used the reconstruction of the basic patterns of circulation for Late Cretaceous to late Eocene of Smith et al. (1981) (Figs. 2.2–2.4). Their maps show the paleogeographic position of land masses, so I superimposed the radiolarian distributions from Figure 2.1 on the maps, and I assumed that the presence of radiolarians indicates oceanic conditions as defined above. Data were also used from the Eurasia lithological maps of Vinogradov, (1967–1969) to obtain the locations of radiolarian biosiliceous sediment. The time frames of Santonian,

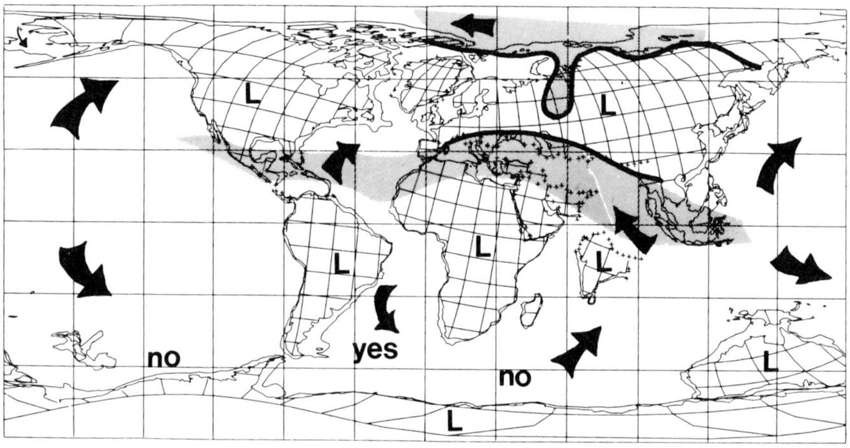

FIGURE 2.3. Paleocene paleogeographic map. Arrows represent major surface circulation patterns. Other designations as in Figure 2.2. "No" means that there is no evidence of siliceous deposition recorded in DSDP cores. Map is adapted from Smith et al. (1981), base map reprinted with the permission of Cambridge University Press.

Paleocene, and the late Eocene were used to correlate with other available data, especially the lithologic maps of Eurasia (Vinogradov, 1967–1969) and the geology of the Russian platform (Naidin et al., 1980).

Santonian – Late Cretaceous

During this time, the distribution of the continents created an almost continuous ocean around the equator (Fig. 2.2). Continental configurations governed the main circulation patterns during the Late Cretaceous. During late Albian to Mastrichtian times, a shallow sea extended southward from the Arctic Sea to the Gulf of Mexico (Western Seaway); evidence for this sea is based on sedimentation patterns (Matsumoto, 1980). Radiolarian deposition in this area was minor. Northwest Europe consisted of small basins separated by structural highs. Tectonic movements on the Russian platform during Late Cretaceous changed the configuration of the continents, with periods of sedimentation in the Turgai Strait. Radiolarian deposits occur throughout the region, but the deposits are not continuous; instead, they provide evidence of regressions and transgressions. In central Asia, transgressive cycles during the Late Cretaceous produced high sedimentation rates. Tectonic movements in Late Cretaceous in southwest Asia make this area difficult to reconstruct. North Africa had a shallow sea that extended into Egypt, Libya, Tunisia, and Algeria (Matsumoto, 1980).

Circulation in the proto-Atlantic probably brought warmer waters to the northern area that mixed with deep seaways on the Russian platform. If monsoon circulation was in effect, part of the year the radiolarian fauna from the southern oceans would have mixed with the fauna from the northern Atlantic, resulting in a fauna that has mixed assemblage characteristics. The overall circulation pattern of the Pacific was basically in the same direction as today, but the connection between the north Pacific and the Arctic may have been a larger seaway than previously thought, especially if radiolarians are found in Cretaceous deposits of the Arctic.

Circulation patterns were much more open during the Cretaceous than today, so radiolarian production along the equatorial belt may have been caused by divergent upwelling as it is today. The intensified upwelling in transgressive seas probably sustained high radiolarian productivity and deposition.

Paleocene

The Paleocene is characterized by a dramatic reduction of biosiliceous deposits containing radiolarians (Fig. 2.3). The deposits of the Paleocene represent a regressive cycle with reduced upwelling. Tectonic movement of the continents may have restricted the

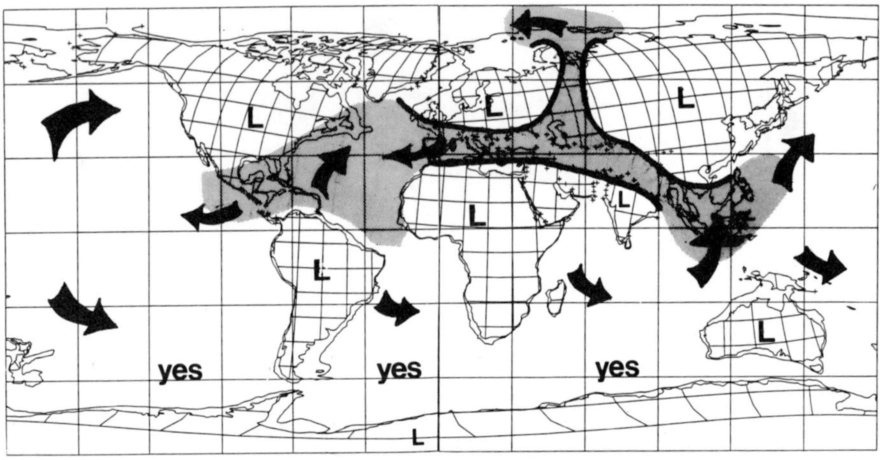

FIGURE 2.4. Eocene paleogeographic map. Arrows represent major surface circulation patterns. Other designations as in Figure 2.2. Map is adapted from Smith et al. (1981), base map reprinted with the permission of Cambridge University Press.

flow of water masses that radiolarians thrive in, or the proliferation of diatoms may have outcompeted the radiolarians.

Paleocene deposits containing radiolarians exist, but assemblage characteristics and high-latitude zonations are unknown. Circulation in the North Atlantic increased because of the opening of the Atlantic, but communications with seas covering the Russian platform and probably the Arctic were lost. Monsoon conditions may not have occurred in the Indian Ocean, because the position of India may have offset the warming and cooling cycles in this area by creating connections between the North Pacific and Arctic. These connections were also much reduced.

Eocene

Circulation patterns were probably similar to those of modern oceans, except that currents flowed through what is now Central America (Fig. 2.2). This similarity also extends to the Eocene radiolarian deposits of the equatorial Pacific and Caribbean. Deposition of siliceous deposits on the Russian platform resumed in the Eocene because of the opening of the Turgai Strait. The radiolarian assemblages of the Eocene are representative of an environment similar to that of the radiolarian assemblages of the Cretaceous. Cooler water conditions are indicated because of the lack of armed spongodiscids and porodiscids; cool-water conditions are also supported by other fossil groups. The Arctic was reconnected with the Russian platform, but the North Pacific connection to the Arctic remained as it was in the Paleocene. The period from middle to late Eocene is noted for its high silica deposition worldwide (Hein and Parrish, 1987; Leinen, 1979). High silica deposition resumed in the southern oceans.

Conclusions

This preliminary study outlines the usefulness of using radiolarians for paleogeographic reconstructions. Radiolarians indicate that during the Late Cretaceous there was upwelling from the Pacific through the present-day Atlantic and into the Caspian Sea to the Ural Mountains region of the Soviet Union. The Arctic Ocean connection to the North Pacific produced circulation patterns strong enough to allow radiolarian deposition in the Arctic, but further work is needed to confirm the intensity of the circulation. Paleocene radiolarians are not well documented, and their absence may indicate a more restrictive circulation pattern that did not allow high productivity of radiolarians. The Eocene radiolarians indicate renewed strong circulation with deposition that reflects a widening of the Atlantic (with conditions not ideal for radiolarian deposition), closing of the north Pacific seaway with the Arctic, and a reopening of the Turgai Strait on the Eurasian continent.

References

Barron, EJ (1986) Physical paleoceanography: A status report. In: Hsü KJ (ed) Mesozoic and Cenozoic Oceans. American Geophysical Union, Washington, Vol 15, pp 1–10.

Barron EJ, Washington WM (1982) Cretaceous climate: A comparison of atmospheric simulations with the geologic record. Palaeogeography, Paleoclimatology, Palaeoecology 40:103–133.

Barron EJ, Harrison CG, Sloan JL, Hay WW (1981) Paleogeography, 180 My to present, Eclog Geology Helv 74:443–470.

Barron JA (1985) Diatom biostratigraphy of the CESAR 6 core, Alpha Ridge. In: Jackson HR, Mudie PJ, Blasco SM (eds) Initial Geological Report of CESAR–the Canadian Expedition to Study the Alpha Ridge, Arctic Ocean, Geological Survey of Canada paper 84-22: 137–148.

Benson RN (1972) Radiolaria, leg 12, Deep Sea Drilling Project. In: Berggren WA et al. (eds) Initial Reports of the Deep Sea Drilling Project 12. U.S. Government Printing Office, Washington, pp 1085–1105.

Berggren WA, Kent DV, Flynn JJ (1985) Jurassic to Paleogene: Part 2. Paleogene geochronology and chronostratigraphy. In: Snelling NJ (ed) The Chronology of the Geological Record, Memoir 10. Geological Society, London, pp 141–195.

Bjørklund K (1976) Radiolaria from the Norwegian Sea, leg 38 of Deep Sea Drilling Project. In: Talwani M, Udintsev G, et al. (eds) Initial Reports of the Deep Sea Drilling Project 38. U.S. Government Printing Office, Washington, pp 1101–1168.

Blueford JR (1986) New radiolarian data from the Orca Group, Prince William Sound Alaska, Circular. U.S. Geological Survey, U.S. Government Printing Office, Washington, pp 21–23.

Blueford JR (1983) Distribution of Quaternary radiolarians in the Navarin Basin geologic province, Bering Sea. Deep Sea Research 30(7):763–781.

Blueford JR (in press) Characteristic radiolarian assemblage of Eocene deposits in California. Micropaleontology.

Blueford JR, Brunner C (1984) Comparison of Eocene radiolarian assemblages of the Sidney Flat and Jameson Shale members of the Markley Formation and the Kellogg Shale, of Northern California. In: Blueford JR (ed) Kreyenhagen Formation and Related Rocks. Society of Economic Paleontologists and Mineralogists Special Publication 37, pp 79–86.

Blueford JR, White LD (1984) Paleoceanographic interpretation of Eocene siliceous deposits from west central California. In: Blueford JR (ed) Kreyenhagen Formation and Related Rocks. Society of Economic Paleontologists and Mineralogists Special Publication 37, pp 67–78.

Borisenko NN (1958) Radiolarii paleobenz Zapadnoi Kubani. Trudy VNII, Krasnodarskii Filial, Vyp. XVII, str. 81–100.

Bukry D (1981) Cretaceous Arctic silicoflagellates. Geo-Marine Letters 1:57–63.

Bukry D (1984) Paleogene paleoceanography of the Arctic Ocean is constrained by the middle or late Eocene age of USGS core F1-422: Evidence from silicoflagellates. Geology 12:199–201.

Bukry D (1985) Correlation of Late Cretaceous Arctic silicoflagellates from Alpha Ridge in Initial Geological Report on CESAR–the Canadian Expedition to Study the Alpha Ridge, Arctic Ocean. In: Jackson HR, Mudie PJ, Blasco SM (eds), Geological Survey of Canada paper 84-22 pp 125–135.

Casey RE (1971) Radiolarians as indicators of past and present water masses. In: Funnell BM, Riedel WR (eds) Micropaleontology of the Oceans. Cambridge University Press, London, pp 331–341.

Chen PH (1975) Antarctic radiolaria. In: Hayes DE, Frakes LA, et al. (eds) Initial Reports of the Deep Sea Drilling Project 28. U.S. Government Printing Office, Washington, pp 437–513.

Clark BL (1935) Tectonics of the Mount Diablo and Coalinga Areas, Middle Coast Ranges of California. Geological Society of America 46(7):1025–1078.

Clark BL, Campbell AS (1942) Eocene radiolarian faunas from the Mt. Diablo Area, California. Geological Society of America special paper 39, 112 pp.

Clark BL, Campbell AS (1944) Radiolaria from Upper Cretaceous of Middle California. Geological Society of America special paper 57 pp 1–61.

Clark BL, Campbell AS (1945) Radiolaria from the Kreyenhagen Formation near Los Banos, California. Geological Society of America Memoir 10(1):51.

Dzinoridze RN, Jouse AP, Koroleva-Golikova GS, Kozlova GE, Nagaeva GS, Petrushevskaya MG, Streinikova NI (1978) Diatoms and radiolarian Cenozoic stratigraphy, Norwegian Basin, DSDP leg 38. In: Supplement of Volumes XXXVIII, XXXIX, XL, and XLI, U.S. Government Printing Office, Washington, pp 289–427.

Foreman H (1968) Upper Maestrichtian Radiolarian of California. Palaeontological Association, London, 82 pp.

Foreman H (1971) Cretaceous Radiolaria, leg 7, DSDP. In: Winterer EL et al. (eds) Initial Reports of the Deep Sea Drilling Project 7. U.S. Government Printing Office, Washington, pp 1673–1693.

Foreman H (1973) Radiolarian. In: Worzel JL, Bryant W, et al. (eds) Initial Reports of the Deep Sea Drilling Project 10. U.S. Government Printing Office, Washington, pp 407–471.

Foreman H (1975) Radiolaria in the N. Pacific, Deep Sea Drilling Project, leg 32. In: Larson RL, Moberly R, et al. (eds) Initial Reports of the Deep Sea Drilling Project, Vol 32. U.S. Government Printing Office, Washington, pp 579–676.

Goll RM, Bjørklund KR (1974) Radiolarian in surface sediments in the South Atlantic. Micropaleontology 20(1):38–75.

Haq BU (1981) Paleogene paleoceanography: Early Cenozoic oceans revisited. Oceanologica Acta SP:71–82.

Hein JR, Parrish JT (1987) Distribution of siliceous deposits in space and time. In: Hein JR (ed) Siliceous Sedimentary Rock - Hosted Ores and Petroleum. Van Nostrand Reinhold, New York, pp 10–57.

Kitchell JA, Clark DL (1982) Cretaceous-Paleogene paleogeography and paleocirculation: Evidence of north polar upwelling. Palaeogeography, Palaeoclimatology, Palaeoecology 40:135–165.

Krasheninnikov VA (1960) Nekotorye radiolarii nizhnego i strednego eotsena zapadnogo predkavkazyaa. Trudy VNIGNI, vyp 16 (Paleontologicheskii Sbornik 3), str 271–307.

Leinen M (1979) Biogenic silica accumulation in the central equatorial Pacific and its implications for Cenozoic paleoceanography: Summary. Geological Society of America Bulletin Part I 90:801–803.

Ling HS (1985) Early Paleocene silicoflagellates and ebridians from the Arctic Oceans. Transaction of Proceedings Palaeontological Society of Japan 138: 79–93.

Lipman RK (1950) Radiolarii eotsena Kyzyl-Kumov. Trudy VSEGEI L, vyp 1, str 51–65

Lipman RK (1952) Materials for the monographic study of Radiolaria of the upper Cretaceous deposits of the Russian Platform. Paleontologiya Stratigrapfiya 24–51 (in Russian).

Lipman RK (1957) Zonalnoe raschlenenie moorskogo paleogena Zapado-Sibirskoi nizmennosti. Gostoptekhizdat L str 201–208.

Lipman RK (1959) Znachenie radiolarii dlya stratai-graficheskogo raschleneniya osadoochynkh porod. Bulletin of the Moscow Society of Naturalist, Department of Geology, 34:67–88.

Lipman RK (1960) Radiolaria in stratigraphy and fauna of the Cretaceous deposits of the Western Siberian Plain. Translations of the All-Union Geological Scientific Research Institute. 29:124–134 (in Russian).

Lipman RK (1971) Paleogene Radiolaria of USSR and their stratigraphic and paleogeographic significance. Proceedings of the Planktonic Conference, Rome, pp 731–738.

Lipman RK (1972) Paleogenovye radiolarii SSSR. Avtoreferat. PhD dissertation, 55 pp.

Lisitzin AP (1972) Sedimentation in the World Ocean Society of Economic Paleontologists and Mineralogists. Special Publication 17:218.

Long JA, Fuge DP, Smith J (1946) Diatoms of the Moreno Shale. Journal of Paleontology 20:89–118.

Luyendyk BP, Forsyth D, Phillips JD (1972) Experimental approach to the paleocirculation of the oceanic surface waters. Geological Society of America Bulletin 83: 2649–2664.

Mamedov NA (1970) Radiolarii paleogenovyeh otlozhenii Schemakhino-Kobystanskoi i Apscherosnkoi oblastei Azerbaidzhana i ich stratigraficheskoe znachenie. PhD dissertation, Baku, 23 pp.

Matsumoto T (1980) Inter-regional correlation of transgressions and regressions in the Cretaceous period. Cretaceous Research 1:359–373.

Molina-Cruz A (1977) Radiolarian assemblages and their relationship to the oceanography of the subtropical southeastern Pacific. Marine Micropaleontology 2: 315–352.

Moore TC (1973) Radiolaria from leg 17 of the Deep Sea Drilling Project. In: Winterer EL, Ewing JL, et al. (eds) Initial Reports of the Deep Sea Drilling Project, Vol XVII. U.S. Government Printing Office, Washington, pp 797–865.

Naidin DI, Sasonova IG, Pojarkova SN, Djalilov MR, Papulov GN, Senkovsky Y, Benjamovsky VN, Kopaevich LF (1980) Cretaceous transgressions and regressions on the Russian Platform in Crimea and Central Asia. Cretaceous Research 1:375–387.

Nigrini C (1967) Radiolaria in pelagic sediments from the Indian and Atlantic Oceans. Scripps Institute of Oceanography Bulletin 11:1–125.

Nigrini C (1968) Radiolaria from eastern tropical Pacific sediments. Micropaleontology 14(1):51–53.

Nigrini C (1970) Radiolarian assemblages in the North Pacific and their application to a study of Quaternary sediments in core V20–130. In: Hays JD (ed) Geological Investigations of the North Pacific. Geological Society of America Memoir 126:139–183.

Oberhansli H Hsü KJ (1986) Paleocene-Eocene paleogeography. In: Hsü KJ (ed) Mesozoic and Cenozoic Oceans. American Geophysical Union, Washington, Vol 15 pp 85–100.

Payne MB (1951) Type Moreno Formation and overlying Eocene strata on the west side of the San Joaquin Valley, Fresno and Merced Counties, California. State of California Division of Mines Special Report 9:1–19.

Pessagno EA Jr (1973) Age and geologic significiance of radiolarian cherts in the California coast ranges. Geology 1:153–156.

Pessagno EA Jr (1974) Upper Creteceous Radiolaria from DSDP site 275. In: Kennett JP, Houtz RE, et al. (eds) Initial Reports of the Deep Sea Drilling Project 29. U.S. Government Printing Office, Washington, pp 1011–1030.

Petrushevskaya MG (1967) Radiolyarii otryadov Spumellaria Nassellaria antarkticheskoi oblasti. In: Povlovskii EN (ed) Issledovaniya Fauni Morei, 3(12). Zoologica Institute Akad SSSR, Leningrad, 186 pp.

Petrushevskaya MG (1971a) Spumellarian and nassellarian Radiolaria in the plankton and bottom sediments of the central Pacific. In: Funnel BM, Riedel WR (eds) The Micropaleontology of the Oceans. Cambridge University Press, London, pp 319–329.

Petrushevskaya MG (1971b) Radiolarian in the plankton and recent sediments from the Indian Ocean and Antarctic. In: Funnel BM, Riedel WR (eds) The Micropaleontology of the Oceans. Cambridge University Press, London, pp 309–317.

Petrushevskaya MG (1974) Cenozoic radiolarians of the Antarctic, leg 29 DSDP. In: Kenneth JP, Houtz RE, et al. (eds) Initial Report of the Deep Sea Drilling Project 29. U.S. Government Printing Office, Washington, pp 541–676.

Petrushevskaya MG, Kozlova GE (1972) Radiolaria: Leg 14. In: Hayes DE, Pimm AC, et al. (eds) Initial Reports of the Deep Sea Drilling Project 14. U.S. Government Printing Office, Washington, pp 495–648.

Plafker G, Keller G, Barron JA, Blueford JR (1985) Paleontologic data on the age of the Orca Group, Alaska. U.S. Geological Survey Open File Report 85-429:1–23.

Renz N (1973) The distribution and ecology of Radiolaria in the central Pacific—plankton and surface sediments. Thesis, University of California, Los Angeles, 251 pp.

Riedel WR, Sanfilippo A (1971) Cenozoic Radiolaria from the Western Tropical Pacific, leg 7. In: Winterer EL et al. (eds) Initial Reports of the Deep Sea Drilling Project 7. U.S. Government Printing Office, Washington, pp 1673–1693.

Riedel WR, Sanfilippo A (1974) Radiolaria from the southern Indian Ocean, DSDP leg 26. In: Davies TA, Luyendyk BP, et al. (eds) Initial Reports of the Deep Sea Drilling Project. U.S. Government Printing Office, Washington, pp 797–865.

Riedel WR, Sanfilippo A (1978) Stratigraphy and evolution of tropical Cenozoic radiolarians. Micropaleontology 24(1):61–96.

Runeva NP (1975) Kompleksy tretichnych radiolarii poluostrova Kamchatka. Trudy VSEGEI, nova seriya, 226:87–92.

Sanfilippo A, Riedel WR (1973) Cenozoic Radiolaria (exclusive of theoperids, artostrobiids, and amphipyndacids) from the Gulf of Mexico, Deep Sea Drilling Project Leg 10. In: Worzel JL, Bryant W, et al. (eds) Initial Reports of the Deep Sea Drilling Project 10. U.S. Government Printing Office, Washington, pp 475–610.

Sanfilippo A, Riedel WR (1974) Radiolarian occurrences in the Caribbean region. Publication de la VII Conference Geologique des Caraibes, Cayenne, pp 145–168.

Seidov DG (1986) Numerical modeling of the Ocean circulation and paleocirculation. In: Hsü KJ (ed) Mesozoic and Cenozoic Oceans. American Geophysical Union, Washington Vol 15, pp 11–26.

Smith AG, Hurley AM, Briden JC (1981) Phanerozoic Paleocontinental World Maps. Cambridge University Press, Cambridge, U.K., 102 pp.

Stewart R (1946) Geology of Reef Ridge, Coalinga District, California. U.S. Geological Survey professional paper 205 C:81–115.

Tolmazin D (1985) Elements of Dynamic Oceanography. Allen and Unwin, London, 181 pp.

Vinogradov AP (1967–69) Atlas of the Lithological-Paleographical maps of the USSR, Ministry Geology USSR and Academy Sciences, USSR, Moscow, 255 plates.

Von Arx WS (1952) A laboratory study of the wind-driven ocean circulation. Tellus 4:311–318.

Weaver F (1976) Antarctica Radiolaria from the SE Pacific Basin. In: Hollister CD, Campbell C, et al. (eds) Initial Reports of the Deep Sea Drilling Project 35. U.S. Government Printing Office, Washington, pp 569–603.

3
Radiolarians, Radiolarites, and Mesozoic Paleogeography of the Circum-Mediterranean Alpine Belts

Patrick De Wever

Overview

Radiolarites are of great importance as bathymetric indicators for paleogeographic reconstructions and geodynamic models. Only recently have we been able to date radiolarites and many age dates are scattered through the specialized geologic literature. The inventory of available data from the Mesozoic of the circum-Mediterranean Alpine fold belts reveals two general periods of radiolarite sedimentation, both associated with sedimentary and ophiolitic sequences: one period occurred during Triassic and/or Liassic time, the other during Dogger and/or Malm time. The Triassic-Liassic sections are allochtonous. Radiolarites associated with extrusive rocks are considered to be the sedimentary cover of an oceanic crust, either of an open ocean, back-arc basin or small ocean basin. Radiolarites intercalated with other sedimentary rocks belong to nappes with unknown basement rocks, probably of thinned continental crust. Radiolarites deposited during Dogger and Malm time are of three types and characterize three environments whose common attribute is a basal diachronism and a synchronous top. Radiolarites intercalated with sedimentary sequences occurred in basins that received deposits since Triassic time (i.e., Lagonegro, Pindos-Olonos zone) or in regions newly invaded by water masses (i.e., Austro-Alpine zone). Detailed local studies suggest deposition on faulted blocks of a rifting margin. Although the base of radiolarites is not synchronous in different units, it nevertheless starts around the Dogger and the maximum development was generally during Oxfordian time. The base of radiolarites associated with ophiolites is everywhere dated as Malm but the bases are not exactly synchronous; data are still too few to analyze the details. The sudden disappearance of all the radiolarites in the Uppermost Jurassic may be attributed to a drastic change in circulation from gyres producing upwelling in the Tethyan basin, to latitudinal, east-to-west circulation through Central America which broke down the upwelling regime in much of the tethyan area.

Introduction

Radiolarites (rhythmically centimeter-bedded cherts alternating with millimeter-bedded shale) are of great importance for paleogeographic reconstructions and geodynamic models because of their bathymetric significance and their frequent association with ophiolites. They are a very good marker for the study of margin formation, subsidence, and evolution and are important for dating the oceanic crust.

However, direct dating of radiolarite was not possible until recently, since these sedimentary rocks do not bear the fossils commonly used for stratigraphy (e.g., foraminifers, nannofossils). Locally they had been indirectly dated by reworked microfauna included in rare beds of microbreccia (i.e., in the Pindos-Olonos zone, Greece—Dercourt et al., 1973; Thiébault et al., 1981; De Wever and Thiébault, 1981). Moreover, the petrographic composition of these rocks did not permit use of the usual etching techniques for extracting fossils. In the 1970s, this difficulty was overcome by the discovery of Dumitricà (1970). His etching technique has been described and improved by several authors (Pessagno and Newport, 1972; De Wever et al., 1979a; De Wever, 1982). Radiolarites of circum-Alpine orogenic belts have since been analyzed paleontologically (Baumgartner et al., 1980; Baumgartner, 1984; Kocher, 1981; Thiébault et al., 1981; De Wever and

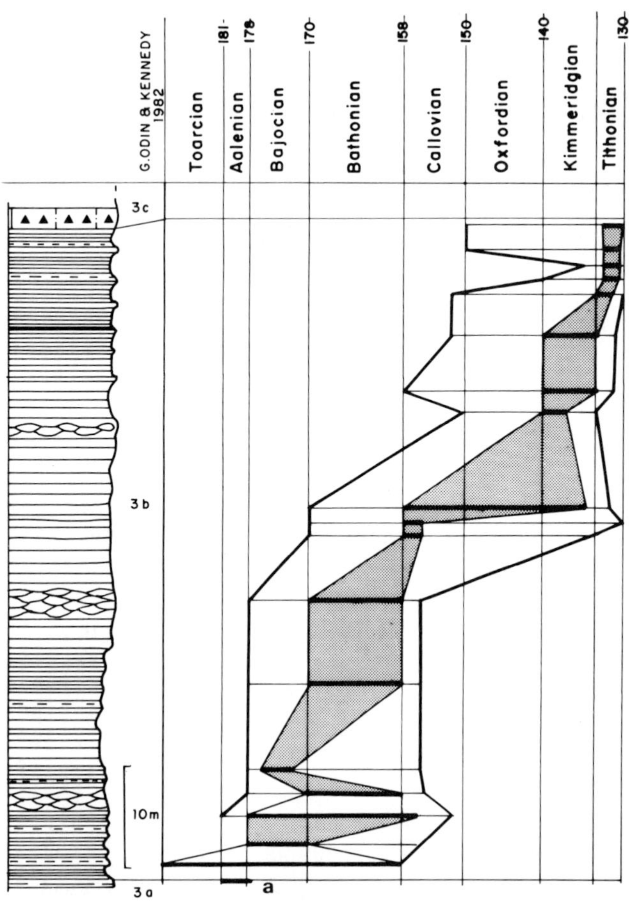

FIGURE 3.1. Synthesis of radiolarites sensu stricto (level 3b of Radiolarite Formation in Fleury, 1980) age dates. From left to right: synthetic lithologic column, radiolarite formation. 3a: Pélites de Kasteli member; 3b: radiolarite member (= radiolarites s.s.); 3c: Calcaires à Calpionelles member. Age assignments: thick horizontal line represents the well-documented age range; thin horizontal line represents the less well documented age range. The narrow age range for each sample is joined to its neighbor (thin line, stippled area); the same is done for wide age ranges (thick line, blank area). a: Age assignments of Dercourt et al. (1973) on the basis of *Lucasella cayeuxi* (Lucas). Time scale is from Odin and Kennedy (1982). Sedimentation rates (of lithified sediment, equivalent to the sedimentation rate of the silica component – SiO_2 representing 95–99% of the total rock). Lower part: 35 m/20 Ma = 1.7 m/Ma. Middle part: 10 m/18 Ma = 0.5 m/Ma. Upper part: 18 m/8 Ma = 2.2 m/Ma. Average: 65 m/46 Ma = 1.4 m/Ma.

Caby, 1981; De Wever and Thiébault, 1981; De Wever, 1982, 1984; De Wever and Origlia-Devos, 1982a,b; Dumitricà and Mello, 1982; Origilia-Devos, 1983; Baumgartner, 1984; El Kadiri, 1984; De Wever and Miconnet, 1985; De Wever et al., 1985, 1986b; De Wever and Cordey, 1986.) As this list of references reveals, age information is recent and dispersed throughout the specialized scientific literature.

Age-Dating Capabilities with Radiolarians and Determination of Sedimentation Rates of Radiolarite

During the 1980s, our knowledge of Mesozoic radiolarian stratigraphy increased a great deal, especially with respect to low-latitude belts. Radiolarian stratigraphy now permits precise dating of otherwise unfossiliferous siliceous rocks.

As an example, in the Pindos-Olonos zone (Greece) (Fig. 3.1), it is possible to record 18 assemblages in a sequence about 60 m thick (Fig. 3.1) representing Bajocian to Tithonian. Age determinations in different sections allow calculation of the average sedimentation rate (for lithified sediment) for the entire formation (1.4 m/Ma). The lower part of the Radiolarite Formation shows a relatively high sedimentation rate of 3 m/Ma; the upper part reveals a lower rate. Therefore, using radiolarians, variation in the sedimentation rate is detectable in a sequence. Calculation of sedimentation rates assumes that the sedimentary environment remained closed to the input or release of silica during diagenetic processes.

The sedimentation rates determined with radiolarians for the Pindos-Olonos zone are compatible with rates proposed by other authors for other Mesozoic bedded chert sections:

2 m/Ma for Franciscan chert (S. Karl, unpublished data in Hein and Karl, 1983)

4 m/Ma as average for different Tethyan radiolarites according to Bernoulli (in Garrision and Fischer, 1969)

0.7 to 1.0 m/Ma for East Alpine radiolarite (Garrison and Fischer, 1969)
1.0 to 5.3 m/Ma (McBride and Thompson, 1970) for partially lithified chert (novaculites from Texas)
3 to 9 m/Ma for East Alpine radiolarite (Schlager and Schlager, 1973)
27 to 34 m/Ma for radiolarite in Japan (Iijima et al., 1978)
2.8 m/Ma for bedded cherts from Inuyama region, Japan (Matsuda et al., 1980)
3.3 to 6.2 m/Ma for radiolarite in Northern Italy (Kocher, 1981)
7.5 m/Ma in the Chichibu Chert Formation in Japan (Matsumoto and Iijima, 1983) for the chert beds and 40 m/Ma for the shale partings

These rates are approximate, because they were not determined with the same chronostratigraphic scale. Different time length variations used for the Jurassic are as follows: Van Hinte (1976), 20 Ma; Odin and Kennedy (1982), 28 Ma; Harland et al. (1982), 25 Ma; Kent and Gradstein (1985), 25 Ma; Westermann (1984), 24 Ma. The maximum difference represents 40%, and the proposed sedimentation rates have to be considered to be within a 40% range.

More recent (unconsolidated) sediments have high rates of sedimentation; for example, siliceous sediments (carbonate-free) found in Antarctic zones show rates of up to 4.8 m/Ma (Brewster, 1980). Some quoted sedimentation rates are up to 40 to 50 m/Ma in the east Pacific and even 90 to 235 m/Ma in diatomaceous sediment from the Miocene of Japan (Iijima et al., 1982; Matsumoto and Iijima, 1983), but these correspond to the bulk sedimentation rate (all phases) and not to the sedimentation rate of only the silica phase.

According to studies of the variation of porosity and density in diatomaceous rocks (Isaacs et al., 1983), a compaction of 60% was proposed.

Studying a silicified wood within radiolarite of the Adoyama Chert Formation (Honshu, Japan), Iijima et al. (this volume) also proposed a compaction of 60%. If we apply this factor to the radiolarites of the Pindos-Olonos zone, the uncompacted rate would range from 3.5 to 7.5 m/Ma (average 6.2 m/Ma). This rate is the same as the average rate proposed for the Adoyama Chert Formation (6 m/Ma; Iijima et al., this volume) and to 7.5 m/Ma proposed by Matsumoto and Iijima (1983) for the Chichibu Chert Formation.

Because many factors influence the original sedimentation rate (turbidites, redeposition, input of silica, and dissolution of the silica tests), Barrett (1981) found it more meaningful to report the number of events (turbidites) per time unit. If we accept that a bed represents one turbidite, this calculation reveals an average of 47 events per Ma (thickness of the series is 60 m, average bed thickness is 4 cm, duration is 32 Ma) for the Pindos-Olonos series (Greece) in contrast to the 100 to 400 events calculated by Barrett for Ligurian cherts of northern Italy. One cycle in the Greek Radiolarite Formation represents around 21,000 years. This cyclicity corresponds well to the period of precession in the Milankovitch theory of climate variation (De Wever, 1987). Alternatively, it has been proposed that the bedding results from diagenesis. This is suggested by the observation that diatomites that are massive when they are still opal-A acquired a pronounced rhythmic bedding when they underwent diagenesis (transformed in opal-CT) (Pisciotto and Garrison, 1981, p 111). The origin of the bedding is not always clear, and both primary and diagenetic bedding no doubt occur. The bedding in the rhythmically bedded Mesozoic cherts of Austria are clearly primary (Vescei et al., this volume). One would not expect the bedding in diatomaceous and radiolarian-rich deposits to be the same because of the different hydrodynamic properties of the two microfossil groups (Hein and Karl, 1983).

Ages of Tethyan Radiolarite

The inventory of available data from Tethyan radiolarites reveals, in general, two periods of pronounced biosiliceous sedimentation, one during Triassic and/or Liassic time, the other during Dogger and/or Malm time (Fig. 3.2). Both time intervals contain radiolarites associated with sedimentary and ophiolite sequences. I will now present a brief lithologic description and ages of the Alpine belts where radiolarites have been dated.

Triassic-Liassic Radiolarite

Radiolarite Interbedded with Sedimentary Rocks

Pindos-Olonos Zone, Greece

In the Pindos-Olonos zone (one of the external Hellenides nappes, Greece) there exist two series of radiolarites. The best known is of Jurassic age and is several tens of meters thick. The other one, only a few meters thick, belongs to the Drimos Limestone

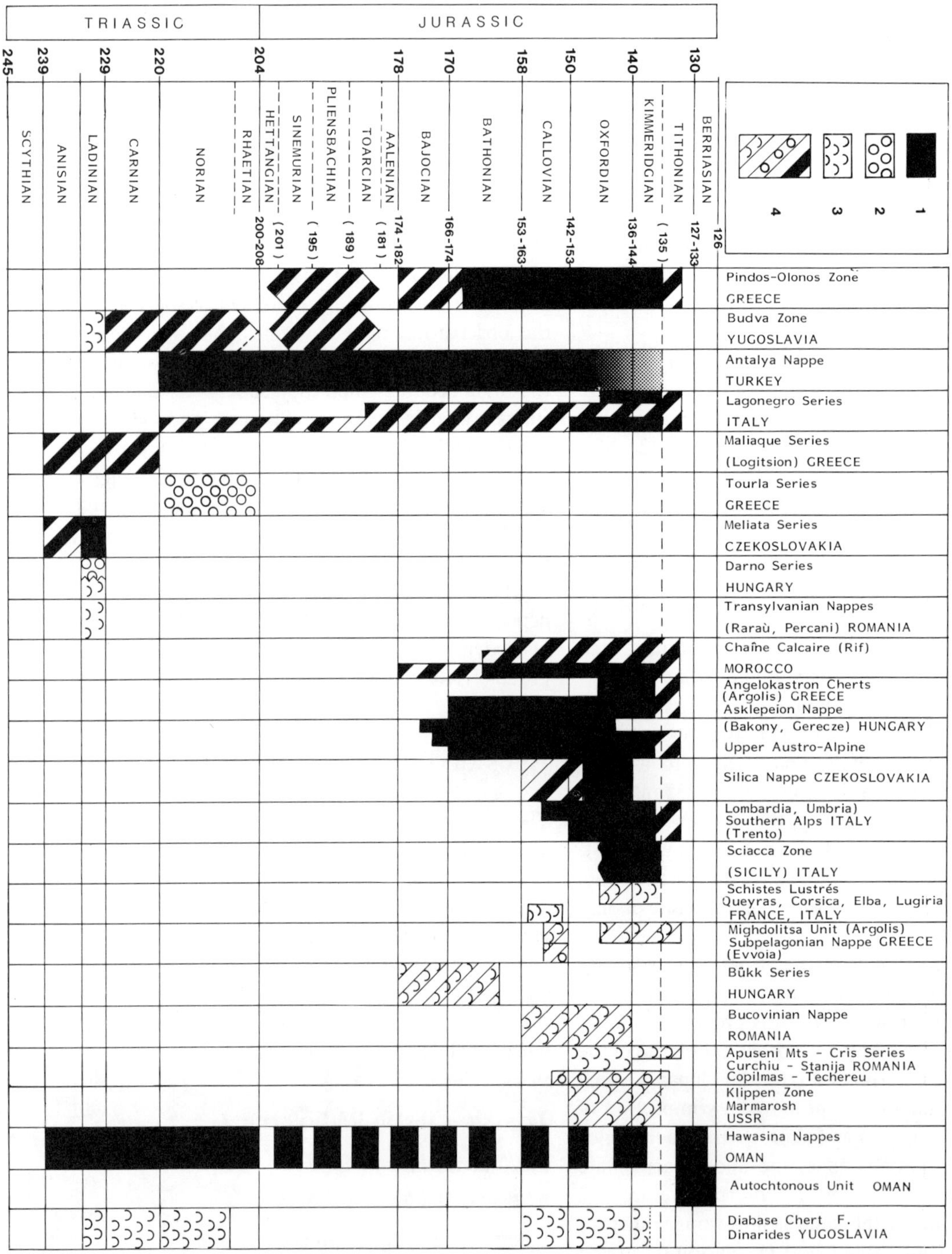

FIGURE 3.2. Summary table of radiolarite ages based on radiolarian data presented in this paper. (1) Paleontological data for radiolarite associated only with sedimentary sequences. (2) Paleontological data for chert from interpillow lava spaces. (3) Paleontological data for radiolarite directly overlying ophiolites. (4) Hatching represents an uncertainty about the age either due to a lack of precision of the paleontological data or due to the structural and/or lithostratigraphic relations.

Formation (Dercourt et al., 1973; Fleury, 1980). It has long been considered to be Triassic but has recently been dated as Liassic (De Wever and Origlia-Devos, 1982b)

Budva Zone (Yugoslavia)

This zone in Yugoslavia is equivalent to the Pindos-Olonos zone in Greece and contains the same two levels of radiolarites. The Budva radiolarite-bearing sequences have recently been dated as Liassic (Obradović and Goričan, this volume; Goričan, 1987).

Maliac Series (Greece)

This sequence (one of the internal Hellenides nappes, Greece) has an unknown basement (presently subducted under the internal zones). Triassic chert has been dated from different sections. Some are associated with only sedimentary rocks (Logistion series, lower Ladinian-Carnian; Ferrière, 1982). According to Vrielynck (1982), the Maliac series may be considered to have been deposited on the lower part of the Pelagonian margin facing toward the Neo-Tethyan ocean. These data are currently the only paleontological evidence of an oceanic area existing in Triassic time between Eurasia and Africa-Arabia.

Antalya Nappe Series (Turkey)

In the Isparta-çay series, radiolarite-bearing intervals are interbedded with limestone yielding halobians, conodonts (Poisson, 1977), and radiolarians (De Wever et al., 1979; De Wever, 1982) of Late Triassic (Norian) age.

Lagonegro Series (Italy)

In southern Italy (Lucany, Lagonegro region), radiolarite overlies a cherty limestone sequence (Triassic) which is in turn overlain by the clastic Galestri Formation (Neocomian). The first deposition of these cherts is diachronous, from Triassic to the south (Lagonegro unit) to Upper Jurassic to the north (San Fele unit) (De Wever and Miconnet, 1985) (Fig. 3.3).

Meliata Unit (Czechoslovakia)

The Meliata unit (Czechoslovakia) belongs to the Slovac edifice. It is an extension of the Bukk unit (Hungary) and represents the termination of the Vardar zone (Dercourt et al., 1985a,b). It has been dated as Triassic using radiolarians (Middle Triassic: latest Anisian or early Ladinian; Dumitricà and Mello, 1982) and conodonts (Kozur and Mock, 1973).

Hawasina Nappes (Oman)

In the Hamrat Duru Group, the Wahra Formation and the Lower Zulla and Upper Sid'r Formations all contain thick sequences of radiolarian-bearing chert interbedded with clastic sedimentary rocks. Samples from Al Jil cherts (= lower part of the Zulla formation in Glennie et al., 1974) yielded radiolarian fauna denoting late Murghabian and Late Triassic (Carnian-Norian) ages (Bourdillon et al., 1987; De Wever, unpublished data). Additional chert samples from the Matbat Formation (upper part of the Zulla + lower part of Guwayza in Glennie et al., 1974) indicate an Early Jurassic age (Bourdillon et al. 1987; De Wever, unpublished data). Several dates mentioned by Blome et al. (1983) are well in accordance with those given here. However, the precise lithostratigraphic position of their samples was not given, so it is impossible to ascertain whether they actually correspond to the formations listed here.

Radiolarites Overlying Ophiolites (Triassic-Liassic)

Darno Series (Hungary)

In the Lower Austro-Alpine unit, chert associated with ophiolite pillow basalt has been dated as Middle Triassic (De Wever, 1984).

Raraù Series (Romania)

In transylvanian nappes (Raraù and Persani Mountains, eastern Carpathian Mountains, Romania), radiolarite has been dated as Middle Triassic (Ladinian; Dumitricà and Mello, 1982).

Diabase-Chert Unit (Yugoslavia)

In the Diabase-Chert Formation from Serbia (Yugoslavia), radiolarian chert (near pillow basalts) has been dated as Late Triassic (Carnian-Norian) (Obradović and Goričan, this volume; Goričan, 1986).

Porphyrite-Chert Unit (Yugoslavia)

In the Porphyrite-Chert Unit from Serbia (Yugoslavia), radiolarian cherts near volcanic rocks have been dated as Middle Triassic (Anisian-Carnian) (Obradović and Goričan, this volume; Goričan, 1986).

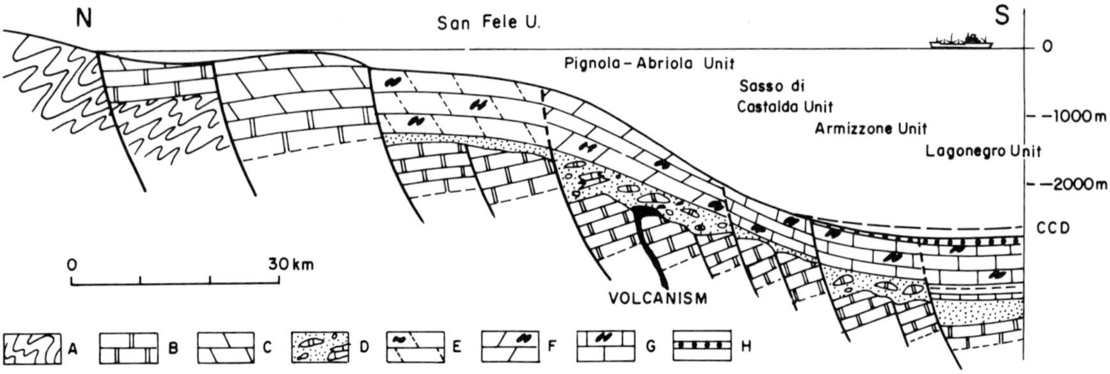

FIGURE 3.3. Reconstruction of Lagonegro (southern Italy) for Norian time (from De Wever and Miconnet, 1985). (A) Hypothetical rock basement (partly provided detritus to the basin). (B) Basement cover and Scythian platform. (C) Primary dolomite (partly provided detritus to the basin). (D) Monte Facito Formation (probable thrusting level). (E–G) Calcaires à Silex Formation (cherty limestone). E–San Fele Unit, massive dolomite; F–Pignola-Abriola Unit, bedded dolomite and dolomitic limestone; G–Sasso di Castalda, Armizzone and Lagonegro Units. (H) Radiolarite belonging to Armizzone and Lagonegro (and Sasso di Castalda ?) Units.

Maliac Series (Greece)

Triassic cherts associated with basalts from the Maliac series (internal Hellenides nappes, Greece) have unknown basement rocks (presently subducted under the inner zones). Chert occurs between pillow lavas and has been dated as Late Triassic (Tourla series, Norian; De Wever, 1982; Ferrière, 1982).

Mid-Upper Jurassic Radiolarites

During this time period, radiolarite is common. As in the previous examples, it is associated with sedimentary and volcanic sequences.

Radiolarites Within Sedimentary Sequences

"Chaînes Calcaires" (Rif, Morocco)

The base of this formation is diachronous based on radiolarian dates, from Bajocian or Bathonian to Callovian or Oxfordian within different structural units (Fig. 3.4), the oldest being to the south (external zones) (De Wever et al., 1985). The top of this chert formation is synchronous in all structural units, Kimmeridgian or early Tithonian.

Southern Alps (Lombardy, Tuscany, and Umbria, Italy)

Radiolarite sections are well known in northern Italy, where their bases are diachronous. Bases are older in basinal deposits (from Bajocian or Bathonian, Lombardy Basin, Umbria) and younger on plateau deposits (mid-Oxfordian, Trento plateau) (Baumgartner et al., 1980; Kocher, 1981; Baumgartner, 1984; Conti et al., 1985; Conti, 1986). Radiolarites are synchronously capped by Maiolica limestone (Govi, 1965; Luthi, 1973).

Lagonegro Series (Lucany, Italy)

As noted above (Fig. 3.3), these series have a diachronous base (from Triassic to Oxfordian), but their top is synchronous (Before Neocomian time) (De Wever and Miconnet, 1985).

Sciacca Series (Sicily, Italy)

The base of this series is unknown, because it is cut by an overthrust fault, but the top has been precisely dated with ammonites, calpionellids, nannofossils, and radiolarians as the exact boundary between Kimmeridgian and Tithonian (De Wever et al., 1986b).

Pindos-Olonos Series (Greece)

Radiolarite has been studied in several sections from northern and southern Greece (Fig. 3.1). The bases of the sections are dated as Bajocian or Bathonian (Thiébault et al., 1981; De Wever and Thiébault, 1981; De Wever and Cordey, 1986), and the top is Tithonian (De Wever and Origlia-Devos, 1982a; De Wever and Cordey, 1986) except in southernmost

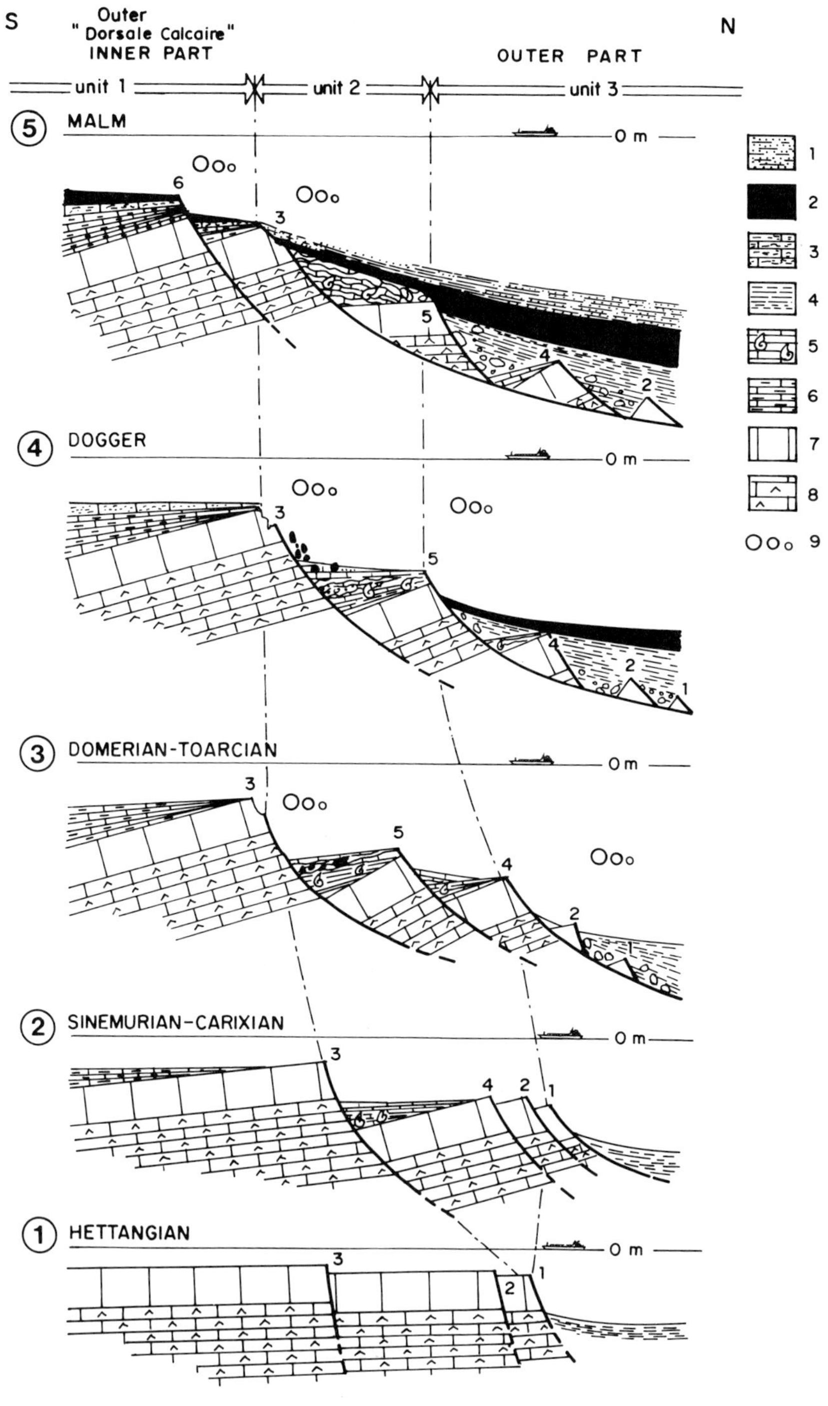

FIGURE 3.4. Paleogeographic evolution of the Chrafate klippes and the external "Dorsale Calcaire" (Rif, Morocco) zone during Jurassic time (from De Wever et al., 1985). (1) Turbidite; (2) radiolarites; (3) "calcaires à filaments"; (4) black shale; (5) Ammonitico Rosso; (6) cherty limestone; (7) massive limestone; (8) dolomitic massive limestone; (9) horizontal grading.

Peloponnesus, where a few outer units contain radiolarites as young as the Late Cretaceous. However, during Tithonian time, some thick, coarse clastic intervals are interbedded (Thiébault et al., 1981; De Wever and Thiébault, 1981).

Budva Series (Yugoslavia)

A series equivalent to the Pindos-Olonos series exists in Montenegro (Yugoslavia), where it has been dated as Middle Jurassic to Late Jurassic (Obradović and Goričan, this volume; Goričan, 1987).

Pelagonian Series (Internal Hellenides, Strimbes, Greece)

Radiolarites from Pelagonian series (Othrys, continental Greece) have been dated as Kimmeridgian-Tithonian with radiolarians (det. Origlia-Devos in Ferrière, 1982, p 930)

Pelagonian Series (Asklepeion Nappe, Angelokastron Cherts, Internal Hellenides, Greece)

Radiolarites have been dated in the Peloponnesus. Their first occurrence is Bathonian (Asklepeion Nappe – Baumgartner, 1984) or late Oxfordian (Angelokastron Chert and Ayios Nikolaos Formation – Baumgartner, 1984; they correspond to the Epidaure zone and Trapezona zone of Vrielynck, 1978, 1980, 1982 at this time period).

Pelagonian Series (Evvoia, Greece)

Radiolarite from northern Evvoia has been dated with radiolarians as Oxfordian or Kimmeridgian, except the latest Kimmeridgian (Baumgartner, 1984; Ferrière et al., 1986). This radiolarite was dated as Malm or Neocomian (Baumgartner and Bernoulli, 1976), but more recent information permits us to restrict the age as mentioned above.

Upper Austro-Alpine Series in Hungary (Bakony and Gerecze Nappes)

In the upper Austro-Alpine nappes of Hungary (Bakony and Gerecze Hills), the base of the radiolarite section is late Bajocian or Bathonian (in Közöskùt, Lökùt, Bakonycsernye sections – Fülöp, 1969, 1976; De Wever, unpublished datà) but elsewhere is Bathonian (Gyenespusztà – Galácz, 1970, 1976, 1980; De Wever, unpublished data). The radiolarite-bearing formation is overlain in some places by a coarse clastic section of Oxfordian age (Lökùt and Tata Hills), but in other places radiolarite persists to the Kimmeridgian (Cszernye Hills) or even to the Kimmeridgian-Tithonian boundary (Margit Hill – De Wever, unpublished data).

Upper Austro-Alpine Series Czechoslovakia (Silica Nappe)

Radiolarite (equivalent to the series of the Slovak Karst) is dated as Callovian or Oxfordian (Dumitricà and Mello, 1982).

Hawasina Nappes (Oman)

In the Hamrat Duru Group, the Wahra Formation, the Lower Zulla and Upper Sid'r Formations all contain thick sequences of radiolarian-bearing chert interbedded with clastic sedimentary rocks (turbidites). Radiolarian fauna obtained from the Sid'r chert indicate a Late Jurassic to Neocomian age (Bourdillon et al., 1987; De Wever, unpublished data). Several dates have been mentioned by Blome et al. (1983) and by Baumgartner (1984,) for the Al Aridh Formation. Their dates are in accordance with those given here. However, they did not provide precise lithostratigraphic positions for their samples, so it is impossible to ascertain whether they correspond to the stratigraphic levels analyzed here.

Autochtonous Unit (Oman)

Overlying the Sahtan Group (shallow-water Jurassic limestone) is the Kahmah Group, which yielded radiolarians in chert at the base of the section interbedded with limestone (with Calpionellids). The chert was dated as late Tithonian or early Berriasian (Bourdillon et al., 1987; De Wever, unpublished data).

Radiolarite Overlying Oceanic Crust

Good outcrops of ophiolite-hosted radiolarite are rare because of metamorphism. Nevertheless, some significant data already exist.

Series of the Ligurian Ophiolitic Nappe (Queyras and Corsica, France; Liguria and Elba, Italy)

The metaradiolarite, which overlies the ophiolite and ophiolitic breccia of the Chabrière series (base of the "Schistes Lustrés"), has been dated in sections from France as late Oxfordian or early Kimmeridgian (De Wever and Caby, 1981) and on rocks from Italy (Traversiera) as middle-upper Callovian (De Wever et al., 1987a).

Nonmetamorphosed radiolarite from Elba overlying pillow lava (Monte Campannelo-Nisporta, Volterraio) or ophicalcite (San Felo Namia) has been dated as middle Oxfordian or early Kimmeridgian (Baumgartner, 1984; Conti and Marcucci, 1986). This age is the same as that determined in other sections (Murlo, Southern Toscany–Conti and Marcucci, 1986; San Colombano and km 59, Corsica–De Wever et al., 1987b). The shale between ophicalcite and radiolarite has been dated at Rocchetta di Vara (eastern Ligurides) (Baumgartner, 1984) as middle Callovian or basal Oxfordian, and the basal part of the chert sequences is dated as late Oxfordian or early Kimmeridgian (Conti et al., 1985; Conti and Marcucci, 1986). In the section at Il Conventino (Monti Rignosi, eastern Tuscany), the radiolarian assemblage gives a late Callovian or early Kimmeridgian age (Conti and Marcucci, 1986). The onset of deposition of the Monte Alpe Chert gives an upper age limit for basalt extrusion in each section. The age of sedimentation overlying Ligurian ocean crust was therefore not synchronous.

Subpelagonian Nappe Series (Migdhalitsa Unit, Argolis, Greece)

Radiolarite from the Migdhalitsa unit (Baumgartner, 1984; Fanary Wildflysch of Vrielynck, 1982) overlying pillow basalt of the Ophiolite Suite (Akros Hill) is dated near its contact with basalt as middle Callovian or early Oxfordian (Baumgartner, 1984)

Subpelagonian Nappe Series (Evvoia, Greece)

Chert samples from interpillow space of the ophiolite in northern Evvoia are dated as Oxfordian (Baumgartner, 1984; Ferrière et al., 1986).

Subpelagonian Nappe Series (Thessaly, Greece)

Oxfordian and/or Tithonian age was obtained for radiolarite overlying ophiolite at Theopetra (De Wever in Ardaens, 1978; Ferrière, 1982), but the rocks dated were not from the base of the section.

Diabase-Chert Formation, Serbian Units (Internal Dinarides, Yugoslavia)

In the Diabase-Chert Formation from western Serbia (Yugoslavia), Upper Jurassic chert (Callovian and/or early Kimmeridgian) containing radiolarians has been identified (Obradović and Goričan, this volume; Goričan, 1986, p 58).

Bukk Ophiolitic Nappe (Bukk Massif, Hungary)

The Bukk may be considered as a Neo-Tethyan sphenochasm between Apulia and Europe (Dercourt et al., 1984a,b, 1985a,b). A Bajocian and/or Bathonian age was obtained (Kozur, 1984) for the upper part of the radiolarite occurring in the wildflysch overlying ophiolites in the Bukk Massif (northeast Hungary) (Geyssant and Lepvrier, 1984).

Bucovinian Nappe (Romania)

Pojorità radiolarite (crystalline Mesozoic zone of southeastern Carpathian Mountains) has long been considered to be of Middle Triassic age (Bancilà and Papiu, 1953) but is now known to be of Callovian and/or Oxfordian age (Mutihac, 1968; Cioflica et al., 1981; Dumitricà, unpublished data), as are most radiolarites of this zone.

Metalliferous Mountains (Southern Apuseni Mountains, Romania)

The Droçea-Cris Unit (of the Cris Nappe) contains radiolarite dated as Oxfordian (Dumitricà in Bleahu et al., 1981) overlying basalt. The Curechiu-Stànija unit has radiolarite dated as Oxfordian-Tithonian (Dumitricà in Bleahu et al., 1981) overlying ophiolite and tuffs.

The Càpîlnas-Techereu Nappe (well developed in Metalliferous Mountains) has radiolarite dated as late Callovian to late Malm (Dumitricà, unpublished data) interlayered with volcanic rocks, both of which overlie ophiolite with a radiometric age of 180 Ma. This ophiolite volcanism was thus probably active before the Jurassic until the Callovian.

Marmaros Unit (Klippen Zone, USSR)

Radiolarite from the Marmaros unit associated with ophiolites of the northeast Eastern Carpathians has been dated as Oxfordian and Kimmeridgian (Tikhomirova, 1983, 1984).

Summary

All Triassic and Liassic radiolarite sections associated with sedimentary rocks are parts of allochtonous sequences. These radiolarites belong to nappes with unknown basement rocks, probably thinned continental crust (Dercourt et al., 1985a,b; Ricou et al., 1985)(Fig. 3.5).

Radiolarite interbedded with sedimentary sequences exists in basins that have received deposits since Triassic time (Lagonegro, Pindos-Olonos

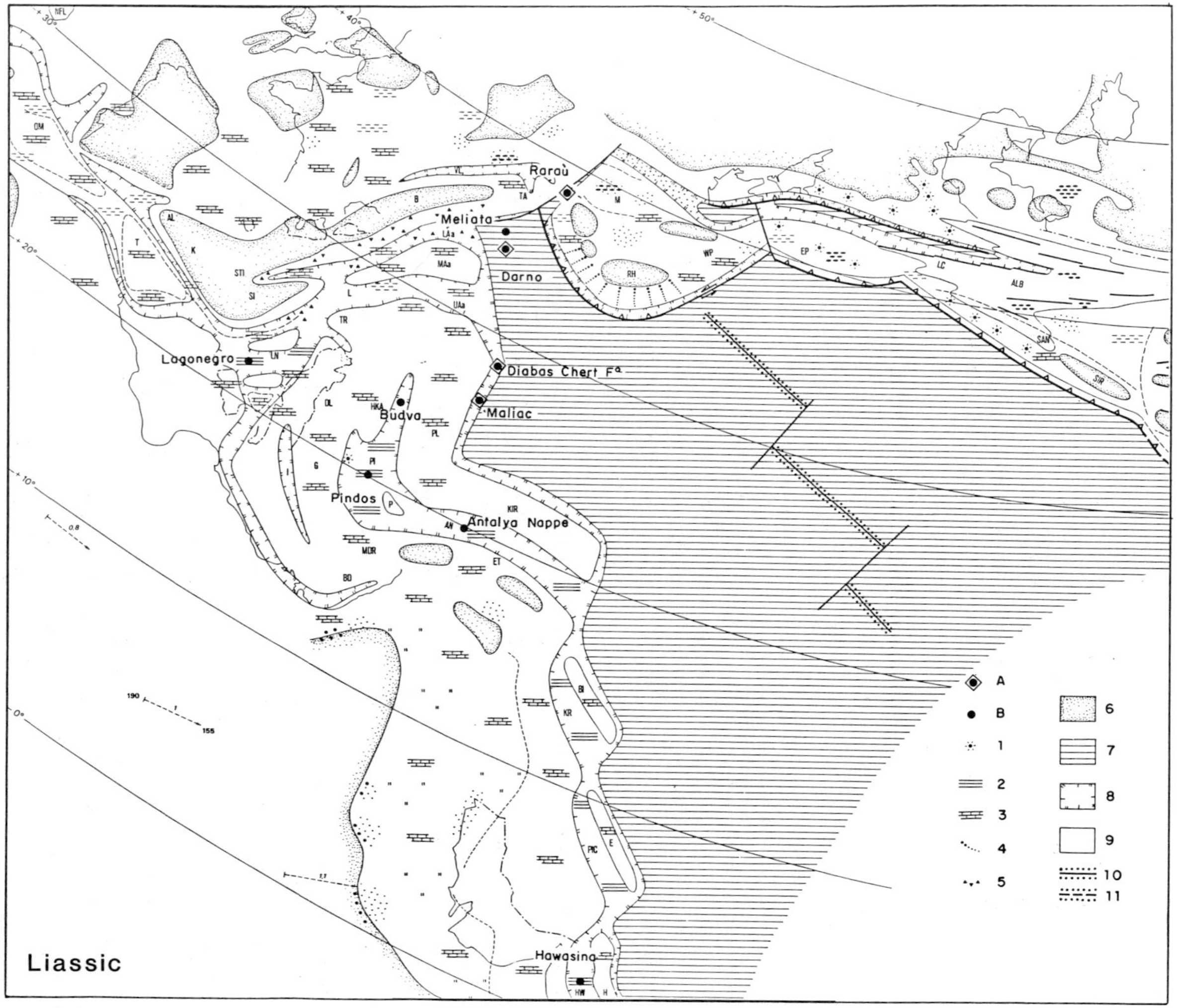

FIGURE 3.5. Location of radiolarite outcrops that were dated by radiolarian assemblages. Paleogeographic map from Dercourt et al. (1985a) map No. 1 (Pliensbachian) is used for this report. (A) Radiolarite associated with sedimentary sequences; (B) radiolarite associated with ophiolites. (1) Volcanism; (2) radiolarite; (3) limestone; (4) flysch; (5) breccia; (6) emergent land (on any type of crust); (7) oceanic crust; (8) thin continental crust (basin); (9) thick continental crust (platform); (10) active spreading ridge; (11) spreading ridge when dying out.

zone, . . .) or in regions newly invaded by major transgressive water masses (i.e., Austro-Alpine zone, Inner Maghrebides). Detailed local studies (Rif, Morocco; Lagonegro, southern Italy; southern Alps, northern Italy) suggest deposition on faulted blocks of a rifting margin. Although the base of the radiolarite is not synchronous in different units, it nevertheless generally starts around the Dogger. The maximum development of biosiliceous sedimentation occurred during Oxfordian time. Their disappearance occurs synchronously near the Tithonian (Fig. 3.6).

All allochtonous radiolarites associated with volcanic rocks have been presumed to be the sedimentary cover of oceanic crust, either of an open-ocean basin crust or of a smaller ocean basin.

The base of radiolarite-bearing sequences (excluding mudstone locally known as umbers in Cyprus and Oman) associated with ophiolite is everywhere dated as Malm, but it is not exactly synchronous; data are still too few for a precise analysis.

The above descriptions consider only true radiolarites, as defined in the Introduction. Other radiolarian dates have recently been obtained from chert nodules in circum-Mediterranean Alpine fold belts but are not considered in this paper because radiolarians were extracted from rock types other than radiolarite sensu stricto. Two sets of results are nevertheless mentioned because of their geodynamic implications. They are as follows: radiolarian-bearing rocks associated with the Samail ophiolite, Oman, and the Troodos ophiolite, Cyprus, yielded a radiolarian fauna dated as Cretaceous: Turonian in Cyprus (Blome and Irwin, 1985): Campanian in Oman (Schaaf and Thomas, 1986; De Wever et al., 1988); and Albian-Cenomanian, Cenomanian-Turonian, and Santonian-Campanian in Oman (Beurrier et al., 1987; Bourdillon et al., 1987). Dated rocks are red mudstone or umbers (not radiolarite).

Work is in progress on the radiolarian stratigraphy of other regions: (1) to the east in the Samkhet Karabagh (Lesser Caucasus, Georgia, USSR), where the Jurassic Lesser Caucasus radiolarite sequences have been dated as Callovian (to Barremian?) with the Kimmeridgian missing (Vishnevskaya, 1984), and (2) in the Klippen Belt of the Pienninnic sequences of the Csorstyn and Kysuca regions (Birkenmayer, 1977). With these new data it will be possible to date the distension of this margin.

Diachronism/synchronism of Radiolarites

Diachronism characterized the earliest radiolarite sedimentation, whereas synchronism characterizes the latest radiolarite sedimentation. In early studies of radiolarites, it appeared that the main radiolarite sequences belonged to the Jurassic, confirming the classic hypothesis on the age of these deposits. Succeeding work dated the base and the top of radiolarite sections more precisely and revealed that the bases were not always of the same age. Detailed studies of the Dogger and Malm radiolarites in different regions documented a diachronism for the earliest radiolarite sedimentation and a synchronism for the top of the sections.

The age variations correspond to different tilted blocks of a continental margin. In some places, age variations correspond to a stretching from the ocean toward the craton (Rif, Morocco; Lagonegro, Italy). Such a margin may have corresponded to the development of a transform fault zone (Lemoine, 1985; Dercourt et al., 1984a, b) between the opening Atlantic Ocean and the subducting Neo-Tethys to the south of the Alboran-Kabyly-Calabria block (De Wever et al., 1985). Radiolarite basins in northern Morocco migrated northward during Middle to Late Jurassic time, suggesting sedimentation on tilted fault blocks (Fig. 3.4) with progressive northward formation of listric faults (Lemoine, 1982, 1985; Lemoine et al., 1981; De Wever et al., 1985).

The diachronism of the first radiolarite deposits in Lagonegro (De Wever and Miconnet, 1985) and Argolis (Baumgartner, 1985) represent the same kind of phenomenon—deposition on tilted blocks of an actively forming margin.

In other places, age variations correspond to the difference noted between basins and plateaus or seamounts such as the Lombardian basin and Trento plateau, northern Italy (Baumgartner, 1984) and Bakony Hills, Hungary (Galacz, 1984; Galacz et al., 1985; De Wever, unpublished data). The first radiolarite deposits are not synchronous, but their maximum extent of development is generally during the Oxfordian.

Disappearance of Radiolarites

It is now accepted that most radiolarites were not deposited in wide-open basins or on continental

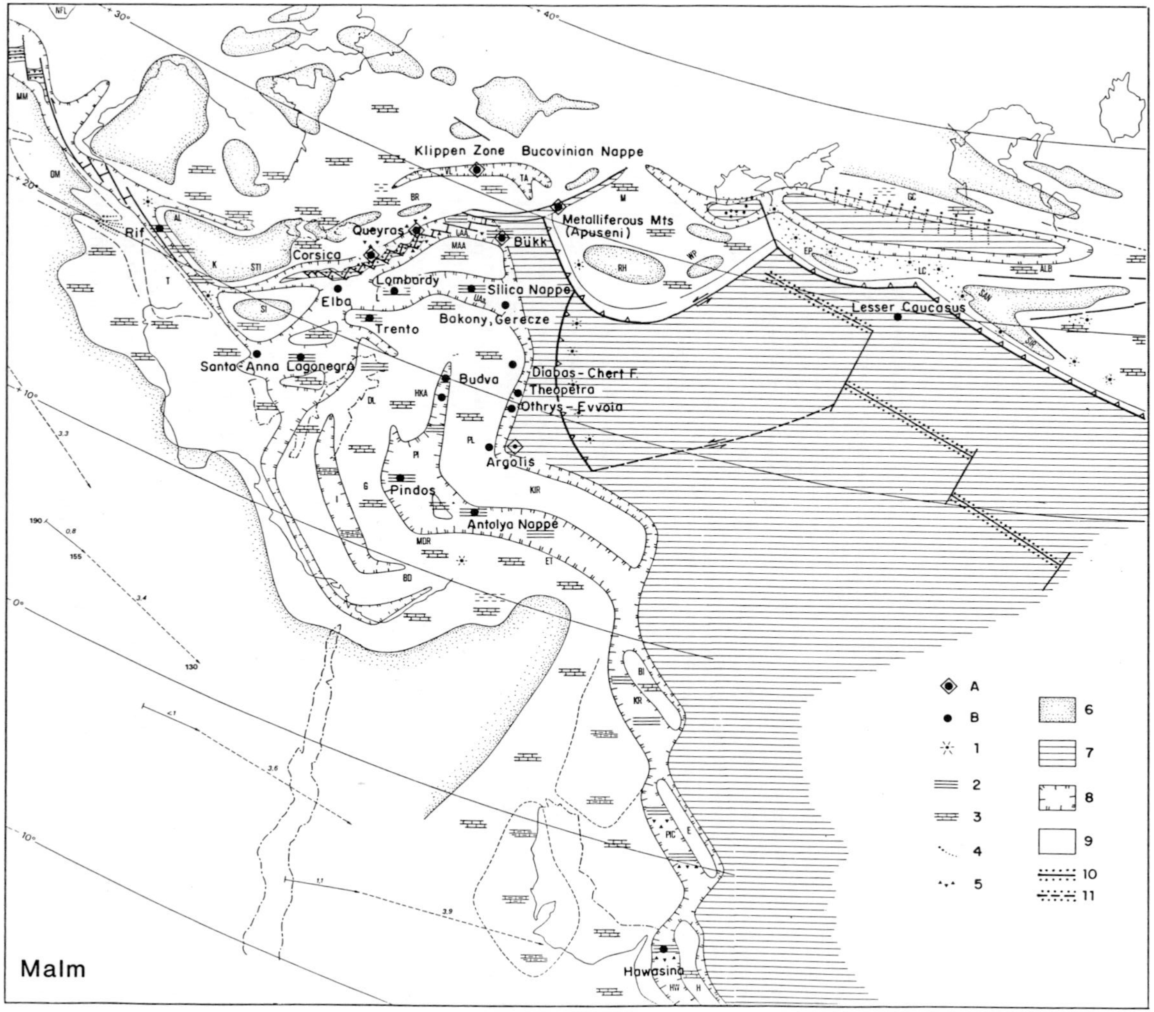

FIGURE 3.6. Location of radiolarite outcrops that were dated by their radiolarian assemblage as Malm. Paleogeographic map from Dercourt et al. (1985a) map No. 2 (Callovian) is used for this report. (A) Radiolarite associated with sedimentary sequences; (B) radiolarite associated with ophiolites. Legend is same as for Figure 3.5.

slopes but in elongate basins (gutters) or small basins with restricted oceanic circulation (Steinberg et al., 1977; Marcoux and Ricou, 1979; Jenkyns and Winterer, 1982; Iijima and Utada, 1983). These relatively confined basins represent gulfs rich in organic matter such as in the modern Gulf of California (Calvert, 1966; Nelson and Goering, 1978; Schrader et al., 1980; De Wever and Thiébault, 1981), Red Sea (Goll, 1969), and marginal seas such as the Sea of Japan (Steinberg et al., 1977), Sea of Okhotsk, and Bering Sea (Jenkyns and Winterer, 1982).

Areas of intense upwelling provide three conditions which enhance the deposition and preservation of radiolarian-rich sediments: (1) high nutrient levels in upwelling regimes promote higher phytoplankton productivity, in turn causing higher rates of radiolarian productivity and deposition of radiolarian-rich deposits. Increased rates of radiolarian deposition decrease the rate of dissolution of siliceous skeletons on burial: (2) increased rates of production of organic matter beneath upwelling zones elevate the CCD, which removes calcareous organisms that dilute the radiolarian-rich sediment: and (3) high organic content prevents the dissolution of silica by coating the tests (Kastner, Scripps Institution of Oceanography, personal communication, 1986).

The general geometry of the Tethys during Jurassic time (an eastwardly open triangle) promoted upwelling on the western side and led to the development of a clockwise gyre (De Wever and Thiébault, 1981; Thiébault et al., 1986). This triangular basin was present from the creation of the neo-Tethys in Triassic time. During the Jurassic, the basin opened and enlarged or spread to the west. This basin evolution fits well with the progressive onset of radiolarite sedimentation from east to west from Triassic to Jurassic time. In addition to the cause cited by Jenkyns and Winterer (1982), the absence of radiolarites in the Atlantic can be explained by the lack of upwelling in this basin and the large distance of this basin from areas of upwelling.

In Tithonian times, radiolarian sedimentation suddenly stopped and was replaced by carbonate sedimentation (e.g., Oberalm in Northern Alps, Maiolica in Southern Alps (Lombardian zone), Biancona in Venetian Alps, Calpionellids Limestone in Elba, Calcare Rupestre in Appennines (Umbria, Marches), Vigla limestone in Greece (Ionian zone), "Calcaires à Calpionelles" in Greece (Pindos-Olonos zone), Lattimusa in Sicily).

Although this change was widespread, its cause need not have been a major one, because only slight biophysical changes in the seas can produce strong modifications in the planktonic life (see El Niño effects; 5- to 20-fold reduction of primary productivity–McGowan, 1984; Barber and Chavez, 1983; Pisias et al., 1986), and because slight changes in production of silica have an exaggerated result in silica sedimentation (Fig. 3.7) (Renz, 1976; De Wever, 1982).

This abrupt change in sedimentation may have occurred worldwide, but outside Tethys (Japan, California, Mexico), radiolarite sedimentation gradually gave way to terrigenous clastic sedimentation.

Modifications in oceanic circulation could have resulted from tectonism associated with the Inner Hellenide zones (De Wever and Thiébault, 1981), but this mountain range did not totally close the Tethyan triangular basin. Moreover, such a closure would have increased the quantity of anoxic sediment, which is not found; rather, anoxic deposits accumulated in isolated basins (Valaisan, Caucasus) since the Malm. In any case, there were shallow seas in this region from Triassic time. This tectonism was probably not extensive enough to modify the oceanographic conditions as far away as Morocco.

Several characteristics we know about the Tethys help to explain the disappearance of radiolarite sedimentation.

The Tethys opened in a westward direction since Triassic time (Aubouin and Tardy, 1980).

East-west exchanges of waters occurred between North and South America from Callovian time, as is attested to by the migration of ammonites (*Reineckeidae* from Pacific to Tethys–Cariou, 1984; Cariou et al., 1985; or *Parasenia* from Tethys to Pacific–Enay and Mangold, 1982). These exchanges involved surface water masses, never deep water masses.

In the Latest Jurassic, a new current formed in the Atlantic, as is evidenced by the arrival of *Pygope* in Greenland (Enay, 1980).

Oceanic conditions existed from Pamir to Mexico (Dercourt et al., 1984a, 1985a).

The exchange of deep water masses began during Tithonian or Berriasian time somewhere between North and South America (Berggren and Hollister, 1974; Thiede, 1979; Kennett, 1982). An oceanic communication has been illustrated by Pindell (1985) between Tethys and Pacific (between the Yucatan block and Mexico) which significantly affected circulation.

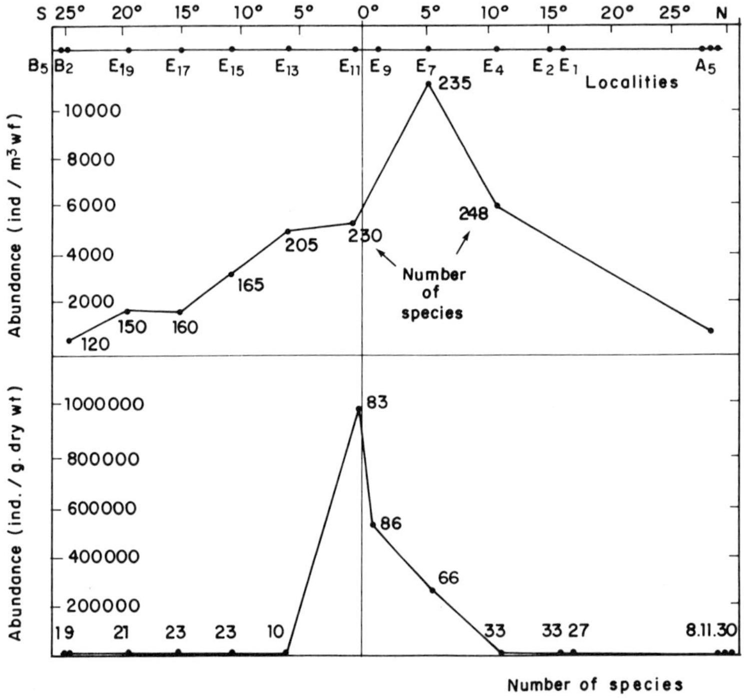

FIGURE 3.7. Comparison between biocenosis and thantocenosis. (A) Radiolarian fauna caught on both sides of the equator in the Pacific Ocean. (B) Radiolarian fauna dredged on the sea floor in the same localities. The shift of the maximum abundance is a result of equatorial gyres.

The sudden disappearance of all the radiolarites in the Latest Jurassic could be explained by a drastic change in circulation due to a latitudinal flow through Central America. Such a new communication drastically changed the current pattern from a gyre in the Tethyan triangular basin (and its associated upwelling) to a latitudinal current and can therefore explain the disappearance of radiolarite sedimentation (De Wever et al., 1986a). This latitudinal current could have, at least locally, flowed from east to west, as suggested by Thiébault et al. (1986).

This hypothesis accords well with the long duration of radiolarite sedimentation of the east Arabian margin of the Tethys Sea (Oman). In Oman, radiolarite sedimentation continued until the Late Cretaceous, because this area remained under the influence of the upwelling conditions (De Wever and Bourdillon, 1987; De Wever et al., 1988; De Wever, unpublished data). When the current pattern changed to latitudinal circulation rather than a gyre, there was a sudden deepening of the basin recorded in the sedimentary rocks of the Autochtonous Unit in Oman. Sedimentation became progressively shallower again during Neocomian time. This initial deepening could have resulted from a "flush" effect at the beginning of the latitudinal current. In Japan, being farther east, where radiolarites were not diluted by clastic sedimentation, this modification in currents had no effect, and radiolarite sedimentation persisted until Neocomian time.

The new circulation pattern affected most of the zones: the oldest rifted platforms (Austro-Alpine, southern Alps) as well as the troughs (Lagonegro, Pindos-Olonos zone, Budva zone). We have therefore to interpret the paleogeography and/or paleobathymetry of the areas where radiolarite deposition persisted after the installation of latitudinal currents (i.e., southernmost Pindos zone, Greece) as confined basins. The regions where radiolarites existed through the Cretaceous were regions sheltered from the new latitudinal current (Oman, Japan).

Acknowledgments. This work has been supported by the CNRS (UA 319, ATP GGO No. 98 1039) and UNESCO IUGS (project IGCP 187). I am indebted to A. Kemp (University of Southampton, U.K.) and J. Tauxe for their help with the English and to R. Garrison (University of California at Santa Cruz), J.

Dercourt (University of Paris), J. Obradović (University of Belgrade and J.R. Hein (USGS, Menlo Park, CA) for reading the manuscript.

References

Ardaens E (1978) Géologie de la chaîne du Vardoussia, comparaison avec le Massif du Koziakas (Grèce continentale). Thèse 3è cycle, Lille, France, 234 pp.

Aubouin J, Tardy M (1980) L'Amérique alpine: Le domaine Caraïbe et ses liaisons avec les cordillères nord et sud américaines: Introduction. In: Géologie des Chaînes Alpines Issues de la Téthys, Colloque C5, 26ème Congrès Géologique International, Paris, pp 14–17.

Bancilà I, Papiu VC (1953) Jaspurile triasice de la Pojorita. Academia Republicii Popular Romine, Buletin Stiinific Sectia de Stiinte Biologice, Agronomice, Geologice si Geografice, Bucharest, 54:665–694.

Barber RT, Chavez FP (1983) Biological consequences of El Niño. Science 222:1203-1210.

Barrett TJ (1981) Stratigraphy and sedimentology of Jurassic bedded chert overlying ophiolites in the north Appennines, Italy. Sedimentology 29:353–373.

Bassoullet JP, Fourcade E, Peybernes B (1985) Paléobiogéographie des "grands" Foraminifères benthiques des marges néotéthysiennes au Jurassique et au Crétacé inférieur. Bulletin de la Société Géologique de France, Paris, 8:699–713.

Baumgartner PO (1984) A middle Jurassic–early Cretaceous low-lattitude radiolarian zonation based on unitary associations and age of Tethyan radiolarites. Eclogae Geologicae Helveticae, Bâle, 77(3):729–837.

Baumgartner PO (1985) Jurassic Sedimentary Evolution and Nappe Emplacement in the Argolis Peninsula (Peloponnesus, Greece). Mémoire de la Société Helvétique des Sciences Naturelles, 99, Bâle, 111 pp.

Baumgartner PO, Bernoulli D (1976) Stratigraphy and radiolarian fauna in a late Jurassic–early Cretaceous section near Achladi (Evvoia, eastern Greece). Eclogae Geologicae Helveticae, 69(3):1027–1061.

Baumgartner PO, De Wever P, Kocher RN (1980) Correlation of Tethyan late Jurassic–early Cretaceous events. Cahiers de Micropaléontologie, Paris, 2:23–85.

Berggren WA, Hollister CD (1974) Paleogeography, paleobiogeography and the history of the circulation in the Atlantic Ocean. In: Hay WW (ed) Studies in Paleo-Oceanography. Section of Economic Paleontologists and Mineralogists, Tulsa, OK, 20:126–186.

Bertrand J, Delaloye M (1976) Datation par la méthode K-Ar de diverses ophiolites du flysch des Gêts (Haute Savoie, France). Eclogae Geologicae Helveticae, Bâle, 69:335-341

Beurrier M, Bourdillon–de Grissac C, De Wever, P Lescuyer JL (1987) Biostratigraphie des radiolarites associées aux volcanites ophiolitiques de la nappe de Samail (Sultanat d'Oman): Conséquences tectogénétiques. Comptes Rendus de l'Académie des Sciences, Paris 304:907–910.

Birkenmayer K (1977) Jurassic and Cretaceous lithostratigraphic units of the Pieniny Klippen belt Carpathians, Poland. Studia Geologica Polonica, Wydawnictwa Geologiczne, Varsovie, XLV:2–115.

Bleahu M, Lupu M, Patrulius D, Bordea S, Stefan A, Panin S (1981) The structure of the Apuseni Mountains. Institut de Géologie et Géophysique, Guidebook Series, Bucharest, 23, 107 pp.

Blome CD Irwin WP (1985) Equivalent radiolarian ages from ophiolitic terranes of Cyprus and Oman. Geology 13:401–404.

Blome CD, Tippit P, Garrison RE, Bernoulli D, Smewing D (1983) Radiolarian Biostratigraphy of Hawasina Complex, Northern Oman. American Association of Petroleum Geologists, abstracts of Annual Convention, Dallas, p. 37.

Bourdillon- de Grissac C, De Wever P, Bechennec F (1987) Nouvelles données biostratigraphiques des Montagnes d'Oman du Permien au Crétacé, à partir des Radiolaires et Foraminifères. Séance spécial. Société Geologique de France, Paris. "Géologie de l'Oman", Paris, abstracts, p 10.

Brewster NA (1980) Cenozoic biogenic silica sedimentation in the Antarctic Ocean. Bulletin of the Geological Society of America, Boulder, Part 1, 91:337–347.

Calvert SE (1966) Accumulation of diatomaceous silica in the sediments of the Gulf of California. Bulletin of the Geological Society of America, Boulder, 77:569–596.

Cariou E (1984) Les Reineckeiidae (Ammonitina, Callovien) de la Téthys occidentale: Systématique, dimorphisme et évolution. Etude à partir des gisements de Centre-Ouest de la France. Documents du Laboratoire de Géologie, Lyon, France, h.s. 8, 559 pp.

Cariou E, Contini D, Dommergues JL, Enay R, Geyssant JR, Mangold C, Thierry J (1985) Biogéographie des Ammonites et évolution structurale de la Téthys au cours du Jurassique. Bulletin de la Société Géologique de France, Paris, 8:679–697.

Cioflica G, Savu H, Nicolae I, Lupu M, Vlad S (1981) Alpine Ophiolitic Complexes in South Carpathians and South Apuseni Mountains. Institut de Géologie et Géophysique, Guidebook Series, Bucharest, 18, 80 pp.

Conti M (1986) New data on the biostratigraphy of the Tuscan cherts at Monte Cetona (southern Tuscany, Italy). In: De Wever P (ed), Eurorad IV, Marine Micropaleontology, Amsterdam, Vol 11, No. 1-3, pp.107–112.

Conti M, Marcucci M (1986) The onset of radiolarian deposition in the ophiolite sequences of the northern Apennines. In: De Wever P (ed) Eurorad IV, Marine Micropaleontology, Amsterdam, Vol 11, No. 1-3, pp 129–138.

Conti M, Marcucci M, Passerini P (1985) Radiolarian cherts and ophiolites in the northern Apennine and Corsica: Age, correlations and tectonic frame of siliceous deposition. Ofioliti (Florence) 10(2/3):203–224.

Danelian T, De Wever P, Cordey F, Kito N (1987) Radiolaires jurassiques de Corse (France). Géobios (Lyon) (submitted).

Dercourt J, Flament JM, Fleury JJ, Meilliez F (1973) Stratigraphie des couches situées sous les Radiolarites de la zone du Pinde-Olonos (Grèce): Le Trias supérieur et le Jurassique inférieur. Annales Géologiques des Pays Helléniques (Athènes) XXV:397–406.

Dercourt J, Geyssant J, Lepvrier C, Bergerat F, Bignot G, Cros P, De Wever P, Geczy B, Guernet C, Lachkar G (1984a) Hungarian mountains in Alpine framework. Acta Geologica Hungarica, Akademia Kiadô Budapest 3-4:213–221.

Dercourt J, Ricou LE, Le Pichon X, Boulin J, Biju-Duval B, Lepvrier C, Geyssant J, Zonenshain LP, Kazwin VG, Knipper AL, Sborshchikov IM, Sorokhtin OG (1984b) Geologic evolution of the Tethys convergence zone from Liassic to present between the Atlantic and Pamir. 27ème Congrès Géologique International, Moscow, (1984), 204pp.

Dercourt J, Zonenshain LP, Ricou LE, Kazmin VG, Le Pichon X, Knipper A, Grandjacquet C, Sorocktin O, Geyssant J, Lepvrier C, Sborshchikov IV, Boulin J, Biju-Duval B, Sibuet JC, Savostin, Westphal M, Lauer JP (1985a) Présentation de 9 cartes paléogéographiques au 1/20,000,000 s'étendant de l'Atlantique au Pamir pour la période du Lias à l'actuel. Bulletin de la Société Géologique de France, Paris, 8:637–652.

Dercourt J, Zonenshain LP, Ricou LE, Kazmin VG, Le Pichon X, Knipper A, Grandjacquet C, Sborshchikov IM, Geyssant J, Lepvrier C, Pechersky DH, Boulin J, Sibuet JC, Savostin, Sorocktin O, Westphal M, Bazhenov ML, Lauer JP, Biju-Duval B (1985b) Geological evolution of the Tethys belt from the Atlantic to the Pamirs since the Lias. Tectonophysics (Amsterdam) 123:241–315.

De Wever P (1982) Radiolaires du Trias et du Lias de la Tethys (Systématique, Stratigraphie). Société Géologique du Nord, Lille, Publication 7, 599 pp.

De Wever P (1984) Triassic radiolarians from the Darno area, Hungary. Acta Geologica Hungarica, Akademia Kiadô (Budapest) 3-4:295–306.

De Wever P (1987) Radiolarites rubanées et variations de l'orbite terrestre. Bulletin de la Société Géologique de France, Paris, 8(III):5, pp 957–960.

De Wever P, Bourdillon-de Grissac C (1987) Nouvelles données biostratigraphiques des Montagnes d'Oman, du Permien au Crétacé, à partir des Radiolaires et Foraminifères. Séance Spécialisée de la Société Géologique de France, "Oman", Paris, abstract, p. 10.

De Wever P, Bourdillon-de Grissac C, Beurrier M (1988) Radiolaires Sénoniens de la Nappe de Samail (Oman). Revue de Micropaléontologie, Paris, (In press).

De Wever P, Caby R (1981) Datation de la base des schistes lustrés postophiolithiques par des radiolaires (Oxfordien supérieur–Kimmeridgien moyen) dans les Alpes Cottiennes (Saint Véran, France). Comptes Rendus de l'Académie des Sciences, (Paris) 292(2):467–472.

De Wever P, Cordey F (1986) Datation directe par les radiolaires de la base et de la pertie médiane des radiolarites s.s. de la série du Pinde-Olonos (en Grèce continentale). In: De Wever P (ed) Eurorad IV. Marine Micropaleontology, Amsterdam, Vol 11, n1-3, pp 113–127.

De Wever P, Dercourt J (1985) Les radiolaires triassico-jurassiques marqueurs stratigraphiques et paléogéographiques dans les chaînes alpines périméditerranéennes: une revue. Bulletin de la Société Géologique de France, Paris, 8:653–662.

De Wever P, Miconnet P (1985) Datations directes des radiolarites du Bassin du Lagonegro (Lucanie, Italie Méridionale). Implications et conséquences. Revista Espanola de Micropaleontologia, (Madrid) XXII(3): 373–402.

De Wever P, Origlia-Devos I (1982a) Datations nouvelles par les radiolaires de la série des radiolarites s.l. du Pinde-Olonos, Grèce. Comptes Rendus de l'Académie des Sciences, (Paris) 294(2):399–404.

De Wever P, Origlia-Devos I (1982b) Datation par les radiolaires des niveaux siliceux du Lias de la série du Pinde-Onolos (Formation de Drimos, Péloponnèse et Grèce continentale). Comptes Rendus de l'Académie des Sciences, (Paris) 294(2):1191–1198

De Wever P, Thiébault F (1981) Les radiolaires d'âge Jurassique supérieur à Crétacé supérieur dans les radiolarites du Pinde-Olonos (Presqu'île de Koroni, Péloponnèse méridional, Grèce). Géobios (Lyon) 14(5): 577–609.

De Wever P, Riedel WR, et al. (1979a) Eurorad 1978: Recherches actuelles sur les radiolaires en Europe. Annales de la Société Géologique du Nord (Lille) XVCIII(4):205–222.

De Wever P, Sanfilippo A, Riedel WR, Grüber B (1979b) Triassic radiolarians from Greece, Sicily, and Turkey. Micropaleontology 25(1):75–110.

De Wever P, Duée G, El Kadiri KH (1985) Les séries stratigraphiques des klippes de Chrafate (Rif septentrional, Maroc). Témoins d'une marge continentale subsidente au cours du Jurassique-Crétacé. Bulletin de la Société Géologique de France, Paris, 8:363–379.

De Wever P, Fourcade E, Ricou LE (1986a) La disparition des radiolarites téthysiennes au Jurassique terminal: L'effet d'un bouleversement de la circulation océanique. Comptes Rendus de l'Académie des Sciences, (Paris) 302:665–670.

De Wever P, Geyssant JR, Azema J, Devos I, Duée G, Manivit H, Vrielynck B (1986b) La coupe de Santa Anna (zone de Sciacca, Sicile): Une synthèse biostratigraphique des apports des macro-, micro-, et nannofossiles du Jurassique supérieur et Crétacé inférieur. Revue de Micropaléontologie, (Paris) 29(3):141–186.

De Wever P, Baumgartner PO, Polino R (1987a) Datations des métaradiolarites des Schistes Lustrés (Alpes

Occidentales). Implications. Comptes Rendus de l'Académie des Sciences (Paris) 305:487–491.

De Wever P, Danelian T, Durand Delga M, Cordey F, Kito N (1987b) Datations des radiolarites post-ophiolitiques de Corse alpine par les radiolaires. Comptes Rendus de l'Académie des Sciences (Paris) 305:893–900.

Dumitricà P (1970) Cryptocephalic and cryptothoracic Nassellaria in some Mesozoic deposits of Romania. Revue Roumaine de Géologie, de Géophysique et de Géographie, Série Géologie (Bucharest) 14:45–124.

Dumitricà P, Mello J (1982) On the age of the Meliata Group and the Silica Nappe radiolarites (localities Drzkovce and Bohunövo, Slovak Karst, CSSR). Geol Prace Spravy (Bratislava) 77:17–28.

El Kadiri K (1984) Les radiolarites jurassiques des klippes de Chrafate (Rif septentrional, Maroc): Stratigraphie, taxonomie. Thèse 3è cycle, Pau, Vol 1, 112 pp : Vol 2, 347 pp.

Enay R (1980) Paléobiogéographie et Ammonites jurassiques: "Rythmes fauniques" et variations du niveau marin; voies d'échanges, migrations et domaines biogéographiques. In: Livre Jubilaire de la Société Géologique de France, 1830–1980. Mémoire hors série, Société géologique de France, Paris, Vol 10, pp 261–281.

Enay R, Mangold C (1982) Dynamique biogéographique et évolution des faunes d'Ammonites du Jurassique. Bulletin de la Société Géologique de France, Paris, 7:1025- 1046.

Ferrière J (1982) Paléogéographie et tectoniques superposées dans les Hellénides internes: Les massifs de l'Othrys et du Pélion (Grèce continentale). Société Géologique du Nord, Lille, publication 8, 970 pp.

Ferrière J, Bertrand J, Simantov J, De Wever P (1986) Comparaison entre des formations volcano-détritiques ("Mélanges") du Malm des Hellénides internes (Othrys, Eubée); implications géodynamiques. Annales Géologiques des Pays Helléniques, (Athènes) (in press).

Fleury JJ (1980) Les zones de Gavrovo-Tripolitza et du Pinde-Olonos (Grèce continentale et Péloponnèse du Nord). Evolution d'une plateforme et d'un bassin dans leur cadre alpin. Société Géologique du Nord, Lille, publication 4, 651 pp.

Fülöp J (1969) Excursion Géologique dans les Montagnes Centrales de Transdanubie, Mecsek et de Villàny. Colloque sur le Jurassique Méditerranéen. Institut Géologique de Hongrie, Budapest, 63 pp.

Fülöp J (1976) The Mesozoic Basemant Horst Blocks of Tata. Geologica Hungarica, Serie Geologica, Budapest, Vol 16, 228 pp.

Galácz A (1970) Biostratigraphic investigation of the middle Jurassic of Gyenespuszta northern Bakony, Transdanubian Central Mountains, Hungary. Annales de l'Université des Sciences de Rolando Eôtvôs; Section Géologie (Budapest) XIII:109–128.

Galácz A (1976) Bajocian (middle Jurassic) sections from the northern Bakony (Hungary). Annales de l'Université des Sciences de Rolando Eôtvôs; Section Géologie (Budapest) XVIII:177–191.

Galácz A (1980) Gyenespusztai Bajoci és Bath Ammonitesek (Bakony Hegység). Geologica Hungarica, Serie Palaeontologica, Budapest, Vol 9, 227 pp.

Galácz A (1984) Jurassic of Hungary: A review. Acta Geologica Hungarica, Akademia Kiadô (Budapest) 27(3–4):359–377.

Galácz A, Horvath F, Vörös A (1985) Sedimentary and structural evolution of the Bakony Mountains (Transdanubian Central Range, Hungary): Paleogeographic implication. Acta Geologica Hungarica, Akademia Kiadô 28(1–2):85–100.

Garrison RE, Fischer AG (1969) Deep-water limestones of the Alpine Jurassic. In: Friedman GM (ed) Depositional Environments in Carbonate Rocks. Society of Economic Paleontologists and Mineralogists, Tulsa, OK, Spec. Pub. 14, pp. 20–54.

Geyssant J Lepvrier C (1984) Microtectonic features of the Bükk Mountains (Hungary). Acta Geologica Hungarica, Akademia Kiadô (Budapest) 3-4:403–408.

Glennie KW, Boeuf MGA, Hughes-Clarke MW, Moody Stuart M, Pilaar WFH, Reinhardt BM (1974) Geology of the Oman mountains. Verhandelingen van het Koninklijk Nederlands Geologisch Mijnbouwkundig Genootschap, Gravenhage, Transaction 31, 3, 430 pp.

Goll RM (1969) Radiolaria: The history of a brief invasion. In: Degens ET, Ross DA (eds) Hot Brines and Recent Heavy Metal Deposits in the Red Sea. Springer, New York, pp. 306–312.

Goričan S. (1986) In: Yugoslav Committee For IGCP (1986) Guide-Book with abstracts for the Third International Conference on Siliceous Deposits, IGCP 187 Beograd, Yugoslavia. 112 pp.

Goričan S (1987) Jurassic and Cretaceous radiolarians from the Budva Zone (Montenegro, SW Yugoslavia). Revue de Micropaléontologie, (Paris) 30(3):177–196.

Govi M (1965) Osservazioni stratigrafiche sul Titoniano di M. Marzio e di Laveno (Varese). Bolotino de la Societa Geologica dell Italiana, Rome, 84:151–168.

Harland WB, Cox AV, Llewellyn PG, Picton CAG, Smith AG, Walters R (1982) A Geologic Time Scale, Cambridge University Press, Cambridge, U.K., 128 pp.

Hein JR, Karl SM (1983) Comparisons between open-ocean and continental margin chert sequences. In: Iijima A, Hein JR, Siever R (eds) Siliceous Deposits in the Pacific Region. Developments in Sedimentology, 36, Elsevier, Amsterdam, pp 25–43.

Iijima A, Utada N (1983) Recent development in the sedimentology of siliceous deposits in Japan. In: Iijima A, Hein JR, Siever R (eds) Siliceous Deposits in the Pacific Region. Developments in Sedimentology, 36, Elsevier, Amsterdam, pp 45–64.

Iijima A, Kakuwa Y, Yamazaki K, Yanagimoto Y (1978) Shallow-sea organic origin of the Triassic bedded chert in central Japan. Journal of the Faculty of Science, University of Tokyo II(19):369–400.

Iijima A, Tada R, Matsumoto R (1982) Sedimentary petrographic study of the Neogene System in Hokkaido. In: Tanai T (ed) Biostratigraphy of the Neogene System in Hokkaido. Hokkaido University Press, Sapporo, pp 67–74.

Iijima A, Kakuwa Y, Matsuda H (1988) Silicified wood from the Adoyama Chert, Kuzuh, Central Honshu and its bearing on compaction and depositional environment of radiolarian bedded chert. In: Hein JR, Obradović J (eds) Siliceous Deposits of the Tethys and Pacific Regions, Springer-Verlag, New York.

Isaacs CM, Pisciotto KA, Garrison RE (1983) Facies and diagenesis of the Miocene Monterey Formation, California: A summary. In: Iijima A, Hein JR, Siever R (eds) Siliceous Deposits in the Pacific Region. Developments in Sedimentology, 36, Elsevier, Amsterdam, pp 247–282.

Jenkyns HG, Winterer EL (1982) Palaeoceanography of Mesozoic ribbon radiolarites. Earth and Planetary Sciences Letters, (Amsterdam) 60:351–375.

Kennett J (1982) Marine Geology. Prentice-Hall, London 813 pp.

Kent DV, Gradstein FM (1985) A Cretaceous and Jurassic geochronology. Geological Society of American Bulletin, (Boulder) 96:1419–1427.

Kocher RN (1981) Biochronostratigraphische untersuchungen oberjurassischer radiolarienfuehrender gesteine, insbesondere der sudalpen. Dissertation of Eidgenoessischen Technische Hochschule Nr 6809, Zürich, 185pp.

Kozur H (1984) Jurassic radiolarians from the Bükk Mountains and some remarks on the tectonic position of the area. Acta Geologica Hungarica Akademia Kaidô (Budapest) 3-4:307–319.

Kozur H, Mock R (1973) Zum Alter und zur tektonischen Stellung der Meliata-Serie des Slowakischen Karstes. Geologicky Zbornik Geologica Carpathica (Bratislava) 24(2):365–374.

Lemoine M (1982) Rifting and early drifting: Mesozoic central Atlantic and Ligurian Tethys. In: Sheridan RE, Gradstein FM, et al., (eds) Initial Reports of the Deep Sea Drilling Project 76. U.S. Government Printing Office, Washington, pp 885–894.

Lemoine M (1985) Structuration jurassique des Alpes occidentales et palinspastique de la Téthys ligure. Bulletin de la Société Géologique de France, Paris, 8(1):127–138.

Lemoine M, Gidon M, Barfety JC (1981) Les massifs cristallins externes des Alpes occidentales: D'anciens blocs basculés nés au Lias lors du rifting téthysien. Comptes Rendus de l'Académie des Sciences, (Paris) 292(II):917–920.

Lemoine M, Bas T, Arnaud-Vanneau A, Arnaud H, Dumont T, Gidon M, Bourbon M, Graciansky (de) PC, Rudkiewicz JL, Megard-Galli J, Tricart P (1986) The continental margin of the Mesozoic Tethys in the western Alps. Marine and Petroleum Geologists, 3:179–199.

Luthi S (1973) Stratigraphie und Sedimentologie des Mte Nudo Beckens (Vol I). Unpublished Diplomarbeit, Eidgenoessischen Technische Hochschule, Zürich.

Marcoux J, Ricou LE (1979) Classification des ophiolites et radiolarites alpino-méditerranéennes d'après leur contexte paléogéographique et structural. Implications sur leur signification géodynamique. Bulletin de la Société Géologique de France, Paris, 7:643–652.

Matsuda T, Isozaki Y, Yao A (1980) Stratigraphic relation of Triassic-Jurassic system in Inuyama district, Mino Belt. Abstract Paper, 87th Annual Meeting of the Geological Society of Japan, p. 105.

Matsumoto R, Iijima A (1983) Chemical sedimentology of some Permo-Jurassic and Tertiary bedded cherts in Central Honshu, Japan. In: Iijima A, Hein JR, Siever R (eds) Siliceous Deposits in the Pacific Region. Developments in Sedimentology. Elsevier, Amsterdam, Vol 36, pp 175–191.

Mc Bride EF, Thompson A (1970) The Caballos Novaculite, Marathon Region, Texas. Geological Society of America, Boulder, Special Paper 122, 129 pp.

McGowan JA (1984) The California El Niño 1983. Oceanus (Woods Hole) 27(2)48–51.

Mutihac V (1968) Structura Geologicà a Compartimentului Nordic din Sinclinalul Marginal Extern (Carpatii Orientali). Edition Academiei, Bucharest, 127 pp.

Nelson DM, Goering JJ (1978) Assimilation of silicic acid by phytoplankton in the Baja California and northwest Africa upwelling systems. Limnology and Oceanography, (Lawrence) 23(3):508–517.

Odin SG, Kennedy WJ (1982) Mise à jour de l'échelle des temps mésozoïques. Comptes Rendus de l'Académie des Sciences, Paris, 294(II):383–386.

Obradović J, Goričan S (1988) Siliceous deposits in Yugoslavia: Occurrences, types and ages. In: Hein JR, Obradović J (eds) Siliceous Deposits of the Tethys and Pacific Regions. Springer-Verlag, New York, pp 51–64.

Origlia-Devos I (1983) Radiolaires du Jurassique supérieur–Crétacé inférieur: Taxonomie et révision stratigraphique (Zone du Pinde-Olonos, Grèce; Zone de Sciacca, Italie; Complexe de Nicoya, Costa Rica, et Forages du DSDP). Thèse 3ème cycle, Paris, 83–53, 328 pp.

Pessagno EA Jr, Newport RL (1972) A new technique for extracting radiolaria from radiolarian cherts. Micropaleontology 18(2):231–234.

Pichon JF (1977) Une transversale dans la zone pélagonienne, depuis les collines de Krapa (SW) jusqu'au massif du Vermion (NE). Les premières séries transgressives sur les ophiolites. VIème Colloquium on the Geology of the Aegean Region, Athènes (1977), Vol 1, pp 163–171.

Pindell JL (1985) Alleghenian reconstruction and subsequent evolution of the Gulf of Mexico, Bahamas and proto-Caribbean. Tectonics (Amsterdam) 4(1):1–39.

Pisciotto KA, Garrison RE (1981) Lithofacies and depositional environments of the Monterey Formation, California. In: Garrison RE et al. (eds) The Monterey Formation and Related Siliceous Rocks of California. Pacific Section of the Society of Economic Paleontogists and Mineralogists, Tulsa, OK, pp 97–122.

Pisias NG, Murray DW, Roelofs AK (1986) Radiolarian and silicoflagellate response to oceanographic changes associated with the 1983 El Niño. Nature 320:259–262.

Poisson A (1977) Recherches géologiques dans les Taurides occidentales (Turquie). Thèse, Orsay, France, 795 pp.

Renz GW (1976) The distribution and ecology of radiolaria in the Central Pacific plankton and surface sediments. Bulletin of Scripps Institution of Oceanography, La Jolla, University of California Press, Berkeley, 22, 267 pp.

Ricou LE, Zonenshain LP, Dercourt J, Kazmin VG, Le Pichon X, Knipper AL, Grandjacquet C, Sborshchikov IM, Geyssant J, Lepvrier C, Pechersky DM, Boulin J, Sibuet JC, Savostin LA, Sorokhtin O, Westphal M, Bazhenov ML, Lauer JP, Biju-Duval B (1985) Méthodes pour l'établissement de neuf cartes paléogéographiques de l'Atlantique au Pamir depuis le Lias. Bulletin de la Société Géologique de France, Paris, 8:625–635.

Ricou LE, Dercourt J, Geyssant J, Grandjacquet C, Lepvrier C, Biju-Duval B (1986) Geological constraints on the Alpine evolution of the Mediterranean Tethys. Tectonophysics (Amsterdam) 123:83–122.

Schaaf A, Thomas V (1986) Les radiolaires campaniens du Wadi Ragmi (nappe de Semail, Oman): Un nouveau repère chronologique de l'obduction omanaise. Comptes Rendus de l'Académie des Sciences, Paris, 303(II):1593–1598.

Schlager W, Schlager M (1973) Clastic sediments associated with radiolarites (Tauglboden-Schichten, upper Jurassic, eastern Alps). Sedimentology (Amsterdam) 20:65–89.

Schrader HW, Kelts K, Curray J, Moore D (1980) Laminated diatomaceous sediments from the Guayamas Basin slope (Central Gulf of California): 250,000 year climatic record. Science 207:1207–1209.

Steinberg M, Desprairies A, Fogelgesang JF, Martin A, Caron D, Blanchet R (1977) Radiolarites et sédiments hypersiliceux océaniques: Une comparaison. Sedimentology (Amsterdam) 24(4):547–563.

Thiébault F, De Wever P, Fleury JJ, Bassoullet JP (1981) Précisions sur la série stratigraphique de la nappe du Pinde de la presqu'île de Koroni (Péloponnèse méridionale, Grèce): L'âge des radiolarites (Dogger-Crétacé supérieur). Annales de la Société Géologique du Nord (Lille) C:91–105.

Thiébault F, De Wever P, Raoult JF (1986) Marqueurs minéralogiques et géochimiques au passage Jurassique-Crétacé dans la série du Pinde-Olonos en Péloponnèse méridional (Grèce). Revue de Géologie Dynamique et de Géographie Physique, (Paris) 27(5):351–362.

Thiede J (1979) Paleogeography and paleobathymetry of the Mesozoic and Cenozoic North Atlantic Ocean. Geojournal (Wiesbaden) 3(3):263–272.

Tikhomirova LB (1983) Pozdneyurskiye-rannemelovyie radiolyarii Karpat (na territorii SSSR). Voprosy Mikropaleontologii, (Moscow) 26:72–86.

Tikhomirova LB (1984) The upper Jurassic and Cretaceous radiolarian of the Mediterranean paleogeography region (the Carpathians and the Lesser Caucasus). Morfologii, Ekologii i Evolutioni Radiolarii, Eurorad IV. Academia Nauk SSR, Leningrad, pp 159–171.

Tippit PR, Pessagno EA Jr, Smewing JD (1983) The biostratigraphy of sediments in the volcanic unit of the Semail ophiolites. Journal of Geophysical Research, 86(B4):2756–2762.

Van Hinte JE (1976) Jurassic time scale. Bulletin of American Association of Petroleum Geologists, Tulsa, 60: 489–497.

Vecsei A, Frisch W, Pirzer M, Wetzel A (1988) Origin and tectonic significance of radiolarian chert in the Austroalpine rifted continental margin. In: Hein JR, Obradović J (eds) Siliceous Deposits of the Tethys and Pacific Regions. Springer-Verlag, New York, pp 65–80.

Vishnevskaya V (1984) Radiolariti kak Analogy Sovremenih Radiolarievih Ilov. Nauka, Moscow, 120 pp.

Vrielynck B (1978) Données nouvelles sur les zones internes du Péloponnèse. Grèce. Les massifs à l'est de la plaine d'Argos. Thèse 3è Cycle, Lille, 134 pp.

Vrielynck B (1980) Données nouvelles sur les zones internes du Péloponnèse. Les massifs à l'est de la plaine d'Argos; Grèce. Annales Géologiques des Pays Helléniques (Athênes) XXIX:440–462.

Vrielynck B (1982) Evolution paléogéographique et structurale de la presqu'île d'Argolide (Grèce) Revue de Géologie Dynamique et de Géographie Physique, (Paris) 23:277–288.

Westermann GE (1984) Gauging the duration of stages: A new approach for the Jurassic. Episodes (Ottawa) 7:2, 26–28.

Yugoslav Committee for IGCP 86 Guide-Book with Abstracts for the Third International Conference on Siliceous Deposits, IGCP 187, Belgrade, Yugoslavia, 112 pp.

4 Siliceous Deposits in Yugoslavia: Occurrences, Types, and Ages

Jelena Obradović and Špela Goričan

Overview

Siliceous deposits are widespread in Yugoslavia especially in Mesozoic sections. They first occur in Middle Triassic deposits and continue through Jurassic and Cretaceous sections. All of these Mesozoic siliceous deposits are of marine origin. Generally three lithologic associations with chert are recognized in the Mesozoic sections. 1. the Porphyrite-chert assemblage of Middle-Late Triassic age, 2. the Diabase-chert Formation of Middle-Late Jurassic age, and 3. carbonate complexes of Middle Triassic to Danian ages.

The first two lithologic associations are connected with the breakup of the Dinaridic plate at the end of Early Triassic, which was followed by hybrid volcanic activity with the greatest production of volcanic rocks in the Ladinian. At that time an oceanic region developed, and existed to the Late Jurassic. Bedded chert formed on transitional and oceanic crust. They formed in different parts of this oceanic basin by different depositional mechanisms. Commonly, olistoliths of the porphyrite-chert assemblage occur in the Diabase-chert Formation. Siliceous deposits associated with carbonate complexes occur in different parts of Yugoslavia, but as an example of this lithologic association we discuss the occurrences in the Budva zone, Montenegin Littoral. The Budva zone represents a depositional trough on thinned continental crust situated between two carbonate platforms. The sedimentation of carbonate rocks with bedded chert took place mostly on the slopes, beginning with subtidal deposition in the latest Ladinian and evolving later into open-shelf deposition.

Introduction

Siliceous deposits are widespread in Yugoslavia, occurring in formations of different ages and tectonic settings. The oldest siliceous rocks, Precambrian and Cambrian quartzite, occur with other metamorphic rocks, mostly rocks of a volcanic-sedimentary assemblage. In younger Paleozoic sections, minor occurrences of lydite (silicified shale, limestone, tuff, slate) occur with metasedimentary rocks. The most widespread siliceous rocks occur in Mesozoic sections in association with carbonate rocks, a volcanic-sedimentary assemblage (the so-called porphyrite-chert assemblage), and the Diabase-chert Formation. All these siliceous rocks from Mesozoic sections are of marine origin. In the Tertiary, both marine and continental siliceous deposits occur. Here we describe only the Mesozoic siliceous sections occurring in Yugoslavia.

Mesozoic Siliceous Deposits from the Dinarides Mountains

Mesozoic siliceous deposits first occurred in Middle Triassic sections and continued through the Mesozoic. Siliceous rocks occur in different lithologic associations and in different tectonic units. Generally, three lithologic associations are recognized: (1) the porphyrite-chert assemblage; (2) the Diabase-chert Formation, and (3) carbonate complexes. These three lithologic associations will be discussed in the next three subsections.

In all these lithologic associations, chert is the dominant siliceous rock. Chert is commonly interbedded with siliceous shale, forming sections up to 50 m thick, or chert occurs as isolated beds and concretions. The silica content of cherts varies from 77% to 99%. Chert is composed of microcrystalline silica (quartz and chalcedony), pigmented by hematite, and contains minor illite, chlorite, carbonate minerals, and in places, pyrite. Radiolaria occur of different sizes and various amounts.

Jasper is subordinant to chert and occurs locally. Jasper is also composed of microcrystalline quartz and chalcedony, pigmented by iron and manganese minerals, and contains radiolarians.

Siliceous shale occurs in thin beds, up to 5 cm thick. The content of silica varies from 30% to 72%, and is rarely as high as 89%. Besides silica minerals (chalcedony and quartz), siliceous shale contains various amounts of illite, carbonate minerals (mostly calcite), hematite, and rarely chlorite and pyrite. Radiolaria commonly occur.

Siliceous limestone consists of siliceous biomicrite in which the siliceous component includes only Radiolaria floating in a micritic (calcitic) matrix; siliceous limestone also consists of silicified biomicrite, oosparite, and other limestone types, with partial to complete replacement of carbonate by silica.

Some siliceous rocks, silicopelites, are considered separately when the origin of the rocks is uncertain; these lithologies include marl, claystone, and tuffaceous rock. Silicopelite has a high silica content (usually more than 90%) but is less compact and more porous than chert and is commonly referred to as porcelanite.

The Porphyrite-Chert Lithologic Assemblage

The term Porphyrite-chert Formation, which was introduced in the geological literature by Ćirić (1954), distinguishes this volcanic-sedimentary lithologic association of Middle Triassic age from the Diabase-chert Formation. The porphyrite-chert lithologic assemblage is widespread and occurs in Montenegro, Bosnia, Hercegovina, Croatia, Slovenia, and partly in Serbia and Macedonia (Fig. 4.1).

At the beginning of the Triassic, there existed a microplate of continental character where the present day Dinarides Mountains occur. In the Early Triassic, sandstone, siltstone, claystone, and shale were deposited on this microplate. In the latest part of the Early Triassic, the plate began to break, which was followed by hybrid magmatic activity. Mainly during the Middle Triassic, ruptures formed in the microplate, probably owing to the thermal production of an antiform at the base of the crust (Karamata, 1975), along which issued basaltic magmas. These basaltic magmas were contaminated with crustal material and differentiated during penetration along small ruptures. Volcanic rocks formed from the end of the Early Triassic to the Carnian.

Volcaniclastic rocks with chert and rare carbonate rocks were deposited during the Anisian to the Carnian also, but were most widespread in the Ladinian. Because of their characteristic green color, these rocks are called *pietra verde* in the literature.

Keratophyre with or without quartz is the most abundant volcanic rock (Knežević, 1975). Earliest Ladinian spilite underlies volcaniclastic rocks in the coastal part of Montenegro and in Bosnia-Hrčavka. In some places (Croatia-Donje Pazarište), basalt, and in other places, andesite underlie volcaniclastic rocks. At times, volcanic eruptions were explosive and produced huge volumes of volcanogenic rocks, which varied from agglomerates and pyroclastic breccia to pelitic tuff.

Tuff and tuffite, which form the greatest part of the assemblage, are of intermediate silica composition and occur in beds up to 60 cm thick, rarely thicker, with horizontal lamination in places. Textures are psammitic to pelitic, rarely microbrecciated. Tuffs are vitroclastic to vitrocrystalloclastic. Alteration consists of argillization, zeolitization, and silicification, with zonal distributions resulting from burial diagenesis.

In Montenegrin Littoral, siliceous rocks are represented by chert beds, chert concretions, and jasper. The last occurs in the lower parts of the series, but above spilites, as lenses 10 m long and 2 to 3 m thick. Jasper and maybe a part of the chert probably formed locally from extensive submarine volcanic activity. In the middle part of the sequence, chert occurs as beds up to 10 cm thick. The rocks are commonly tuffaceous, producing cherty tuff. Besides silica, cherty tuff consists of plagioclase, celadonite, and zeolites. The last replaced irregular volcanic glass shards.

Chert concretions of different sizes and shapes and cherty tuffs are secondary siliceous rocks. Concretions formed by mineral segregation, diagenetic reorganization of silica entirely of biogenic origin,

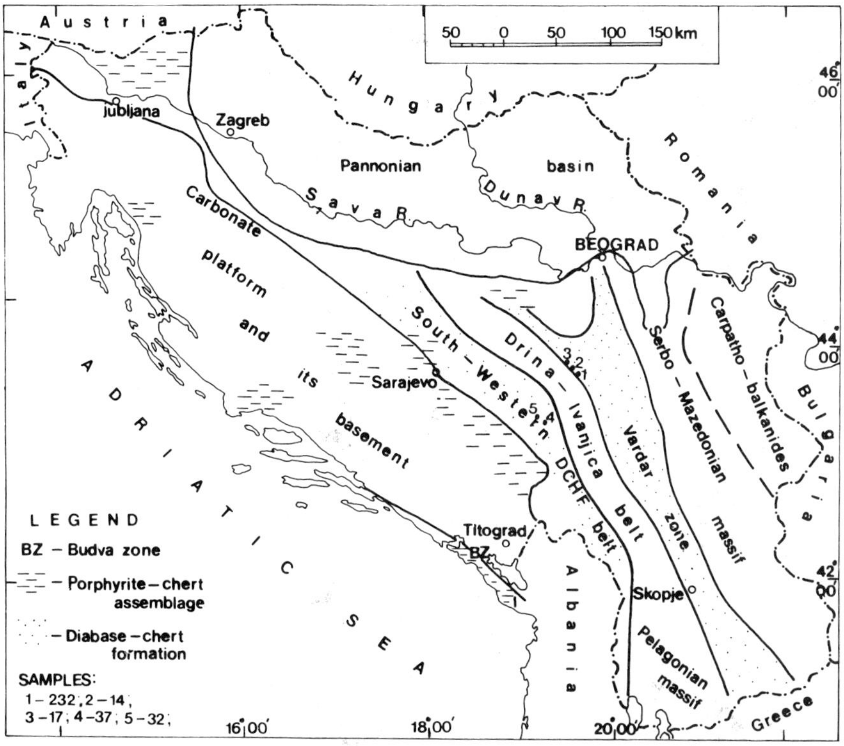

FIGURE 4.1. Distribution of the porphyrite-chert assemblage and the Diabase-Chert Formation (DCHF). Modified from Karamata (1986) and Grubić (1986).

chiefly from Radiolaria. Cherty tuff formed by partial replacement of tuff by silica.

In western Serbia, the porphyrite-chert assemblage occurs in two NW-SE trending belts (Knežević et al., 1978) which are separated by the Drina-Ivanjica Paleozoic belt. The northern belt is well developed in the Podrinje Mountain area in a limestone-dolomite belt of Middle Triassic age. The extent and amount of volcanism in western Serbia are not as great as in the Outer Dinarides (Montenegro). Volcanism is of intermediate composition and includes andesite (porphyrite), keratophyre with or without quartz, and basaltic andesite. Volcanogenic rocks are widespread and include pyroclastic breccia and tuff. Volcaniclastic rocks occur together with siliceous rocks and rarely with carbonate rocks. Siliceous rocks consist of chert (beds and concretions) and cherty tuff. There are not differences between chert in western Serbia and in Montenegro.

In the southern belt, this assemblage, according to some authors (Dimitrijević and Dimitrijević, 1973, 1974; Dimitrijević, 1986), occurs as olistolithes in the assemblage of Jurassic age, the so-called Diabase-chert Formation.

Near Ovčar Banja, olistoliths of the porphyrite-chert-assemblage are composed of tuff and tuffites and chert is interbedded with siliceous shale and limestone. The chert beds are up to 20 cm thick and siliceous shale is up to 5 cm thick. Besides silica minerals, chert consists of sericite, illite, and rarely calcite and pyrite. Radiolarians and sericite are commonly oriented and in places deformed. Siliceous shale has the same mineral composition, only the content of clay minerals is greater and silica less than in chert.

The chert from this locality, based on radiolarians, is latest Illyrian to Langobardian in age. Radiolarians from sample 17 (Fig. 4.2(8)) are abundant but poorly preserved. The assemblage contains only forms with robust tests and single stout spines broken from the radiolarians. The following species were found: *Baumgartneria* sp. (Fig. 4.2(8)), *Eptingium manfredi* Dumitricâ 1978) (Fig. 4.2(11)), *Falcispongus* sp. (Fig. 4.2(7)), *Oertlispongus inaequispinosus* (Dumitricâ, Kozur and Mostler,

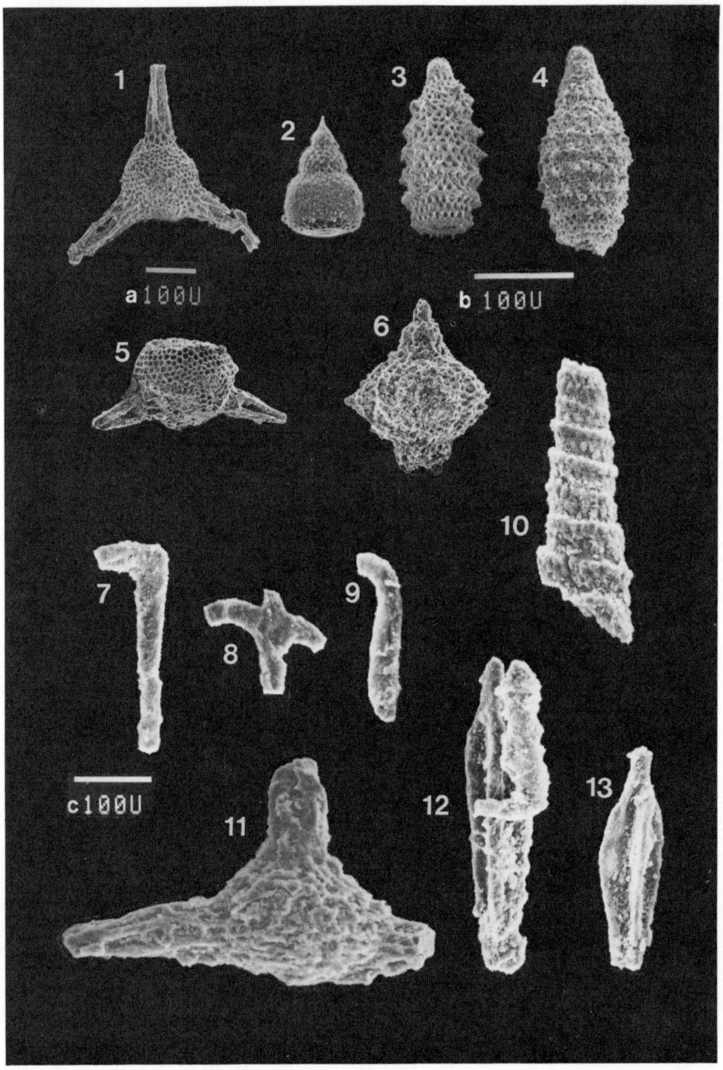

FIGURE 4.2. Radiolaria from an olistolith of the porphyrite chert assemblage; section is located on Nova Varoš to Bistrica Road (sample 37, photographs 1–6) and from the olistolith of the same assemblage in a section located on the road near Ovčar Banja (sample 17, photographs 7–13). SEM negative numbers are listed after each species name. Scale bar (a) for figures 1, 5; bar (b) for 2, 3, 4, 6; and bar (c) for 7–13. (1) *Triactoma jonesi* (Pessagno), 86/147/6. (2) *Eucyrtidiellum unumaensis* (Yao), 86/146/6. (3, 4) *Parvicingula dhimenaensis* Baumgartner, 3: 86/146/4; 4: 86/146/5. (5) *Triactoma blakei* (Pessagno), 86/ 147/5. (6) *Unuma* sp., 86/147/4. (7) *Falcispongus* sp., 85/2640. (8) *Baumgartneria* sp., 85/2640. (9) *Oertlispongus inaequispinosus* Dumitricâ, Kozur and Mostler, 85/2645. (10) *Triassocampe scalaris* Dumitricâ, Kozur and Mostler, 85/ 2645. (11) *Eptingium manfredi* Dumitricâ, 85/2639. (12, 13) Straight four-bladed spines, 85/2642.

1980) (Fig. 4.2(9)), and *Triassocampe scalaris* (Dumitricâ, Kozur and Mostler, 1980) (Fig. 4.2(10)).

On the basis of *Oertlispongus inaequispinosus*, a latest Illyrian to Langobardian age (Dumitricâ, 1982) is assigned to this olistolith. *Eptingium manfredi* and *Triassocampe scalaris* have a longer range.

Another olistolith from the porphyrite-chert assemblage occurs also in the Diabase-chert Formation near Bistrica, on the Bistrica to Priboj road, and it is composed of chert of Ladinian age. Radiolarian fauna in sample 32 are poorly preserved and incomplete. The assemblage includes the following stratigraphically important species: *Oertlispongus inaequispinosus* (Dumitricâ, Kozur and Mostler, 1980) (Fig. 4.3(2)), *Pseudostylosphaera japonica* (Nakaseko and Nishimura, 1981) (Fig. 4.3(1)), and *Triassistephanidium laticornis* (Dumitricâ, 1978) (Fig. 4.3(3)).

Triassistephanidium laticornis has been reported only from the Fassanian (Dumitricâ, 1978; Dumitrikâ et al., 1980; Kozur and Réti, 1986). *Oertlispongus inaequispinosus* is known from the Fassanian and Langobardian (Dumitricâ, 1982). *Pseudostylosphaera japonica* is one of the characteristic species of Yao's (1982, 1984) Ladinian *Triassocampe deweveri* assemblage zone. Thus, a Ladinian age is assigned to this olistolith.

Chemical analysis of jasper and chert from the porphyrite-chert assemblage (Table 4.1) shows high

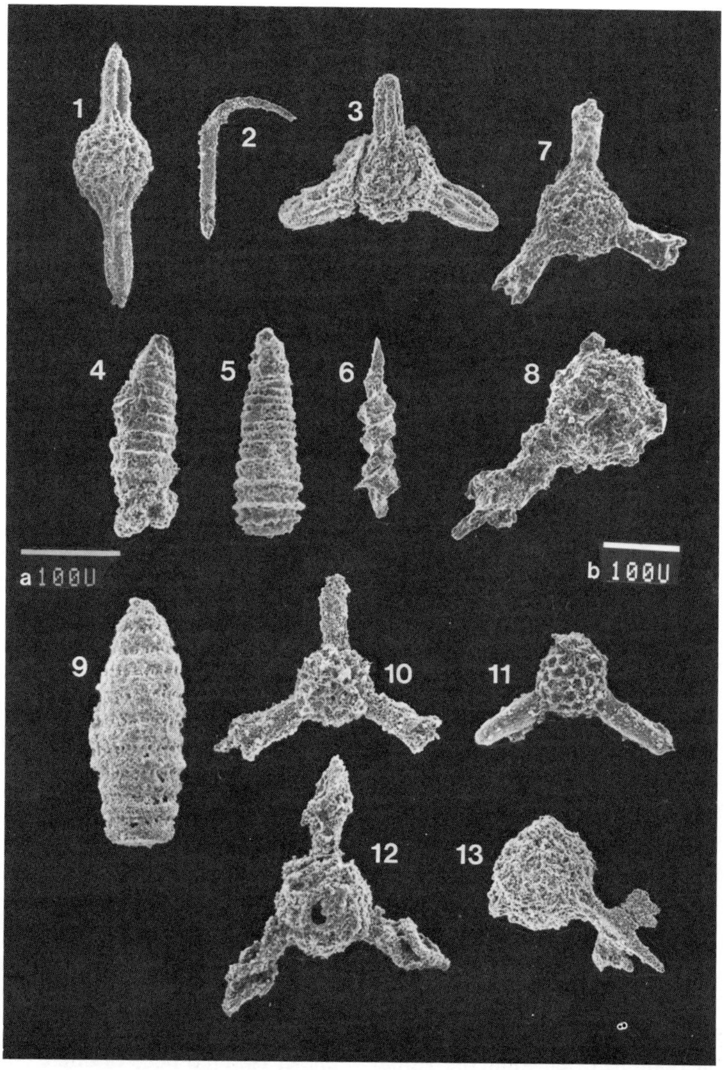

FIGURE 4.3. Radiolaria from an olistolith of the porphyrite-chert assemblage in a section located on the Bistrica to Priboj Road (sample 32, photographs 1–3), from the Diabase-Chert Formation section located on Čačak to Ovčar Banja Road, near the first dam (sample 232/11, photographs 4–8), and from the Diabase-chert Formation section located on Čačak to Ovčar Banja Road (sample 14, photographs 9–13). SEM negative numbers are listed after the species name. Scale bar (a) for Figures 4, 5, 9; bar (b) for the others. (1) *Pseudostylosphaera japonica* (Nakaseko and Nishimura), 86/153/9. (2) *Oertlispongus inaequispinosus* Dumitricâ, Kozur and Motler, 86/149/2. (3) *Triassistephanidium laticornis* Dumitricâ, 86/148/8. (4, 5) *Triassocampe* sp., 4: 86/149/7; 5: 86/146/1. (6) Twisted spine, 86/149/4. (7) *Capnuchosphaera* sp., 86/149/10. (8) *Vinassaspongus transitus* Kozur and Mostler, 86/150/1. (9) *Latium longulum* Blome, 86/153/7. (10) *Capnodoce anapetes* De Wever, 86/148/1. (11) *Capnodoce* sp., 86/148/2. (12) *Capnuchosphaera triassica* De Wever, 86/153/8. (13) *Kahlerosphaera* ? sp., 86/147/10.

TABLE 4.1. Chemical analysis of siliceous rocks from the porphyrite-chert assemblage.

	1	2	3	4
SiO_2	95.73%	83.45%	92.80%	88.25%
TiO_2	0.15	0.21	0.08	0.25
Al_2O_3	0.58	5.87	1.50	3.15
Fe_2O_3	1.48	0.63	0.78	1.24
FeO	0.24	1.57	0.16	0.96
MnO	0.03	0.10	0.07	0.05
MgO	0.38	1.43	0.24	1.30
CaO	0.21	0.24	1.87	1.25
Na_2O	0.20	1.08	0.13	0.27
K_2O	0.22	2.77	0.39	1.77
P_2O_5	–	–	tr	–
H_2O^+	0.84	1.67	2.38	1.37
H_2O^-	0.28	0.36	0.09	0.12
Totals	100.34%	100.38%	100.49%	99.98%

SiO_2, Al_2O_3, MgO, CaO, and H_2O were analyzed by gravimetric methods; Fe_2O_3, FeO, MnO, and P_2O_5, by calorimetric methods; and K_2O and Na_2O, by flame-photometric method.
(1) Jasper, No. 1, Montenegrin Littoral, loc. Bečići, analyst D. Nikolić. (2) Chert tuff, No. 6, Montenegrin Littoral, loc. Bečići. (3) Chert, No. 11, western Serbia, southern belt, loc. Ovčar Banja, analyst P. Ristić. (4) Chert, No. 17, western Serbia, southern belt, loc. Ovčar Banja, analyst S. Šaponja.

silica contents, from 88% to 95%, 83% in cherty tuff. The silica represents mostly the diagenetic products of biogenic silica, except for jasper, which may contain some volcanic or hydrothermal silica. Alumina is greatest in cherty tuffs, as are the alkali elements, especially K_2O. Total iron varied from 0.92% to 2.20%.

The Diabase-Chert Formation

The Diabase-chert Formation composes a part of the Inner Dinarides—the ophiolitic belt—and the Vardar tectonic zone. The distribution of this formation forms one belt that begins to the northwest of Zagreb, crosses Bosnia and Stara Raška to the south, and ends in the Merdita zone in Albania (Fig. 4.1). Another belt crosses western Serbia and the Vardar zone and continues in the direction of Greece.

According to some authors, the Diabase-chert Formation is part of the ophiolite complex of the Dinarides Mountains, which formed between the late Lias and Latest Jurassic, the time of the most intensive ocean basin formation of the Dinaridic eugeosyncline (Ćirić, 1954, 1984). According to other authors, this formation is considered to be an ophiolitic melange with an olistostrome structure and is connected with marginal parts of continental plates, situated over subducted oceanic crust (Dimitrijević and Dimitrijević, 1973, 1974; Dimitrijević, 1986). Yet others have postulated a volcanic-sedimentary lithologic assemblage (Pamić, 1969) or formation during the opening and closing of narrow oceanic basins during Middle Triassic to Latest Jurassic time (Karamata, 1975, 1981, 1986).

Siliceous rocks from this formation are most commonly bedded chert, rhythmically interbedded with siliceous shale, producing sections up to 50 m thick, rarely thicker. In some parts of this formation—for example, in ophiolitic melange and near the boundary with the Vardar zone—there are olistolithes of the porphyrite-chert assemblage containing bedded chert, isolated beds of chert, and chert concretions as well as olistoliths of Triassic and Jurassic limestone with chert. Because of this mixing of different lithologic assemblages, the occurrence of siliceous rocks is complex.

On the road from Čačak to Ovčar Banja, near the first dam, a section of pillow lava and bedded chert occurs. Early workers considered the pillow lavas of spilitic composition in this area to be of Late Jurassic age. However, chert that overlies the spilite is of Triassic age. Thin chert beds, up to 30 cm thick, rhythmically interbedded with siliceous shale (up to 5 cm thick) produce sections up to 30 m thick. Chert is composed of silica minerals and radiolarians, is pigmented by hematite, and contains many veins filled by chalcedony and calcite.

From this section, radiolarians in sample 232 are common but are also etched, compressed, and fragmented. The following taxa were recognized: *Cap-*

nuchosphaera sp. (Fig. 4.3(7)), *Triassocampe* sp. (Fig. 4.3(4,5)), and *Vinassaspongus transitus* (Kozur and Mostler, 1981) (Fig. 4.3(8)).

The genus *Triassocampe* is known worldwide to occur in Ladinian to Norian rocks (Blome, 1984). *Capnuchosphaera* is common in Carnian and Norian strata (De Wever, 1982), and it has recently been reported from Ladinian strata in Hungary (De Wever, 1984). *Vinasspongus transitus* has been assigned only to the lower Norian of the type locality (Kozur and Mostler, 1981). On the basis of radiolarians, a Ladinian to Norian age is assigned to this section. Unfortunately, the poorly preserved faunas do not allow a more exact age assignment, but they nevertheless define a Triassic age for the sedimentary rocks above the pillow lavas.

Near this section, also on the road from Čačak to Ovčar Banja, an outcrop of chert interbedded with siliceous shale is exposed. In the lower part of the section, chert is dominant, and siliceous shale occurs in small amounts. In the upper part of the section, chert is also dominant but the proportion of siliceous shale increases. Chert occurs in beds 5 to 40 cm thick, which are brecciated in places and laced with quartz veins.

In sample 14 from this outcrop, the following radiolarian species were determined: *Capnodoce anapetes* (De Wever et al., 1979) (Fig. 4.3(10)), *Capnodoce* sp. (Fig. 4.3(11)), *Canuchosphaera triassica* (De Wever et al., 1979) (Fig. 4.3(12)), *Kahlerospaera* ? sp. (Fig. 4.3(13)), and *Latium longulum* (Blome, 1984) (Fig. 4.3(9)).

The radiolarian assemblage can be assigned to the Capnodoce Zone (Pessagno et al., 1979; Blome, 1984), which is defined by the first and last occurrences of Capnodoce and ranges from late Carnian to middle Norian. The base of this zone is still questionable. The top of the zone is somewhat younger, because De Wever (1982) found Capnodoce also in late Norian rocks. The Carnian-Norian age is also supported by the occurrences of *Latium longulum* (Blome, 1984) and *Capnuchosphaera triassica* (De Wever, 1982).

A third section of rhythmically bedded chert and siliceous shale, with rare limestone, occurs on the right side of the road from Nova Varoš to Bistrica (Obradović et al., 1986). Chert occurs in thin beds up to 15 cm thick, rarely 20 cm thick. Beds are dark red, rarely green, and are composed of microcrystalline quartz and chalcedony pigmented by hematite. Various amounts of illite are present. In some places calcite replaces silica minerals, including radiolarians. The rocks are laced with veins of different sizes that are filled by quartz and/or calcite. Microstylolites occur in the beds of chert. Siliceous shale beds are 3 to 5 cm thick with characteristic fissility. Besides illite and microcrystalline silica minerals, hematite, calcite, and radiolarians are present. Limestone, calcarenite and biomicrite are rare components of this assemblage. These carbonate rocks are silicified to various degrees, and in some rocks ankerite occurs. In sample 37 from this section, the following radiolarian species were recognized: *Eucyrtidiellum unumaensis* (Ichikawa and Yao, 1976) (Fig. 4.2(2)), *Parvicingula dhimenaensis* (Baumgartner, 1984) (Fig. 4.2(3,4)), *Triactoma blakei* (Pessagno, 1977a) (Fig. 4.2(5)), *Triactoma jonesi* (Pessagno, 1977a) (Fig. 4.2(1)), and *Unuma* sp. (Fig. 4.2(6)).

The age of the assemblage was based on Baumgartner's (1984) stratigraphy. The first occurrence of *Triactoma blakei* is in U.A.2, and the final occurrence of *Eucyrtidiellum unumaensis* is in U.A.8, which allows only a broad age assignment of Callovian to early Kimmeridgian.

Silica content of chert from this assemblage is high, up to 97.5% (Table 4.2). The highest content of alumina occurs in chert 37 of Callovian to early Kimmeridgian age, which is rich with clay minerals. Chert is characterized by enrichments of Mn and Ba, and siliceous shale by enrichments of B, Zr, V, Cu, and Cr (Table 4.3).

Siliceous Rocks Associated with Carbonate Complexes

In many parts of Yugoslavia, siliceous rocks are associated with carbonate complexes, but here we will discuss only the Lastva section, Budva zone, in the Montenegrin Littoral. The Budva zone, which is equivalent to the Pindos-Olonos zone, is a depositional trough on thinned continental crust situated between two carbonate platforms (Fig. 4.4). The sedimentation of carbonate rocks took place mostly on the slopes, beginning with subtidal deposition (in Ladinian) and evolving later into open-shelf deposition toward the basin.

The Lastva section in the Budva zone, Montenegrin Littoral, consists of a Middle Triassic to Latest Cretaceous carbonate complex. Dominant lithologies include biomicrite, intramicrite, biosparite, intrabiosparite, oosparite, dolomitized limestone of different types, and dolomite (Fig. 4.5). Dolomi-

TABLE 4.2. Chemical analysis of siliceous rocks from the diabase-chert assemblage.

	1	2	3	4
SiO_2	93.75%	97.44%	92.80%	85.60%
TiO_2	0.08	–	0.08	0.16
Al_2O_3	0.80	–	1.50	6.85
Fe_2O_3	1.36	0.25	0.78	2.32
FeO	0.84	0.19	0.16	1.13
MnO	0.01	0.06	0.07	0.05
MgO	1.90	–	0.24	1.00
CaO	0.62	0.50	2.87	0.05
Na_2O	0.16	0.20	0.13	0.20
K_2O	0.30	0.14	0.39	0.45
P_2O_5	–	0.01	tr	0.01
H_2O^+	0.92	1.24	1.38	1.30
H_2O^-	0.21	0.18	0.09	0.56
Totals	100.95%	100.21%	100.49%	99.68%

SiO_2, Al_2O_3, MgO, CaO, and H_2O were analyzed by gravimetric methods; Fe_2O_3, FeO, MnO, and P_2O_5, by calorimetric methods; and K_2O and Na_2O, by flame-photometry. Analyst is S. Šaponja. (1) Chert, No. 14, loc. road from Čačak to Ovčar Banja. (2) Chert No. 13, loc. road from Čačak to Ovčar Banja. (3) Chert No. 11(232), loc. road from Čačak to Ovčar Banja, near the first dam. (4) Chert No., 37, loc. road from Nova Varoš to Bistrica.

tization, silicification, and, rarely, dedolomitization were characteristic processes that affected these rocks. Microstylolites and stylolites commonly occur.

The thickness of the Ladinian and Upper Triassic carbonates in the Budva zone varies from 40 to 222 m, and that of the Jurassic carbonates varies from 67 to 422 m. The Lower Cretaceous sedimentary rocks are 72 to 769 m thick, and Upper Cretaceous rocks are 83 to 419 thick. The section we studied (Fig. 4.5) consists of Jurassic carbonates 209 m thick, Lower Cretaceous sedimentary rocks 70 m thick, and Upper Cretaceous rocks 344 m thick. Characteristically, the content of siliceous rocks from the Lastva section is about 14%.

Chert is widespread, and siliceous rocks commonly include siliceous shale, siliceous limestone, marl, and tuffaceous rocks. In chert, quartz and chalcedony are dominant, and clay minerals (mostly illite), carbonate minerals (mostly calcite, rarely dolomite and ankerite), hematite, pyrite, and organic material are minor. Silica content varies between 89% and 97%; very rarely is it 99%. Radiolarians commonly occur.

The silica content of siliceous shale varies from 83% to 89%. The mineral composition is the same as for chert, but the proportion of clay minerals is higher. Some rocks consist of radiolarians floating in a mudstone matrix. Most commonly, however, radiolarians and remains of other fauna occur in a partly silicified mudstone.

Based on origin, siliceous limestone is separated into two groups. The first group includes biomicrite with radiolarians, and the second group includes partial to complete silicification of biomicrite, oosparite, and other lithologies. In the first group, radiolarians were transported by currents and deposited in the carbonate mud. Some siliceous shales with radiolarians were formed in the same way.

TABLE 4.3. Spectrochemical analyses of siliceous rocks from the diabase-chert assemblage.

	1	2	3	4
B	39	230	105	10
Ga	*	10	*	2
Cr	20	100	46	1
V	32	120	42	3
Ni	16.5	130	86	1
Co	*	27	11	1
Cu	37	76	60	1
Zn	*	190	*	10
Mn	155	3,150	5,000	1
Zr	38	180	70	10
Sc	*	21	11	2
Y	*	60	32	3
La	*	56	*	10
Sr	23	290	90	3
Pb	5	19	*	3
Ba	110	230	210	3

*Below the limit of detection. 4, the limit of detection for each element in ppm. Analyst Lj. Ranković.
(1) Chert, No. 14, loc. Čačak to Ovčar Banja road. (2) Siliceous shale, No. 122, loc. Čačak to Ovčar Banja road. (3) Siliceous shale, No. 123, loc. Čačak to Ovčar Banja road.

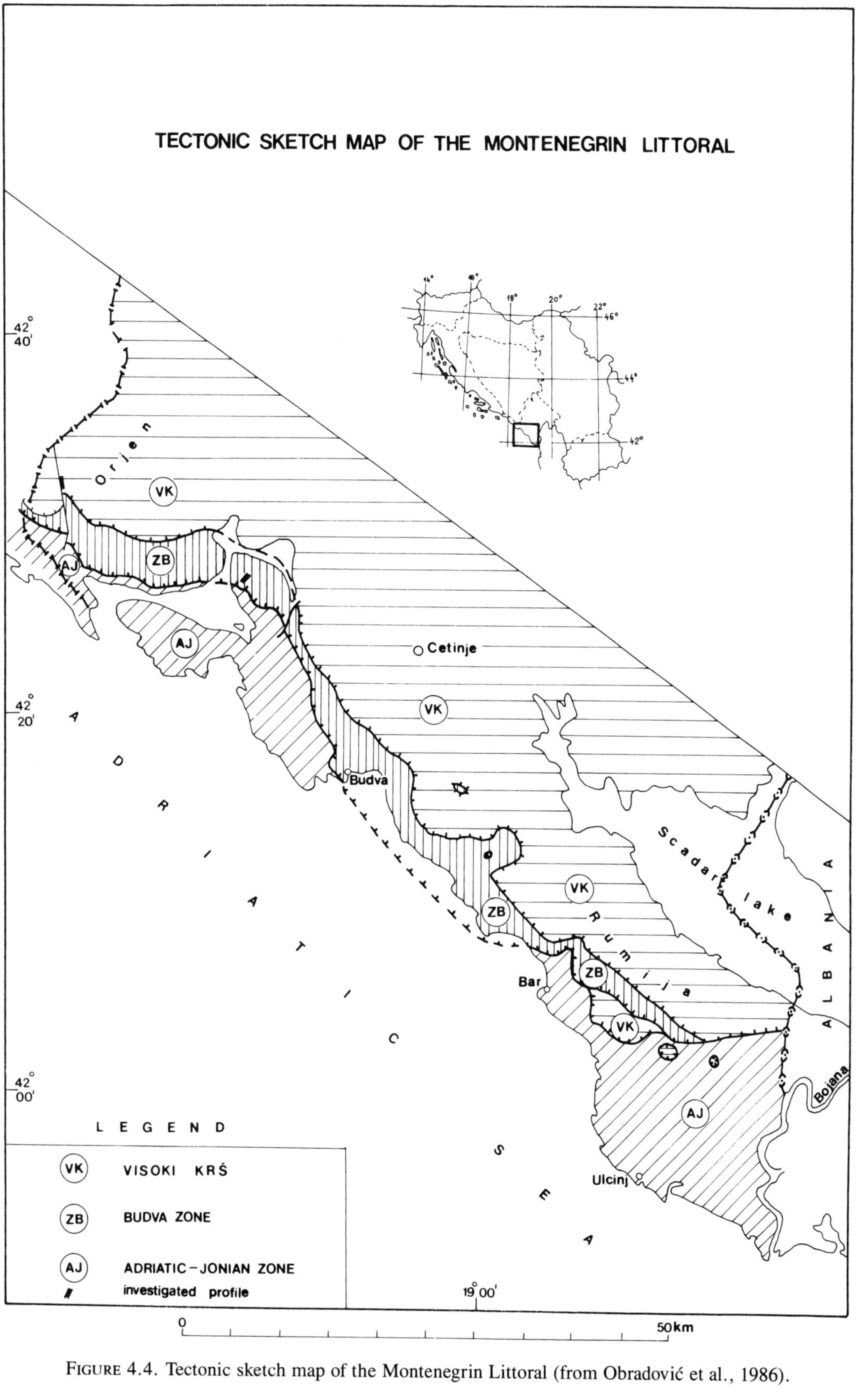

FIGURE 4.4. Tectonic sketch map of the Montenegrin Littoral (from Obradović et al., 1986).

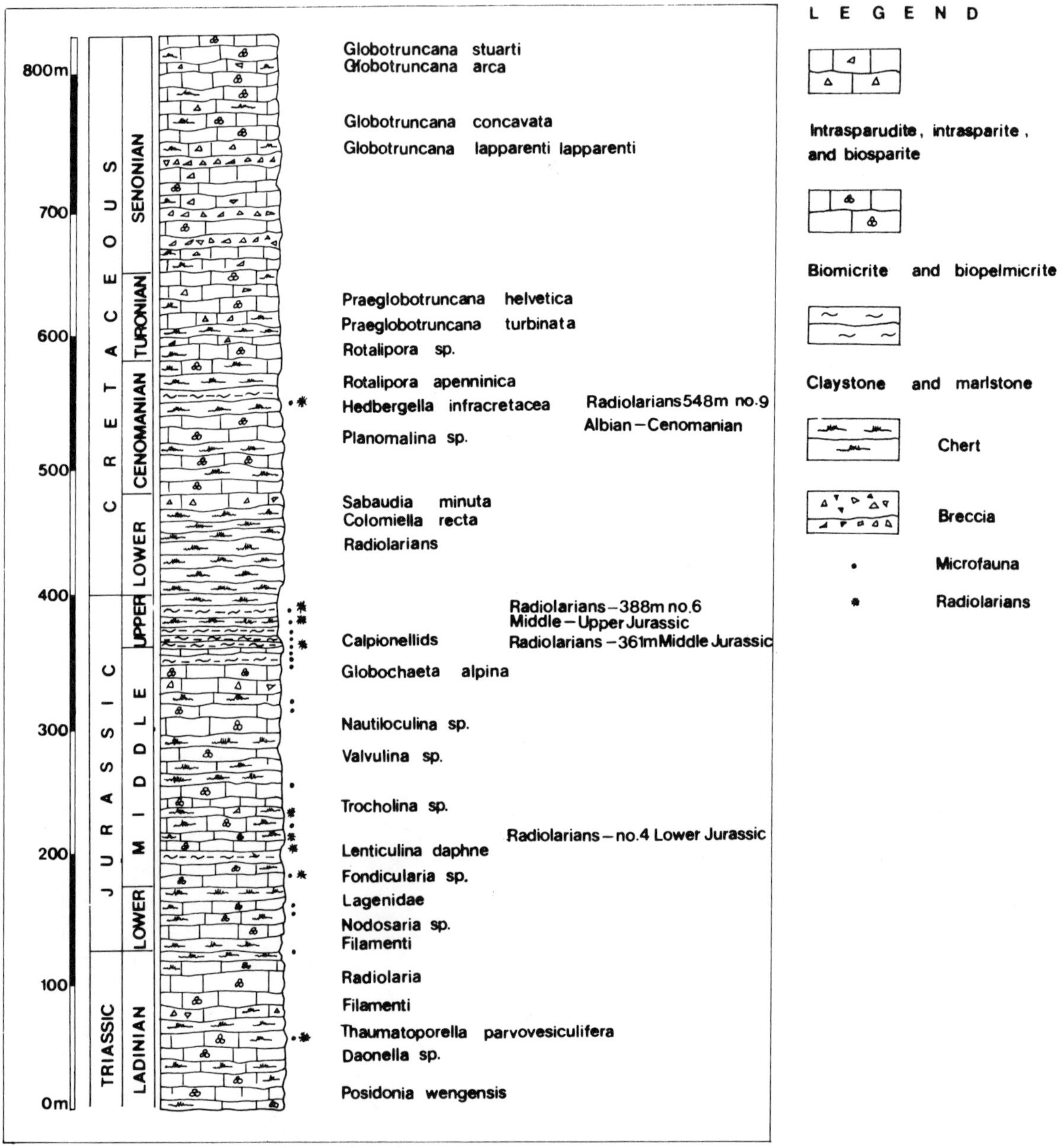

FIGURE 4.5. Geological column of Donja-Gornja Lastva, Budva zone, Tivat (from Obradović et al., 1986).

Some siliceous rocks are called siliceous pelites when the origin of the rocks is uncertain and include marl, claystone, and tuff. These rocks have a high content of silica (more than 90%), but they are less compact and more porous than cherts. They are referred to in other places as porcelanite.

Chert occurs as concretions and bedded chert. Chert concretions of various sizes and shapes formed in all Mesozoic carbonate complexes by mineral segregation and reorganization of the primary silica, which was first (in the Middle and late Triassic) of volcanic origin and later of biogenic origin (Jurassic and Cretaceous).

Bedded chert occurs as isolated beds in carbonate rocks and as groups of beds of chert separated by siliceous shale interbeds. Isolated beds of chert occur in all Mesozoic carbonate complexes and have bed thicknesses up to 20 cm. They were formed by partial to complete replacement of limestone. Sedimentation of chert beds interbedded with shale, which occur mostly in Cretaceous carbonate turbidite sections up to 40 m thick, took place in slope environments.

The radiolarian fauna from the Lastva section (Donja-Gornja Lastva, near Tivat) will be illustrated and described in more detail elsewhere, but data for four samples are presented here. Sample GL4 contains the following radiolarian taxa: *Canoptum* sp., *Katroma neagui* (Pessagno and Poisson, 1981), *Parahsuum simplum* (Yao, 1982), and *Wrangellium* sp.

It is assigned to the Early Jurassic on the basis of stratigraphic ranges of *Parahsuum simplum* (Yao, 1982, 1984) and *Katroma neagui* (Pessagno and Poisson, 1981; De Wever, 1982). The genus *Wrangellium* has also been reported only from the Lower Jurassic so far (Pessagno and Whalen, 1982).

In sample ZB-28 (Fig. 4.5), a rich and well-preserved radiolarian assemblage was found. Some of the taxa include: *Acanthocircus suboblongus* (Yao, 1972), *Andromeda praepopodbielensis* (Baumgartner, 1984), *Cyrtocapsa mastoidea* (Ichikawa and Yao, 1976), *Gorgansium silviense* (Pessagno and Blome, 1980), *Lupherium officerense* (Pessagno and Whalen, 1982), *Mesosaturnalis tetraspinus* (Yao, 1972), *Parvicingula dhimenaensis* (Baumgartner, 1984), *Tetraditryma pseudoplena* (Baumgartner, 1980), *Unuma typicus* (Ichikawa and Yao, 1976), and *Unuma echinatus* (Ichikawa and Yao).

The assemblage can be assigned to U.A.1 (Bathonian) of Baumgartner's (1984) range chart, since *Parvicingula dhimenaensis* and *Tetradytrima pseudoplena* have their first appearance and *Unuma echinatus* has its last appearance in this association. The range of *Andromeda praepopodbielensis*, which has been recorded only from U.A.0, should most probably be extended to U.A.1. Our faunas correspond to the *Unuma echinatus* zone according to the zonation of Yao (1984), which includes three diagnostic species: *Unuma typicus, Cyrtocapsa mastoidea*, and *Unuma echinatus*.

Sample GL6 (Fig. 4.5) was taken 27 m above sample ZB-28 (361 m). It contains the following radiolarian species: *Acanthocircus suboblongus* (Ichikawa and Yao, 1972), *Archaeospongoprunum imlayi* (Pessagno, 1977a), *Emiluvia premyogii* (Baumgartner, 1984), *Eucyrtidiellum unumaensis* (Ichikawa and Yao, 1976), *Parvicingula dhimenaensis* (Baumgartner, 1984), *Stichocapsa japonica* (Ichikawa and Yao, 1976), and *Triactoma jonesi* (Pessagno, 1977a).

Radiolarians are poorly preserved and are represented by rather long-range species. According to Baumgartner (1984), the assemblage could be assigned to the time interval of U.A.1 to U.A.8 (late Bathonian to early Kimmeridgian) on the basis of the first occurrence of *Parvicingula dhimenaesis* in U.A.1 and the last occurrences of *Emiluvia premyogii* and *Eucyrtidiellum unumaensis* in U.A.8.

Sample GL9 (Fig. 4.5) contains the following radiolarians: *Archaeodictyomitra simplex* (Pessagno, 1977b), *Holocryptocanum* sf. barbui (Dumitricâ, 1970) (morphotype with a smoother abdominal surface), *Mita gracilis* (Squinabol, 1903), *Pseudodictyomitra lodogaensis* (Pessagno, 1977b), *Stichomitra communis* (Squinabol, 1903), and *Thanarla praeveneta* ? (Pessagno, 1977b).

According to this fossil assemblage, the age should be Albian or Cenomanian. The assemblage shows most species in common with *Holocryptocanum barbui–H. geysersensis* assemblage zone (Nakaseko and Nishuimura, 1981; Yao, 1984). Baumgartner (1984) found *Holocryptocanium barbui* from the Valanginian on. Our Radiolaria tests show smoother surfaces than described and illustrated by Baumgartner. Forms with a smoother abdominal surface seem to be restricted to a younger age. The lower limit of the age assignment is supported also by the presence of *Mita gracilis*, which starts in the late Albian (Schaaf, 1985). No representative of the *Holocryptocanium barbui* group, either with a smooth abdominal surface or with distinctive pore frames, has been reported from strata younger than Cenomanian.

Mesozoic Siliceous Deposits from the Carpatho-Balkanides

Siliceous deposits in association with carbonate rocks occur in the Carpatho-Balkanides tectonic zone. According to Vasić (1985), chert occurs in carbonates of Jurassic to Berriasian age. Carbonate sedimentation began in the Lias and ended in the Berriasian, with short interruptions at the Lias and Dogger boundary, as indicated by hard grounds.

The Dogger (Bajocian, Bathonian, and Callovian) is characterized by thin, condensed stratigraphic sections. The thickness of Bajocian and Bathonian red bedded limestone is 50 cm. Callovian Ammonitico Rosso contains many molds of ammonites and is 6 to 10 m thick. Red limestone deposited during the Malm is interbedded with gray sparite limestone, and together they form a 30-m-thick section.

From the end of Dogger to the Berriasian, chert occurs as concretions or as isolated beds formed by partial to complete replacement of various types of limestone by silica. Silica is of biogenic origin.

Radiolarians commonly occur. Silicification of limestone occurred during early to late diagenesis.

Discussion

In the carbonate complexes, cherts occur as concretions and bedded chert. Chert concretions of various sizes and shapes formed in all Mesozoic carbonate complexes by mineral segregation and reorganization of primary silica, a minor constituent of the carbonate host rock.

Bedded chert occurs as isolated beds in carbonate rocks and as groups of beds of chert separated by siliceous shale interbeds. Isolated beds of chert occur in all Mesozoic carbonates and are formed by partial to complete replacement of limestone.

Also in the carbonate complexes, up to 50-m-thick sections of rhythmically bedded chert and siliceous shale occur. An example of a rhythmically bedded chert and shale is the Cretaceous carbonate turbidites of the Lastva section in the Budva zone. The formation of this section can be explained in two ways. First, radiolarian-rich turbidity currents (Nisbet and Price, 1974) deposited the chert beds, whereas siliceous shale represents mainly interturbidite hemipelagic sedimentation, which accumulated slowly and which was probably modified by near-surface dissolution of Radiolaria. We did not find any evidence of turbidity current deposition except that the bedded chert occurs within carbonate turbidite sections. It is difficult to explain both the carbonates and the cherts by the mechanism of turbidite deposition.

The second possible depositional mechanism is that deposition of the bedded chert took place on offshore banks located on the slopes of the Budva trough. Episodic production of radiolarians in the surface water produced rapid sedimentation of chert beds and intervening slow sedimentation of siliceous shale. Around the banks, turbidity currents transported and deposited carbonate material from the carbonate platforms.

In the Budva zone, which is equivalent to the Pindos-Olinos zone (De Wever, this volume), chert with radiolarians occurs in sections of Middle Triassic through Cretaceous ages.

The occurrence of chert from the porphyrite-chert assemblage and the Diabase-Chert Formation is more complicated than from the carbonate complexes. Some cherts, isolated beds, and concretions formed by the same mechanisms as the chert from carbonate complexes. However, some cherts in the porphyrite-chert assemblage formed by replacement of tuffs by silica (isolated beds).

Middle-Upper Triassic chert interbedded with siliceous shale in the porphyrite-chert assemblage and its olistoliths in the Diabase-Chert Formation formed in the protected parts of a marine basin, isolated from the influx of clastic material. During the middle-Late Triassic, the sea was shallow, and carbonate sedimentation took place on the margins. The relief of the surrounding land was subdued, producing a low rate of erosion. The influx of clastic material was minimal. Hemipelagic sedimentation of fine-grained detrital material occurred at a constant but slow rate and was intermittently interrupted by rapid accumulation of radiolarians, either from turbidity currents or from increased productivity in surface waters.

The Nova Varoš to Bistrica Road section of the Callovian to early Kimmeridgian age consists of chert interbedded with siliceous shale and rarely, carbonate rocks. Chert from this section in the Diabase-Chert Formation formed from deposition by turbidity currents. Siliceous shale was deposited as hemipelagic mud between intervals of turbidity currents. Alternatively, the sedimentation of the chert and shale may have taken place in the protected parts of the basin by the double accumulation model of Iijima and Utada (1983).

In conclusion, we suggest that the breaking of the Dinaridic plate began at the end of the Early Triassic, followed by hybrid volcanic activity, with the greatest production of volcanic rocks in the Ladinian. At that time, an oceanic region developed and existed to the Late Jurassic, which is now occupied by the Inner Dinarides and Vardar zone (Karamata, 1986). Bedded chert formed in the different parts of the basin by different depositional mechanisms. Conditions for bedded chert formation included separation of coarse clastic material, high productivity of siliceous organisms, the influx of fine-grained clastic material, and persistence of such conditions for a long period of time (Iijima and Utada, 1983). All of those conditions occurred in Middle Triassic to Late Jurassic time.

Acknowledgments. We have benefited from discussions and collaboration with D. Milovanović, N. Vasić, B. Mirković, M. Mirković, and P. Vujisić. We are grateful to S. Karamata and J.R. Hein for their critical review of the manuscript and to P.O. Baumgartner, who reviewed the radiolarian part of the manuscript. Scanning electron micrographs were taken by V. Segalla and J. Werning. Chemical analyses were performed by S. Šaponja. Infrared spectroscopy was done by V. Poharc-Logar, and spectrochemical analysis by Lj. Ranković. V. Mudronja assisted in the preparation of the figures.

References

Baumgartner PO (1980) Late Jurassic Hagiastridae and Patulibracchiidae (Radiolaria) from the Argolis Peninsula (Peloponnessus, Greece). Micropaleontology 26(3):274–322.

Baumgartner PO (1984) A middle Jurassic–early Cretaceous low-latitude radiolarian zonation based on unitary associations and age of Tethyan radiolarites. Ecolgae Geologicae Helveticae 77(3):729–837.

Bešić Z (1975) Geology of Monte Negro (in Serbian). Special Publication Academy Science Monte Negro I:1.

Blome CD (1984) Upper Triassic Radiolaria and radiolarian zonation from western North America. Bulletin American Paleontology 85/318:5–88.

Ćirić B (1954) Einige Beobachtungen uber die Diabase-Hornstein-Formation der Dinariden. Bull Serv Geol XI:31–88.

Ćirić B (1984) Complexe ophiolitique et formations comparables dans les Dinarides. Mem Ser Geol Geophys XXIII:1–44.

D'Argenio B, Radoičić R, Sgrosso J (1971) A paleogeographic section through the Italodinaridic external zones during Jurassic and Cretaceous time. Nafta (Zagreb) 4/5:195–207.

De Wever P (1982) Radiolaires du Trias et du Lias de la Tèthys (Systèmatique, Stratigraphie). Société Géologique Nord 7:1–599.

De Wever P (1984) Triassic radiolarians from the Darnò area (Hungary). Acta Geologica Hungarica 27/34: 295-306.

De Wever P, Sanfilippo A, Riedel WR, Gruber B (1979) Triassic radiolarians from Greece, Sicily and Turkey. Micropaleontology 25(1):75–110.

Dimitrijević MD, Dimitrijević MN (1973) Olistostrome melange in the Yugoslavian Dinarides and late Mesozoic plate tectonics. Journal Geology 81:328–340.

Dimitrijević MD, Dimitrijević MN (1974) Internal fabric of Mesozoic bodies in the melange around Nova Varoš: A case for gravity tectonics. Bull Sci Sec A 19:3–4.

Dimitrijević MD (1986) Siliceous rocks in the ophiolitic melange. Guide book with abstracts. In: Obradović J (ed) Third International Conference on Siliceous deposits, IGCP 187, pp 43–49.

Dumitricâ P (1970) Cryptocephalis and Cryptothoracit nacellaria in some Mesozoic deposit of Romania. Review Románian Geology Geophysics Geography Series in Geology 14:45–124.

Dumitricâ P (1978) Family Eptingiidae n. fam., extinct Nassellaria (Radiolaria) with sagittal ring. DS Inst Geol Geophys 64:27–38.

Dumitricâ P (1982) Triassic Oertlisponinae (Radiolaria) from eastern Carpathians and southern Alps. D S Inst Geol Geophys 67:57–74.

Dumitricâ P, Kozur H, Mostler H (1980) Contribution to the radiolarian fauna of the middle Triassic of the southern Alps. Geol Paleontol Mitt Innsbruck 10(1): 1–46.

Grubić, A (1986) Tectonic sketch of Yugoslavia. In Obradović J. (ed) Third International Conference on Siliceous deposits, IGCP 187, Yugoslavia, pp. 9–31.

Ichikawa K, Yao A (1976) Two new genera of Mesozoic cyrtoid radiolarians from Japan. Special Publication, Progress in Micropaleontology, pp 110–117.

Iijima A, Utada M (1983) Recent developments in the sedimentology of siliceous deposits in Japan. In: Iijima A, Hein JR, Siever R (eds) Siliceous Deposits in the Pacific Region. Elsevier, Amsterdam, pp 45–64.

Karamata S (1975) Geological evolution of Yugoslavian region from Triass to Quaternary. Radovi Inst Geol Rud Istr Nuklearnih i dr Min Sirovina 10:1–16 (in Serbian).

Karamata S (1981) Time and space in the plate tectonic modelling of tectonic, magmatic and metamorphic processes in Tethys-type orogenic belt. *Bull Acad Sci Serb* 21:27–47.

Karamata S (1986) "The Diabase-Chert Formation"– some genetic aspects. Guide-book and abstracts. In: Obradovič J (ed) Third International Conference on Siliceous Deposits, IGCP 187, Yugoslavia, pp 31–43.

Knežević V (1975) Triassic igneous rocks of Montenegro. Acta Geol VIII/1–25, 41:107–147 (in Serbian).

Knežević V, Obradović J, Djordjević P (1978) Triassic volcanic rocks and volcanic-sedimentary formation in western Serbia, in zone northern from Drina-Ivanjica crystalline. Radovi Geol Inst 12:219–228 (in Serbian).

Kozur H, Mostler H (1981) Beiträge zur Erfoschung der mesozoischen Radiolarien. Teil IV: Thalassosphaeracea Haeckel, 1862, Hexastylacea Haeckel, 1882 emend. Petruševskaja, 1979, Sponguracea Haeckel, 1862 emend. und weitere triassische Lithocycliacea, Trematodiscacea, Actinommacea und Nassellaria. Geol Palaont Mitt Innsbruck, Sonderbd, pp 1–208.

Kozur H, Réti Z (1986) The first paleontological evidence of Triassic ophiolites in Hungary. N Jb Geol Palaont Mh 1986(5):284–292.

Nakaseko K, Nishimura A (1981) Upper Jurassic and Cretaceous Radiolaria from the Shimanto group in southwest Japan. Science Report College General Education Osaka University 30(2):133–203.

Nisbet EG, Price I (1974) Siliceous turbidites: Bedded cherts as redeposited, ocean-derived sediments. In: Hsü KJ, Jenkyns HC (eds) Pelagic Sediments on Land and Under the Sea. Special Publication International Association of Sedimentologists 1:351–366.

Obradović J (1972) Siliceous rocks from the volcano-sedimentary formation in Montenegrin Littoral. Radovi Geol Inst 8:25–33 (in Serbian).

Obradović J (1978/79) Middle Triassic volcanoclastites from Dinarides. Geol An Balk Pol XLIII–XLIV:3–71 (in Serbian).

Obradović J, Milovanović D, Vasić N (1986) Siliceous rocks in Mesozoic from Serbia. I. In association with the Porphyrite-chert and Diabase-chert Formation. Glasnik Prir Muzeja Srpske Zemlje 40 (in press) (in Serbian).

Obradović J, Mirković B, Mirković M, Goričan S, Vujisić P (1986) Siliceous rocks from Mesozoic carbonate complex, Budva zone, Montenegrin littoral. Montenegrin Academy Science Collected Scientific Works Academy Zarija Bešić 2:317–343.

Pamić J (1969) Middle Triassic Spilite–Keratophyre association of the Dinarides and its position in the Alpine magmatic-tectonic cycle. Geol Glas 13:205–217.

Pessagno EA, Jr (1977a) Upper Jurassic Radiolaria and radiolarian biostratigraphy of the California Coast Ranges. Micropaleontology 23(1):56–113.

Pessagno EA, Jr (1977b) Lower Cretaceous radiolarian biostratigraphy of the Great Valley Sequence and Franciscan Complex, California Coast Ranges. Cushman Foundation Special Publication 15.

Pessagno EA, Jr, Bloome CD (1980) Upper Triassic and Jurassic Pantanelliinae from California, Oregon and British Columbia. Micropaleontology 26(3):225–273.

Pessagno EA Jr, Poisson A (1981) Lower Jurassic Radiolaria from the Gümüslü Allochton of southern Turkey (Taurides Occidentales). Bull Min Res Explor Inst Turkey 92:42–69.

Pessagno EA, Jr, Whalen PA (1982) Lower and Middle Jurassic Radiolaria (multicyrtid Nassellariina) from California, east-central Oregon and the Queen Charlotte Islands, B.C. Micropaleontology 28(2):111–269.

Pessagno EA Jr, Finch W, Abbot PL (1979) Upper Triassic Radiolaria from the San Hipolito Formation, Baja California. Micropaleontology 25(2):160–197.

Radoičić R (1982) Carbonate platforms of the Dinarides; the example of Montenegro-West Serbian sector. Bull Serb Acad Sci Arts LXXX 22:35–47.

Schaaf A (1985) Un nouveau canevas biochronologue du Crètacè infèrièur et moyen: Les biozones à radiolarres. Sci Geol Bull 38(3):227–269.

Squinabol S (1903) Le Radiolarie dei noduli selciosi nella Scaglia degli Euganei: Contributione Riv. Italiana Paleon. 9:105–150.

Stefanovska D, Antonijević R (1967) Characteristics of development of carbonate series from Cukali zone. VIII Cong KBGA, Beograd, pp 419–423.

Vasić N (1985) Jurassic carbonate and siliceous sediments from the area Pesača-Boljetin, East Serbia. Unpubl MS thesis, Faculty of Mining and Geology, Belgrade.

Yao A (1972) Radiolarian fauna from the Mino belt in the northern part of the Inuyama area, Central Japan. Part I. Spongosaturnalids. Journal of Geoscience of Osaka City University 15:21–64.

Yao A (1982) Middle Triassic to early Jurassic radiolarians from the Inuyama area, central Japan. Journal Geoscience Osaka University 25(4):53–70.

Yao A (1984) Subdivision of the Mesozoic complex in Kü-Yura area, southwest Japan and its bearing on the Mesozoic basin development in the southern Chichibu terrane. Journal Geoscience Osaka University 27(2): 41–103.

5

Origin and Tectonic Significance of Radiolarian Chert in the Austroalpine Rifted Continental Margin

A. Vecsei, W. Frisch, M. Pirzer, and A. Wetzel

Overview

Rhythmically bedded radiolarian chert of Late Jurassic age (Ruhpolding Formation) was deposited by bottom currents and low-density turbidity currents in the Northern Calcareous Alps. The radiolarians were concentrated by sorting of hydrodynamically equivalent particles. High production of radiolarians and low terrigenous input were characteristic of these deposits. Carbonate dissolution was of secondary importance. The radiolarian chert was deposited below the aragonite compensation depth, but for the most part, above the calcite compensation depth.

Deposition of the radiolarian chert marks a change in the tectonic and topographic setting of the Austroalpine rifted continental margin. A horst and graben pattern, formed at the beginning of the Jurassic, was replaced at the time of the Oxfordian/Kimmeridgian boundary by a tilted block array facing the Piemont Ocean. The change in topography enabled an open, low-energy current system to be established, which was responsible for the deposition of the radiolarian chert. Steepening of the topography resulted in sedimentation by high-energy bottom currents and various types of sediment gravity flows along the steep flanks of the asymmetrical tiltblock basins (Tauglboden Formation).

Introduction

Rhythmically bedded radiolarian chert has been the object of several studies in the Northern Calcareous Alps (e.g., Garrison and Fischer, 1969; Schlager and Schlager, 1973; Schlager and Schöllnberger, 1974; Diersche, 1980). Interest has focused on these rocks because they are an important marker horizon in the tectonic and subsidence evolution of the Austroalpine rifted continental margin.

However, understanding of both the origin and the significance of these rocks remains controversial. In this paper, which is based on the detailed studies of Vecsei (1986) and Pirzer (1988), the sedimentology of the rhythmically bedded radiolarian chert and associated rocks is analyzed and interpreted in terms of depositional mechanisms and tectonic environment. Based on these results, we propose a model that relates the origin of the radiolarian chert to the tectonic evolution of the Austroalpine continental margin.

Field Areas

The chert studied is exposed in the Northern Calcareous Alps, a thin-skin thrust complex which belongs to the Austroalpine megaunit and is made up of largely unmetamorphosed Permo-Mesozoic sequences. The field areas are located in the central part of this east-west trending complex: the Osterhorn Mountains southeast of Salzburg, and the Unken syncline north of Lofer in Austria (Fig. 5.1). The two regions comprise a well-developed Jurassic sequence with no major tectonic disturbances.

Geological Framework

During the Jurassic, the Northern Calcareous Alps were part of the continental margin of the Piemont or South Penninic Ocean, and are referred to as the Austroalpine continental margin (Fig. 5.2). The Piemont Ocean was a narrow ocean connecting the

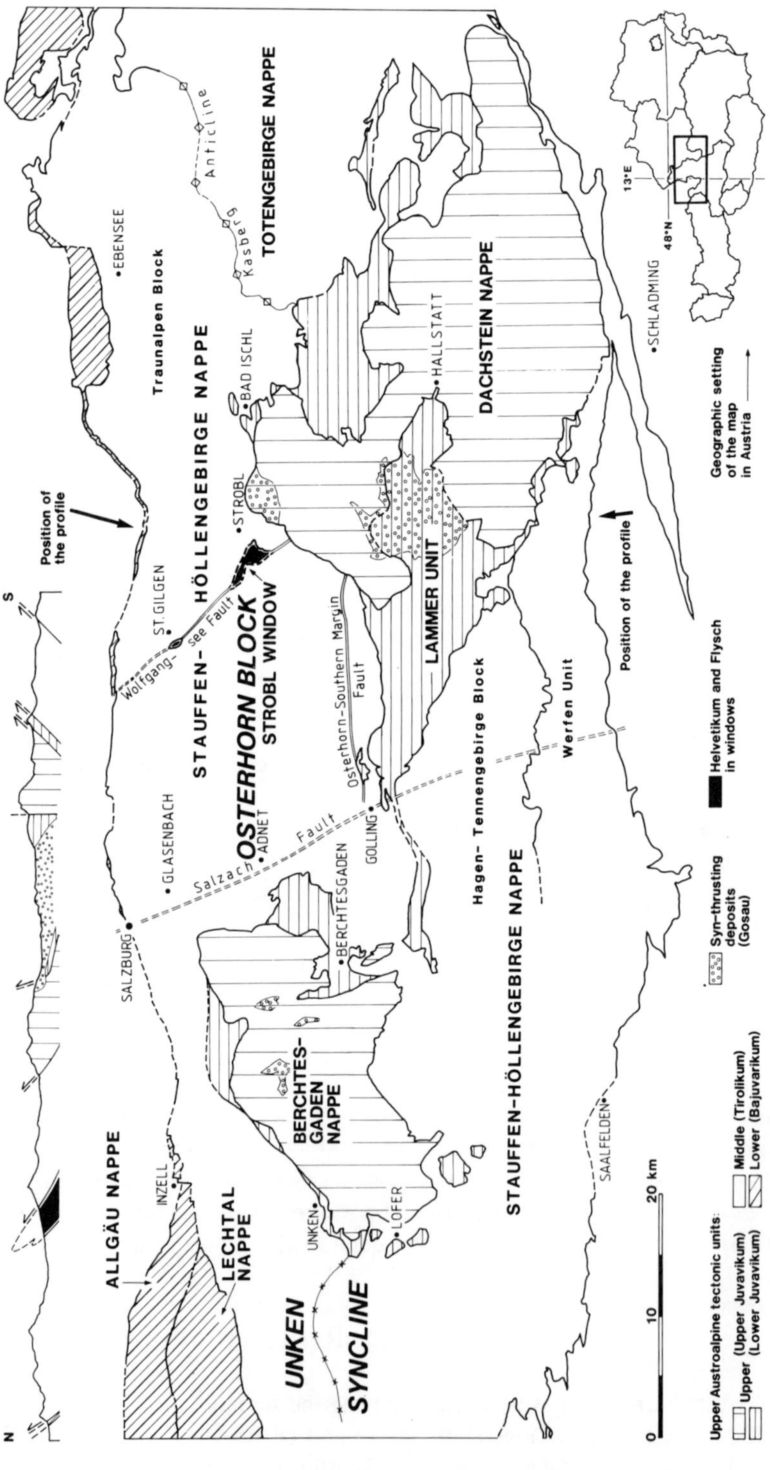

FIGURE 5.1. Sketch map showing the Osterhorn block and the Unken syncline within the central part of the Northern Calcareous Alps. After Tollmann (1976).

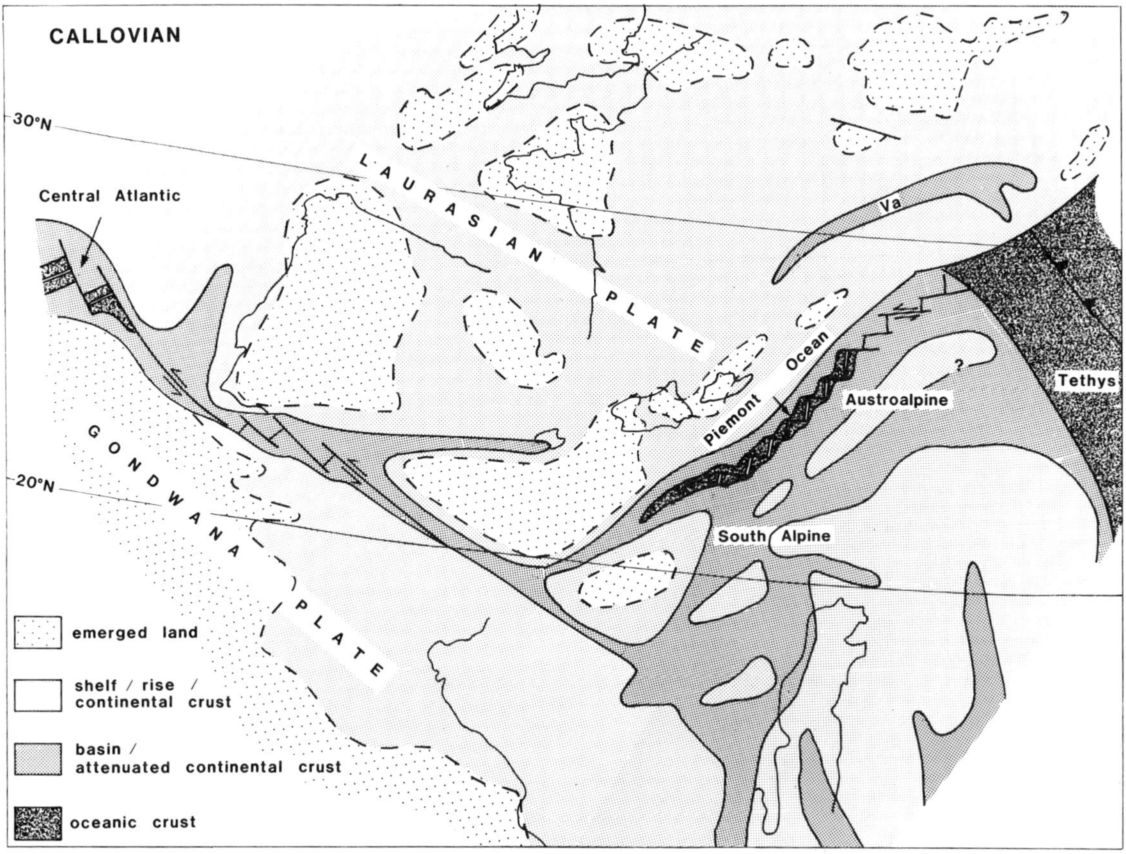

FIGURE 5.2. Paleogeographic reconstruction of the Piemont Ocean and adjoining areas during the Callovian. The Northern Calcareous Alps are part of the Austro-Alpine realm. Va, Valais. After Frisch (1979) and Dercourt et al. (1986).

Tethys to the east and the central Atlantic to the west. Transtension in the Austroalpine continental margin led to the opening of this ocean during the middle-Late Jurassic (Frisch, 1979; Weissert and Bernoulli, 1985; Dercourt et al., 1986).

The rhythmically bedded radiolarian chert occurs in a sequence typical of the Jurassic of the Mediterranean region (Bernoulli and Jenkyns, 1974) (Figs. 5.3, 5.4). In the Middle and Late Triassic, the Northern Calcareous Alps were dominated by a carbonate platform with thick reef buildups, lagoonal deposits, and local basinal sequences. The carbonate platform disintegrated during the Rhaetian and Hettangian. This disintegration coincided with a change to pelagic sedimentation over the entire area.

In the areas studied, the typical sequence reflecting this evolution passes from (1) carbonate platform sedimentary rocks (Hauptdolomit, Norian), through (2) marls with intercalated ahermatypic reefs (Kössen Formation and Oberrhätischer Riffkalk, Rhaetian) and (3) red crinoid limestone (Hierlatzkalk, Hettangian), to (4) red cephalopod limestone (Adnet and Klaus limestones, Hettangian through Oxfordian), which laterally passes into hemipelagic basinal limestone (Allgäu Formation). (5) The radiolarian chert (Ruhpolding Formation) was deposited around the time of the Oxfordian/Kimmeridgian boundary (Huckriede, 1971). The chert is overlain in most sections by (6) radiolarian limestone of varying thickness containing sediment gravity flow deposits (Tauglboden formation, Kimmeridgian; Schlager and Schlager, 1973). The clastic carbonate passes upsection into (7) pelagic to hemipelagic and, in part, redeposited limestone (Oberalm Formation, Tithonian), which resembles the contemporaneous Maiolica facies in the Southern Alps (Weissert, 1981; Stow et al., 1984).

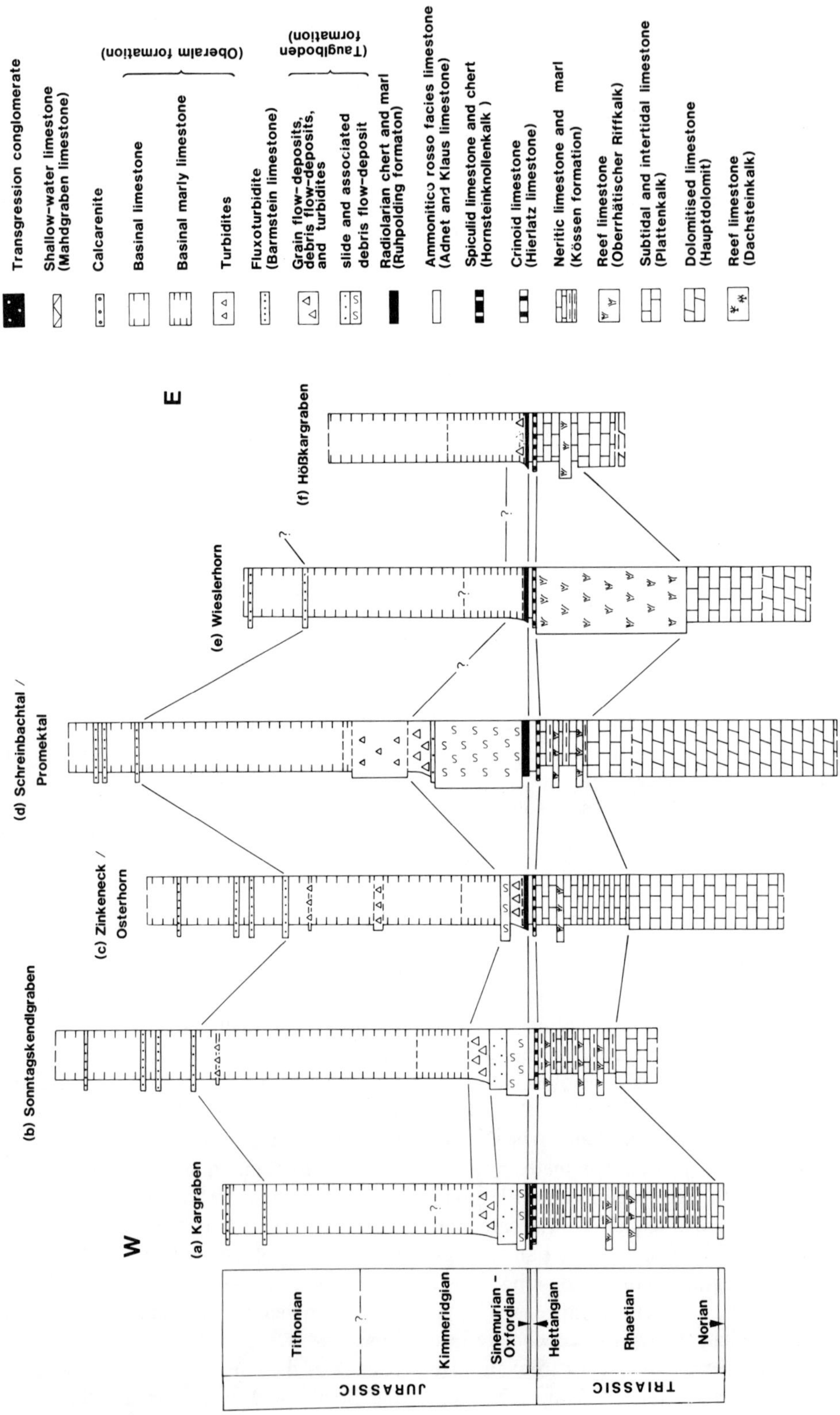
W
E
(a) Kargraben
(b) Sonntagskendlgraben
(c) Zinkeneck / Osterhorn
(d) Schreinbachtal / Promektal
(e) Wieslerhorn
(f) Hößkargraben
JURASSIC
TRIASSIC
Tithonian
Kimmeridgian
Sinemurian - Oxfordian
Hettangian
Rhaetian
Norian
Transgression conglomerate
Shallow-water limestone (Mahdgraben limestone)
Calcarenite
Basinal limestone
Basinal marly limestone
Turbidites
(Oberalm formation)
Fluxoturbidite (Barmstein limestone)
Grain flow-deposits, debris flow-deposits, and turbidites
slide and associated debris flow-deposit
(Tauglboden formation)
Radiolarian chert and marl (Ruhpolding formaton)
Ammonitico rosso facies limestone (Adnet and Klaus limestone)
Spiculid limestone and chert (Hornsteinknollenkalk)
Crinoid limestone (Hierlatz limestone)
Neritic limestone and marl (Kössen formation)
Reef limestone (Oberrhätischer Riffkalk)
Subtidal and intertidal limestone (Plattenkalk)
Dolomitised limestone (Hauptdolomit)
Reef limestone (Dachsteinkalk)

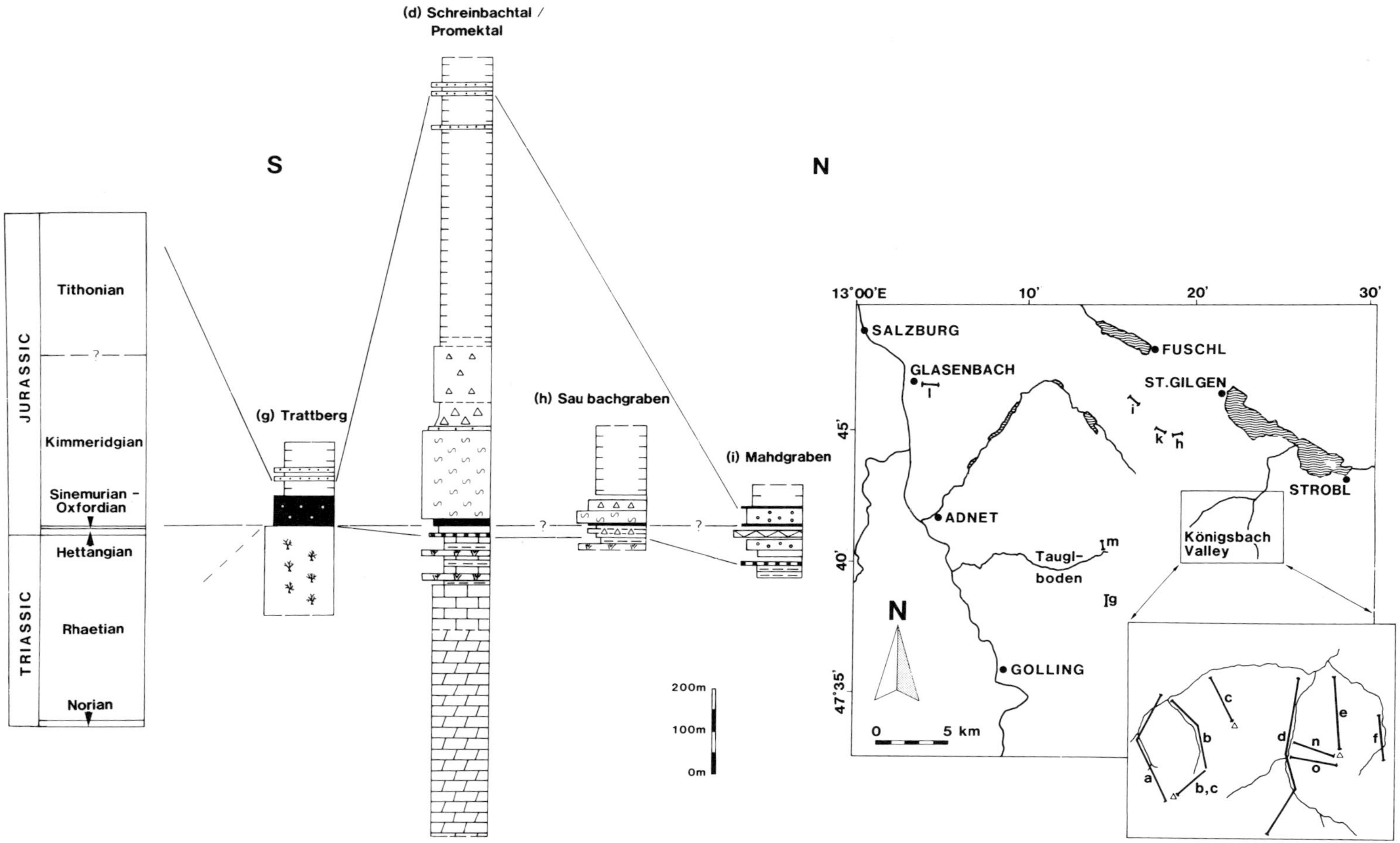

FIGURE 5.3. Geographic position and stratigraphy of the studied sections in the Osterhorn block. Mahd-Graben section after Plöchinger (1964); Saubach-Graben section after Plöchinger (1975); the other sections are from Vecsei (1986). Sections l-o not shown (l: Glasenbach; m: Urban-Graben; n: KuhglückGraben; o: Himmelreich-Graben).

Methods

The Jurassic paleogeography of the central part of the Northern Calcareous Alps was reconstructed on the basis of the distribution and thickness of the basinal and condensed facies (see Tollmann, 1976; Diersche, 1980). The distribution and the transport directions determined from sedimentary structures support this reconstruction (Vecsei, 1986; Pirzer, 1987). Facies analysis was conducted on 29 sections located in the Osterhorn Mountains and the Unken Syncline (Figs. 5.3, 5.4). These were compared with sections form other regions in the Northern Calcareous Alps described in the literature.

Two sections, undisturbed by both synsedimentary erosion and tectonism, were selected for microfacies analysis (Himmelreich-Graben and Kuhglück-Graben in Promektal area, Osterhorn Mountains; Fig. 5.3). Samples were taken at 50-cm intervals; every bed at the base of the chert sections was sampled. Sedimentary structures were studied on polished slabs. X-ray radiography of 3-mm-thick slabs revealed sedimentary structures in detail. Examination of microstructures was carried out by thin-section analysis and SEM studies. Compositional trends were obtained by energy-dispersive absorption analysis (EDAX) attached to the SEM.

Sedimentological Features of the Jurassic Sequence

The Sequence Below the Radiolarian Chert

Between the Hettangian and Oxfordian, the types of sedimentation in the Northern Calcareous Alps were divided into areas of condensed sequences deposited on structural highs (horsts) and areas of extended sequences in basins (for compilation, see Tollmann, 1976).

Condensed sequences are characterized by red cephalopod limestone (Adnet and Klaus limestones) of the typical Rosso Ammonitico facies. According to Wendt (1970), this limestone contains algal remains near the base, testifying to formation at a water depth within the photic zone. The nodular appearance of these rocks is due to intense bioturbation and superimposed differential early lithification (Wendt, 1971; Jenkyns, 1974). The small thickness of the sequence and faunal distribution testify to low accumulation rates. The upper part of the red cephalopod limestone is even more condensed than the lower part, as suggested by the increased number of hardgrounds, which are partially coated with ferromanganese crusts.

The extended sequences are characterized by strongly bioturbated, pelagic and hemipelagic limestone and marl (Allgäu Formation; Jacobshagen, 1965) with intercalated turbidites. The change from crinoidal to pelagic bivalve debris in these turbidites indicates a gradual increase in water depth of the source areas through time (Bernoulli and Jenkyns, 1970).

Because of the contrasting types of sedimentation and the distribution of the sedimentary rocks, the depositional environment of the Hettangian to Oxfordian deposits is best characterized by a horst and graben tectonic setting. The exact geometry of the grabens, however, is not well known. Local sediment gravity flows during the Liassic were triggered along scarps formed by normal faulting (e.g., Glasenbach-Klamm and Saubachgraben in the Osterhorn Mountains; Bernoulli and Jenkyns, 1970; Plöchinger, 1975).

Radiolarian Chert and Associated Rocks

Rhythmically bedded radiolarian chert was deposited either directly on the red cephalopod limestone, in places capped by ferromanganese crusts (Fig. 5.6A), or above a thin member of pelagic bivalve limestone overlying the cephalopod limestone. Outside the field areas studied in detail are sections in which the radiolarian chert overlies basinal limestone (Allgäu Formation).

Radiolarian chert occurs near the base of a characteristic sequence of sedimentary rocks (Fig. 5.5). Deposits related to steep slopes (e.g., grain flow deposits) occur only near the southern margins of the depositional basins. These sedimentary rocks decrease in thickness and frequency from south to north within the basins and wedge out on the southern flanks of the topographic highs. The asymmetry of the basin infill (Diersche, 1980) indicates that the basins were half-grabens or tilted blocks with a polarity toward the Piemont Ocean. These half-grabens came into existence around the time of the Oxfordian/Kimmeridgian boundary and replaced the earlier horst and graben topography. They persisted during the Kimmeridgian and Tithonian. Half-graben margins frequently did not conform to the margins of the earlier-formed horsts and grabens. This evolution resulted in widespread changes in topography so that some rises became basins, and vice versa. The

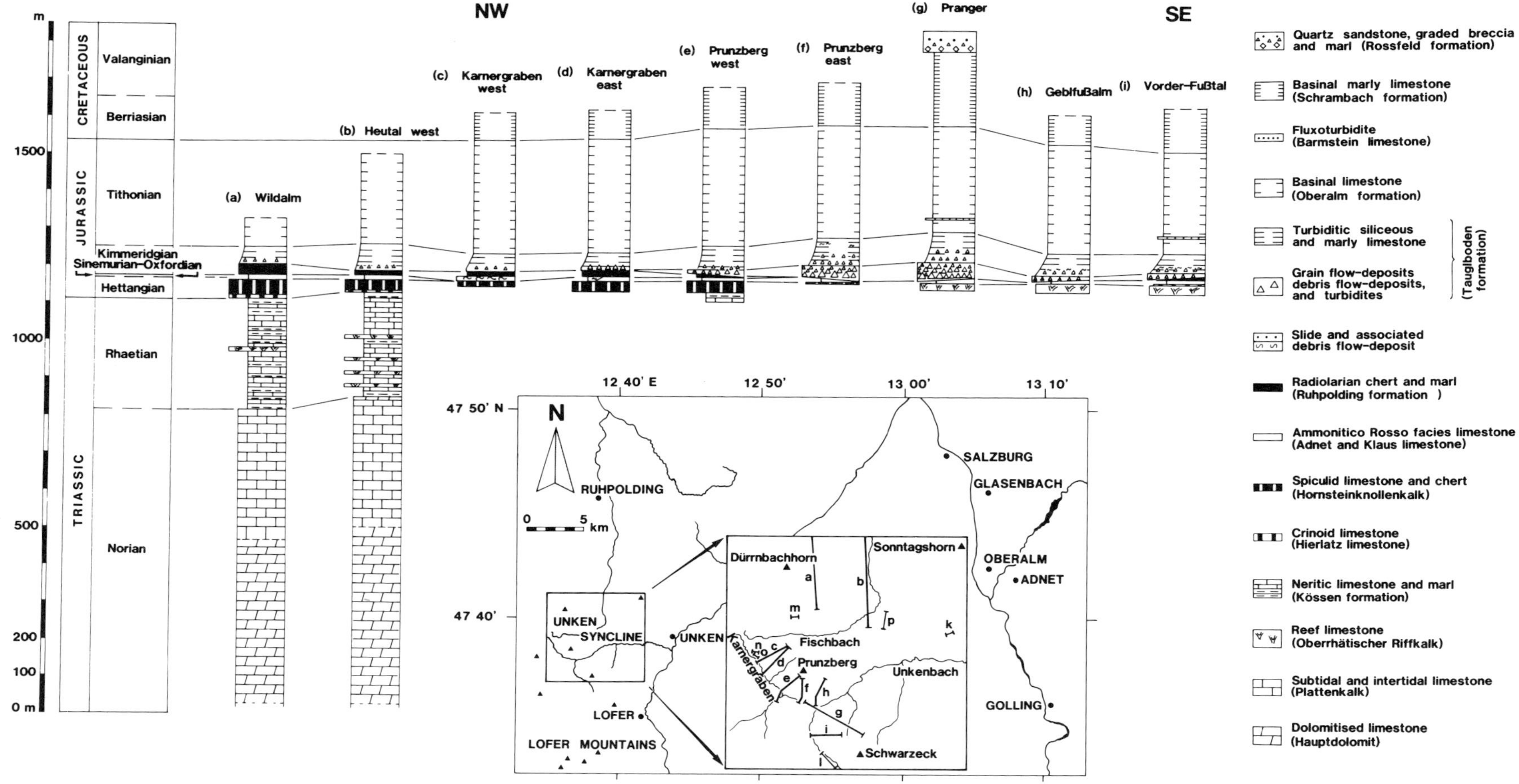

FIGURE 5.4. Geographic position and stratigraphy of the studied sections in the Unken syncline. Sections k–p not shown (k: Talbruck; l: Rottenbach; m: Hochgimpling; n: Muckklause – N; o: Muckklause – S; p: Kasersbach). After Pirzer (1988).

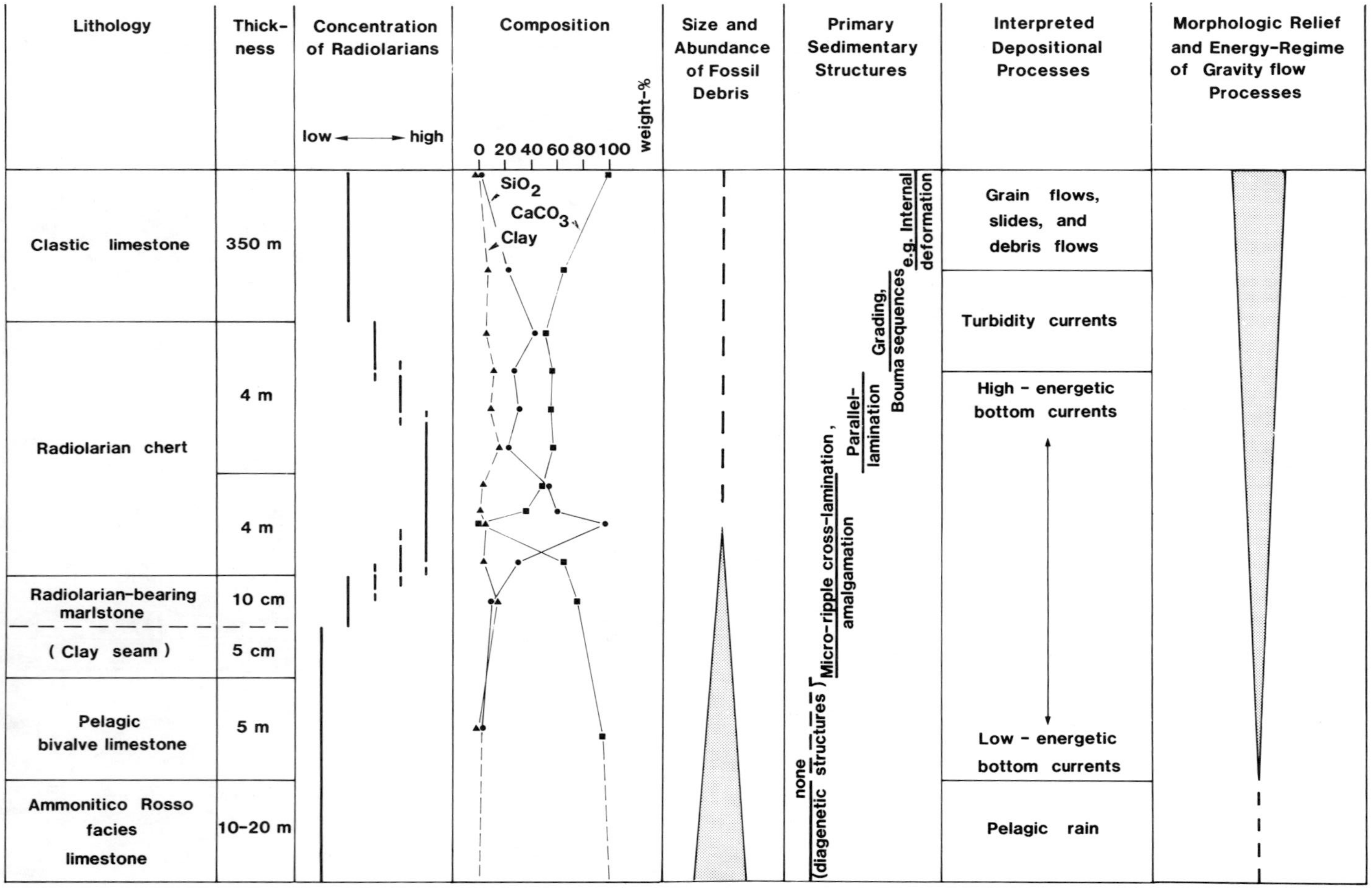

FIGURE 5.5. Properties and interpretation of the sequence containing the radiolarian chert, compiled from the Himmelreich-Graben and Schreinbachtal/Promektal sections in the Osterhorn block. The clay portion was calculated from the Al_2O_3 content.

FIGURE 5.6. (A) Outcrop showing the red cephalopod limestone (l) overlain by rhythmically bedded radiolarian chert (r). Talbruck, Unken syncline. (B) Detail from (A) showing the sequence of red cephalopod limestone (l), ferromanganese crusts (x), claystone (c), and rhythmically bedded radiolarian chert (r). (C) Outcrop of a debris flow deposit consisting of centimeter- to meter-size carbonate clasts in a marly matrix. Note deformation and random arrangement of the components. Tauglboden Formation, Kargraben, Osterhorn block. (D) Outcrop of rhythmically bedded radiolarian chert separated by millimeter-thick claystone seams. The bedding planes are sharp and undulatory, and they persist laterally except where amalgamated. Himmelreich-Graben. (E) Outcrop of redeposited carbonate beds of the Tauglboden Formation. Note sharp bedding planes and intercalated marlstone seams. Promektal, Osterhorn block.

depositional regime drastically changed as a consequence of this change in topography.

Four main facies are recognized within the upper Oxfordian and Kimmeridgian; from bottom to top these are (1) pelagic bivalve limestone, (2) radiolarian-bearing marlstone, (3) radiolarian chert, and (4) radiolarian limestone containing sediment gravity flow deposits.

Pelagic Bivalve Limestone

The pelagic bivalve limestone is deposited on top of the red cephalopod limestone. The up to 5-m-thick lithology may wedge out laterally and is missing in many sections. It probably accumulated in minor depressions on the horsts.

The decimeter-thick beds predominantly consist of conchs of planktic bivalves (Flügel, 1978). Many shells are imbricated (Fig. 5.7A). The matrix is mostly microsparite, but micrite and fine-grained terrigenous material occur as well. The beds are massive and ungraded. Beds are continuous in outcrop and do not show significant lateral or vertical thickness variations. Bedding planes are sharp.

Both the enrichment of mollusks and their preferred orientation are interpreted to result from deposition by currents (Flügel, 1967; Kuhry et al., 1976). Particles were sorted by currents so that hydraulically equivalent components were concentrated and imbricated during deposition.

Radiolarian-Bearing Marlstone

The radiolarian-bearing marlstone is a unit 10 to 20 cm thick. It forms the base of the Ruhpolding Formation. The micritic marlstone also contains radiolar-

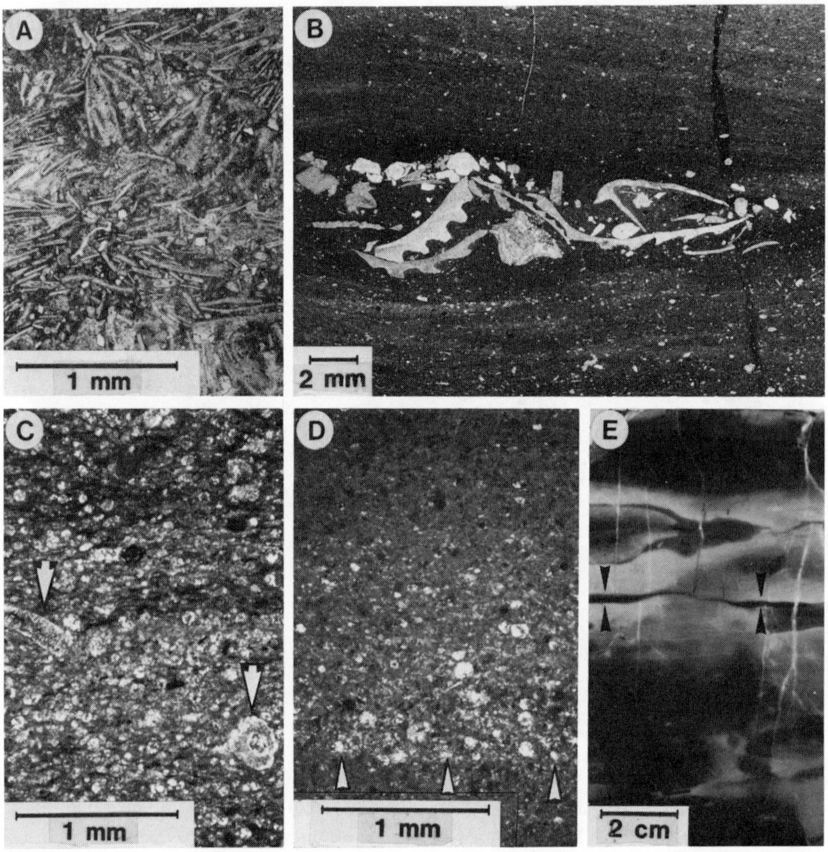

FIGURE 5.7. Thin-section photographs of samples from Himmelreich-Graben, Osterhorn block. (A) Pelagic bivalve limestone. This rock consists mainly of imbricated bivalve shells in a micritic to microsparitic matrix. (B) Radiolarian-bearing marlstone arranged into microscale ripples. Note layer with imbricated debris of aptychi, bivalves, and foraminifera. (C) Micro-cross-laminated radiolarian chert bed cut perpendicular to the bedding. Most radiolarians are not deformed; Nasselaria and some fossil debris (arrows) are oriented approximately parallel to the bedding. The radiolarians float in a fine-grained quartz cement and some fine-grained terrigenous material. (D) Chert deposited from a low-density turbidity current at the transition from calcareous radiolarian chert (Ruhpolding Formation) to clastic limestone (Tauglboden Formation). Radiolarians are normally graded with respect to number. The matrix is fine-grained siliceous limestone. Base of the turbidite is marked by arrows. (E) Polished section of a claystone seam (arrows) partly preserved in an amalgamated radiolarian chert bed. Light shades are due to secondary bleaching of the rock.

ians. Coarse-grained calcitic fossil debris (bivalves, aptychi, and foraminifera; Fig. 5.7B) is arranged in layers parallel to the bedding. No aragonitic fossil remains were observed. Aptychi are oriented with their convex sides down. The content of radiolarians significantly increases compared to the underlying limestone. The bedding planes are uneven. The beds show significant thickness variations, possibly due to diagenetic overprint as described by Ricken (1986). Enrichment of the sand-size and coarser-grained biogenic particles (up to several centimeters in length) in layers indicates deposition by currents. Likewise, the finer particles exhibit the influence of currents in the form of microscale ripples (Fig. 5.7B). The sedimentary structures are strongly deformed by compaction.

In some sections the radiolarian-bearing marlstone is partly or entirely replaced by an up to 5-cm-thick bed of bright-red claystone (Fig. 5.6B); the origin of this bed remains unclear.

Radiolarian Chert

Radiolarian chert forms the major part of the Ruhpolding Formation. Chert occurs in all sections located in the east-west trending half-grabens or on their flanks, and wedges out toward the adjacent paleohighs to the north and south (Figs. 5.3, 5.4). The radiolarian chert shows a systematic change upsection (Fig. 5.5).

The color of the chert is dark red in more siliceous beds and is, rarely, gray or pale red in calcareous beds. Bed thickness is variable (3–35 cm). Bedding planes are sharp (Fig. 5.6D) and undulatory in the lower part of the chert sections and plane in the upper part of the sections. The sharp bedding planes are a result of reworking of the sediment (Price, 1977).

The chert can be classified as radiolarian wackestone to packstone which contains more than 30% radiolarians, more than 2% carbonate, and some fine-grained terrigenous material in a silica matrix (Fig. 5.5). The carbonate content systematically increases upsection. We assume that the chert was derived from radiolarian grainstone by diagenetic dissolution of the more delicate siliceous organisms. Radiolarian spines perpendicular to the bedding planes were removed by pressure solution during diagenesis (Fig. 5.7C), but spines oriented parallel to the bedding planes are abundant. Many radiolarian shells were not deformed. The deformation of bioturbation structures (*Thalassinoides, Zoophycos*, and *Chondrites*) indicates that compaction of the beds did not exceed about 50% (Wetzel and Aigner, 1986).

The chert beds include some fine-grained calcite fossil debris (bivalves, foraminifera, and aptychi; Fig. 5.7C). The amount of fossil debris sharply decreases from the radiolarian-bearing marlstone to the basal portion of the radiolarian chert, owing to a change in the depositional energy regime and hydraulically equivalent sorting of particles. The size and abundance of fossil debris further decreases upsection.

A variety of sedimentary structures were observed: (1) Microscale cross lamination (Fig. 5.8C) is predominantly found in the lower part of the chert sections. The radiolarians form laminae at low angles to the bedding. The silica content fluctuates within a single bed and may attain 98%. (2) Parallel lamination (Fig. 5.8D) is dominant in the middle part of chert sections; these beds are richer in carbonate than the cross-laminated chert. (3) Thin beds of individual depositional events displaying incomplete Bouma sequences frequently occur in the upper part of the sections (Fig. 5.7D); these calcareous cherts are transitional to siliceous carbonates. The vertical succession of primary sedimentary structures coincides with an overall thickening upward in layers representing individual depositional events, which are grouped into beds by amalgamation. We interpret this sequence of sedimentary structures to reflect an increase in energy of transport processes, which resulted from differential subsidence that was not balanced by the rate of accumulation, leading to an increase in relief.

Sorting and deposition of hydraulically equivalent components indicate that the radiolarian chert was deposited by bottom currents. In the basal part of the formation, low-energy bottom currents (<5–10 cm/s; McCave, 1984) produced microscale cross-lamination which coincides with the greatest enrichment in radiolarians. Upsection, parallel laminae were formed by higher-energy bottom currents. The frequency and thickness of beds formed from low-density turbidity currents increase farther upsection. The parallel-laminated chert and turbidites were, in part, later modified by low-energy bottom currents.

Amalgamation of layers frequently occurs, indicating erosion before further deposition. Some shale seams between the cross-laminated chert are discontinuous as the result of this erosion (Fig. 5.7E). Sole marks were not observed and are unlikely to occur because of the relatively low-energy regime of the currents and the fine grain size of the sediment.

The claystone seams of the radiolarian chert beds are low in calcite near the base of the formation. Higher up in the sections, carbonate content and bed thickness increase. Simultaneously, the carbonate content in the chert beds increases.

In a restricted area within the Unken syncline (Muckklause and Karnergraben; Fig. 5.4), debris flow and slide deposits containing components of red cephalopod limestone, gray nodular spiculite limestone (Liassic), and deformed radiolarian cherts occur near the base of the Ruhpolding Formation. Such rocks are otherwise known only from the Tauglboden Formation (see below).

Radiolarian Limestone

Clastic radiolarian limestone (Tauglboden Formation; Schlager and Schlager, 1973) mainly formed by redeposition of calcarenite and calcilutite. The bedding surfaces are plane and sharp; the thickness of the majority of the beds varies from 10 to 30 cm (Fig. 5.6E). Sections of single depositional events (gravity slides) reach 80 m in thickness and can be traced laterally for several kilometers. The beds

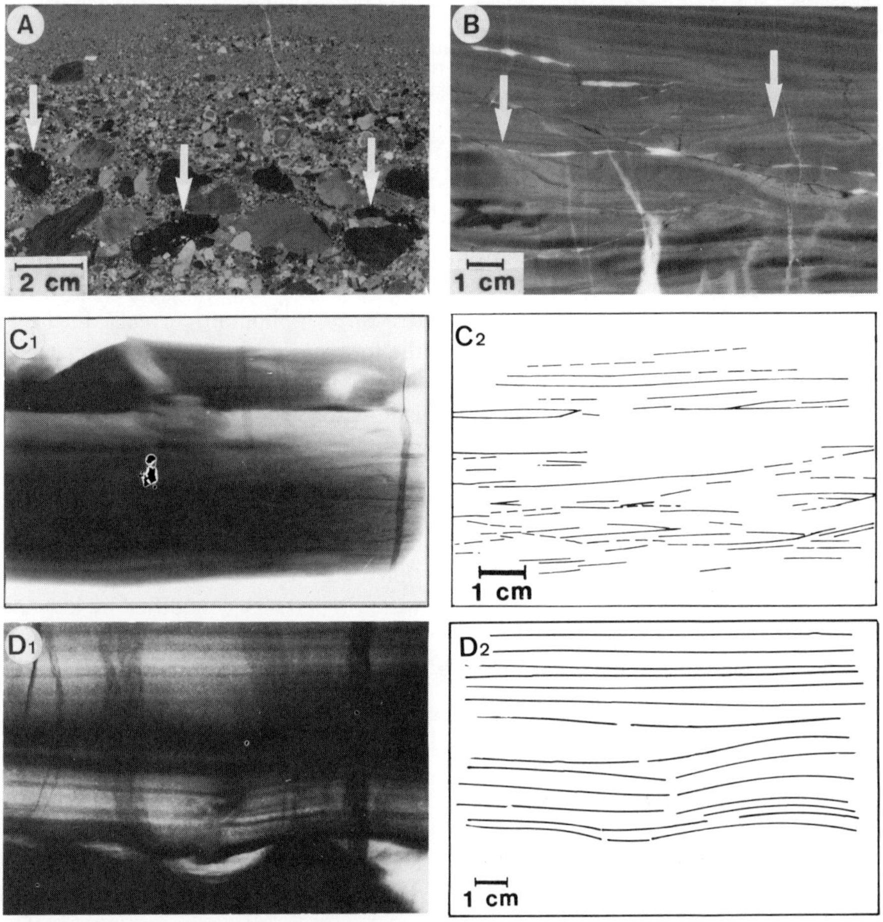

FIGURE 5.8. (A) Polished slab of a rock deposited from a high-density turbidity current, showing grading of lithified carbonate and deformed marlstone clasts (arrows). The matrix is composed of micrite and fine-grained terrigenous material. Tauglboden Formation, Promektal, Osterhorn block. (B) Polished slab of a calcareous radiolarian chert bed within Tauglboden Formation. Note cross-lamination, parallel lamination, and truncations (arrows). The color changes from light gray to dark red in the lower part of the bed. (C_1) Print of X-ray radiograph of micro-cross-laminated and parallel-laminated radiolarian chert. The sedimentary structures are vaguely seen on reproductions. Himmelreich-Graben. (C_2) Line drawing of sedimentary structures from (C_1). (D_1) Print of X-ray radiograph of parallel-laminated radiolarian chert. The basal lamina truncates the surface of an underlying chert bed and fills indentions in the surface of the bed. Redeposited sediment within the Tauglboden Formation, Promektal. (D_2) Line drawing of sedimentary structures from (D_1).

mainly consist of clasts derived from older formations. The coarse grain size and sedimentary structures indicate transportation and deposition by both low- and high-density turbidity currents, debris flows, grain flows, slides, and megaslides (Figs. 5.6C, 5.8A). An example of a megadebris flow is the Schwarzbergklamm Breccia in the Unken Syncline (Garrison, 1964). The Tauglboden Formation attains a thickness of more than 350 m in parts of the Osterhorn Mountains, but wedges out in other areas. The beds alternate with few-centimeter-thick marl layers which consist of micrite and fine-grained terrigenous material; these represent hemipelagic background sedimentation.

Calcareous radiolarian chert is found as individual beds or groups of beds in several horizons within the Tauglboden Formation. They are cross-laminated, parallel-laminated, or internally truncated, and they frequently show a change from gray to red within single beds (Fig. 5.8B).

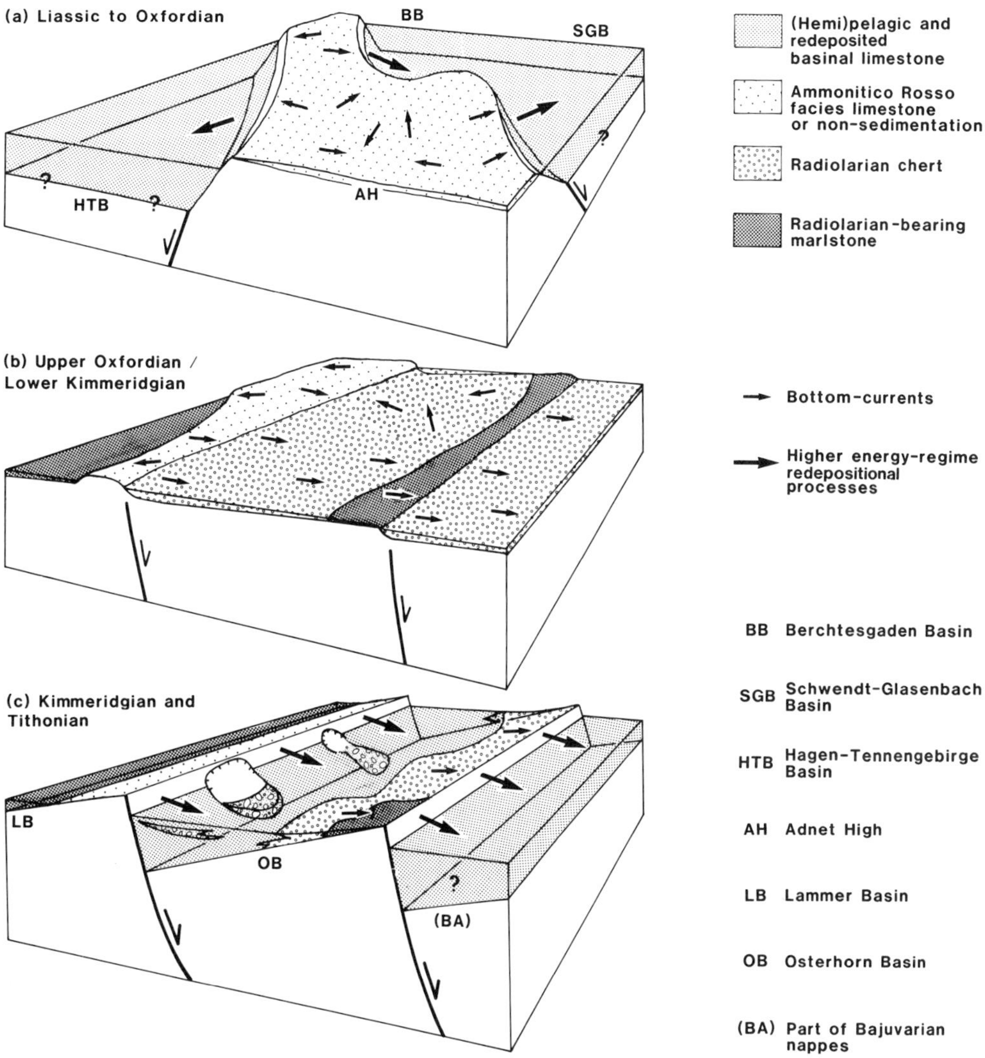

FIGURE 5.9. Model for the development of submarine topography and sedimentary evolution during the Jurassic of the Northern Calcareous Alps. (a) In the lower Liassic, horsts and grabens formed by differential subsidence of a Late Triassic carbonate platform. From the Liassic to the Oxfordian, cephalopod limestone was deposited on the horsts, and marl and limestone were deposited in the grabens. (b) At the time of the Oxfordian/Kimmeridgian boundary, the topography was leveled by the transition from horsts and graben to tilted blocks. Radiolarian chert was deposited by bottom currents over most of the area. (c) During the Kimmeridgian and Tithonian, the continental margin developed into an array of tilted blocks. Coarse carbonate clastics were shed northward from the steep margins of the tilted blocks. The coarse carbonate clastics change to calcareous radiolarian chert in places. During the upper Tithonian, the listric faults became inactive. As a consequence, the half-grabens were filled by hemipelagic and redeposited limestone (Oberalm Formation).

Depositional Environment of the Radiolarian Chert

The sedimentary characteristics of the rhythmically bedded radiolarian chert in the Ruhpolding and Tauglboden Formations indicate that a combination of physical and chemical processes were responsiblc for the accumulation of the radiolarian chert, especially hydrodynamic sorting and carbonate dissolution. The radiolarian chert was deposited during a period characterized by an extensive bottom-water current system. The enrichment in radiolarians was primarily caused by current activity. The bottom current system, in turn, was initiated by the tectonic evolution and the resulting topographic configuration of the area.

The tectonic significance of the radiolarian chert is based on the selective sedimentation of radiolarians by currents due to sorting of hydraulically equivalent particles. This is indicated by the coincidence of microscale cross-lamination and maximum concentration of radiolarians. The change in relief and basin configuration from horst and graben to tilted block topography with its elongate basins (Fig. 5.9), in connection with the opening of the Piemont Ocean, allowed for a better exchange of bottom waters. This permitted bottom currents and low-density turbidity currents to prevail over a large area, resulting in the accumulation of radiolarian ooze.

Low sediment accumulation rates resulted in a longer exposure of carbonate particles on the sea floor, promoting carbonate dissolution. Relatively low sedimentation rates may have resulted from the absence of redeposited lithified carbonate clasts as well as low productivity of calcareous plankton before the Late Jurassic (Bernoulli and Jenkyns, 1970; Jenkyns, 1971).

The absence of aragonite remains in the radiolarian chert and associated rocks indicates deposition of these rocks below the aragonite compensation depth. The occurrence of crushed calcite fossil debris and the redeposited character of the radiolarian chert do not allow for a reliable determination of the depositional depth of the radiolarian chert in relation to the calcite compensation depth (CCD). The calcite-bearing claystone seams between the radiolarian chert beds, however, indicate deposition of these rocks above the CCD.

Discussion and Conclusions

The mechanisms that led to the accumulation of radiolarian chert in the Late Jurassic are certainly complex, but both the distribution pattern of the sediment and its sedimentary structures indicate that the sorting of hydrodynamically equivalent particles was most important. Clastic material, laminations of different kinds, and incomplete Bouma sequences indicate deposition by currents. The energy of the currents was low near the base of the radiolarian chert sections (Ruhpolding Formation) but increased upsection, where deposition resulted from high-energy bottom currents, turbidity currents, and even higher-energy processes in the Tauglboden Formation.

A high production rate of radiolarians during most of the Jurassic is revealed by the abundance of radiolarians in most Jurassic sedimentary rocks of the Northern Calcareous Alps (Tollmann, 1976). Furthermore, the terrigenous input was low in the Late Jurassic owing to subsidence of the entire area—i.e., the Austroalpine and Penninic depositional realms. Carbonate dissolution also favored concentration of the radiolarians.

As local tectonism and topography play a fundamental role in the accumulation process of the radiolarians, it is suggested that the chert should be, to a certain degree, time-transgressive. At present no dates are available to test this hypothesis.

The rhythmically bedded radiolarian chert of the Ruhpolding Formation is considered to represent only a short period of time. This is supported by the relatively small thickness of this formation (4 to 20 m). Moreover, because of the deposition by currents and gravity flow processes, the average sedimentation rate would have greatly exceeded values known for pelagic radiolarian ooze in modern deep-sea environments (often >1 mm/1,000 years; Jenkyns and Winterer, 1982). Rapid sedimentation applies especially to the calcareous radiolarian chert of the Ruhpolding and Tauglboden Formations, which was deposited by high-energy bottom currents and low-density turbidity currents. Schlager and Schlager (1973) calculated a high average rate of sedimentation for all members of the Tauglboden Formation: 0.58 mm/1,000 years in the compacted state.

The change in the tectonic regime, which is considered one of the fundamental causes for the formation of the rhythmically bedded radiolarian chert around the time of the Oxfordian/Kimmeridgian boundary, produced a characteristic sedimentary sequence. The Hettangian through Oxfordian sequence below the Ruhpolding Formation shows an irregular distribution pattern of red cephalopod limestone typical of starved sedimentation on horsts, and basinal marly limestone with intercalated turbidites. The occurrence of coarse clastic sediment along fault scarps in the Liassic and their decrease in the Dogger and Oxfordian testify to decreasing tectonic activity during this period. Around the Oxfordian/Kimmeridgian boundary, renewed tectonic activity drastically changed the topography and depositional environments ("Ruhpolding stratigraphic turning point"; Schlager and Schöllnberger, 1974). The distribution of radiolarian chert and overlying formations indicates a linear arrangement of the basins and a marked basin asymmetry. These features define an array of tilted blocks facing toward the Piemont Ocean.

The deposition of the rhythmically bedded radiolarian chert in the Northern Calcareous Alps was therefore related to the tectonic evolution of the Austroalpine rifted continental margin. The great similarity between the chert-bearing sequences of the Northern Calcareous Alps and other Mediterranean regions (McBride and Folk, 1979; Kälin et al., 1979) suggests that corresponding relations may be valid for these other regions as well. The opening of the Piemont Ocean and the reorganization of the topography in the adjacent Northern Calcareous Alps are considered to be responsible for the establishment of a new current system (see Hsü, 1976, for paleoceanographic considerations). The change in paleotopography from horsts and grabens to tilted blocks resulted in a characteristic pattern of low-energy bottom currents, which led to an enrichment of radiolarians and formation of chert beds. Higher-energy regimes are typical of the period following the radiolarian chert deposition. It is likely, therefore, that topographic relief increased during the Kimmeridgian. This increase was responsible for the changes in the depositional mechanisms from the Ruhpolding Formation to the Tauglboden Formation.

The formation of the tilted tectonic blocks may correlate with a period of strong crustal distension in the Piemont Ocean. The incipient formation of oceanic crust in the northeasternmost part of this ocean, which has been discussed here, may be seen in connection with a regional unconformity ("breakup unconformity") in the limestone sequence of the Oberalm Formation, which overlies the Tauglboden Formation and is essentially of Tithonian age. The unconformity indicates an end to the tectonic activity that created the tilted blocks and a smoothening of the topography. The distension after continental breakup was accomplished by the spreading of oceanic crust in the Piemont Ocean. Continental breakup, however, was likely to have occurred earlier in the western Alps and the Ligurian Apennines, because these regions were more distant to the pole of plate rotation and therefore experienced a greater amount of distension.

Acknowledgments. W. Wetzel prepared the photographs. X-ray radiographs were supplied by Mrs. Kolter, Strahleninstitut, University of Tübingen. D. Bernoulli and J.R. Hein critically read the manuscript and gave valuable comments. S. Borchert and J.R. Hein improved the English text. All this is gratefully acknowledged.

References

Bernoulli D, Jenkyns HC (1970) A Jurassic Basin: The Glasenbach Gorge, Salzburg, Austria. Verhandlungen der Geologischen Bundesanstalt, Wien, 1970:504–531.

Bernoulli D, Jenkyns HC (1974) Alpine, Mediterranean, and central Atlantic Mesozoic facies in relation to the early evolution of the Tethys. In: Dott RH Jr, Shaver RH (eds) Modern and Ancient Geosynclinal Sedimentation. Society of Economic Paleontologists and Mineralogists, Special Publication 19:129–160.

Dercourt J et al. (1986) Geological evolution of the Tethys belt from the Atlantic to the Pamirs since the Lias. In: Aubouin J, LePichon X, Monin AS (eds) Evolution of the Tethys. Tectonophysics 123:241–315.

Diersche V (1980) Die Radiolarite des Oberjura im Mittelabschnitt der Nördlichen Kalkalpen. Geotektonische Forschungen 58:1–217.

Flügel E (1978) Mikrofazielle Untersuchungsmethoden von Kalken. Springer, Heidelberg, pp 1–454.

Flügel HW (1967) Die Lithogenese der Steinmühl-Kalke des Arracher Steinbruchs (Jura, Österreich). Sedimentology 9:23–53.

Frisch W (1979) Tectonic progradation and plate tectonic evolution of the Alps. Tectonophysics 60:121–139.

Garrison RE (1964) Jurassic and early Cretaceous sedimentation in the Unken Valley area, Austria. PhD thesis, Princeton University, Princeton, NJ, pp 1–88.

Garrison RE, Fischer AG (1969) Deep-water limestones and radiolarites of the Alpine Jurassic. In: Friedman GM (ed) Depositional Environments in Carbonate Rocks. Society of Economic Paleontologists and Mineralogists, Special Publication 14:20–56.

Huckriede R (1971) Rhyncholithen-Anreicherung (Oxfordium) an der Basis des Älteren Radiolarits der Salzburger Kalkalpen. Geologica et Palaeontologica 5:131–147.

Hsü KJ (1976) Paleoceanography of the Mesozoic Alpine Tethys. Geological Society of America, Special Paper 170:1–44.

Jacobshagen V (1965) Die Allgäu-Schichten (Jura-Fleckenmergel) zwischen Wettersteingebirge und Rhein. Jahrbuch der Geologischen Bundesanstalt, Wien, 108:1–114.

Jenkyns HC (1971) The genesis of condensed sequences in the Tethyan Jurassic. Lethaia 4:327–352.

Jenkyns HC (1974) Origin of red nodular limestones (Ammonitico Rosso, Knollenkalke) in the Mediterranean Jurassic: A diagenetic model. In: Hsü KJ, Jenkyns HC (eds) Pelagic Sediments: On Land and Under the Sea. International Association of Sedimentologists, Special Publication 1:249–271.

Jenkyns HC, Winterer EL (1982) Palaeoceanography of Mesozoic ribbon radiolarites. Earth and Planetary Science Letters 60:351–375.

Kälin O, Patacca E, Renz O (1979) Jurassic pelagic deposits from southeastern Tuscany; aspects of sedimentation and new biostratigraphic data. Eclogae Geologicae Helvetiae 72:715–762.

Kuhry B, De Clercq SWD, Dekker L (1976) Indications of current action in late Jurassic limestones, radiolarian limestones, saccocoma limestones and associated rocks from the Subbetic of SE Spain. Sedimentary Geology 15:235–258.

McBride EF, Folk RL (1979) Features and origin of Italian Jurassic radiolarites deposited on continental crust. Journal of Sedimentary Petrology 49:837–868.

McCave IN (1984) Erosion, transport and deposition of fine-grained marine sediments. In: Stow DAV, Piper DJW (eds) Fine-Grained Sediments: Deep-Water Processes and Facies. Geological Society of London, Special Publication 15:35–69.

Pirzer M (1988) Die jurassischen Sedimentserien der Unkener Synklinale. Diploma thesis, University of Tübingen, Tübingen, F.R.G. (in preparation).

Plöchinger B (1964) Die tektonischen Fenster von St. Gilgen und Strobl am Wolfgangsee (Salzburg, Österreich). Jahrbuch der Geologischen Bundesanstalt, Wien, 107: 11–69.

Plöchinger B (1975) Das Juraprofil an der Zwölferhorn-Westflanke (Nördliche Osterhorngruppe, Salzburg). Verhandlungen der Geologischen Bundesanstalt, Wien, 1975:27–33.

Price I (1977) Facies distinction and interpretation of primary cherts in a Mesozoic continental margin succession, Othris, Greece. Sedimentary Geology 18:321–335.

Ricken W (1986) Diagenetic bedding: A model for marl-limestone alternations. Lecture notes in Earth Science 6, Springer, Heidelberg, pp 1–210.

Schlager W, Schlager M (1973) Clastic sediments associated with radiolarites (Tauglboden-Schichten, Upper Jurassic, Eastern Alps). Sedimentology 20: 65–89.

Schlager W, Schöllnberger W (1974) Das Prinzip stratigraphischer Wenden in der Schichtfolge der Nördlichen Kalkalpen. Mitteilungen der Geologischen Gesellschaft in Wien 66/67:165–193.

Stow DAV, Wezel FC, Savelli D, Rainey SCR, Angell G (1984) Depositional model for calcilutites: Scaglia Rossa limestones, Umbro-Marchean Appennines. In: Stow DAV, Piper DJW (eds) Fine-Grained Sediments: Deep-Water Processes and Facies. Geological Society of London, Special Publication 15:223–241.

Tollmann A (1976) Analyse des klassischen Nordalpinen Mesozoikums. Deuticke, Wien, pp 1–580.

Vecsei A (1986) Sedimentation und Tektonik im Jura der Osterhorngruppe (Nördliche Kalkalpen, Österreich). Diploma thesis, University of Tübingen, Tübingen, F.R.G., pp 1–90.

Weissert HJ (1981) Depositional processes in an ancient pelagic environment: The lower Cretaceous of the Southern Alps. Eclogae Geologicae Helvetiae 74: 339–352.

Weissert HJ, Bernoulli D (1985) A transform margin in the Mesozoic Tethys: Evidence from the Swiss Alps. Geologische Rundschau 74:665–679.

Wendt J (1970) Stratigraphische Kondensation in triadischen und jurassischen Cephalopodenkalken der Tethys. Neues Jahrbuch für Geologie und Paläontologie Monatshefte 1970:433–448.

Wendt J (1971) Die Typlokalität der Adneter Schichten (Lias, Österreich). Annales Instituti Geologici Publici Hungarici 54:105–116.

Wetzel A, Aigner T (1986) Stratigraphic completeness: Tiered trace fossils provide a measuring stick. Geology 14:234–237.

6
Distribution and Evolution of Siliceous Rocks in Bulgaria

Ivan K. Nachev and Chavdar I. Nachev

Overview

Siliceous rocks are widespread in Bulgaria and span a wide age range, from Cambrian to Quaternary. Lithologies are of different types including diatomite, spongolite, radiolarite, siliceous shale, jasper, silicified limestone, nodular chert, chert pebbles, and chalcedony sand. The mineral composition is also variable including quartz, chalcedony, cristobalite, opal-CT, and opal-A. The siliceous rocks are polygenetic. Their evolution is governed by changes in: (1) dominance of siliceous organism types, radiolarians, sponges, and diatoms; (2) types of volcanic activity and tectonic environment; spreading-center basaltic volcanism, island-arc submarine basaltic volcanism, and subaerial acid volcanism; (3) sedimentary environments, open-ocean metachert; marginal sea siliceous shale and chert; back-arc trough siliceous shale and bedded chert; continental margin siliceous shale; intra-arc trough jasper, siliceous shale, bedded chert, nodular chert, silicified limestone, and vein agate; back-arc trough radiolarite, spongolite, and nodular chert; active continental margin silicified limestone, jasper, nodular chert, and agate; epicontinental sea nodular chert, spongolite, diatomite, and silicified limestone; Neogene lake diatomite; Quaternary fluviatile chert pebbles; Quaternary weathering chalcedony sand.

Consequently, the composition and type of siliceous rocks in Bulgaria correlate well with the development of siliceous organisms, volcanic activity, and changes in the depositional and tectonic environments.

Introduction

Siliceous rocks are widespread in Bulgaria and are of diverse types, ages, and origins. We review the distribution and evolution of siliceous rocks in Bulgaria and discuss the relationships between the mineral composition of the rocks, the evolution of siliceous organisms, volcanism, and paleogeodynamic environments. Regional and specific studies are in progress.

Previous Work

Siliceous rocks have not been well studied in Bulgaria. They are only mentioned in regional stratigraphic and tectonic studies. Published sedimentological studies include Oxfordian (Atanasov, 1954) and other Jurassic siliceous nodules (Nachev, 1962, 1976), Upper Cretaceous siliceous nodules (Nachev and Slavova-Nacheva, 1961; Sultanov, 1982), Aptian chalcedony beds (Goranov, 1965) and siliceous nodules (Nachev and Kunchev, 1984), Paleogene spongolites (Sultanov, 1980), Paleogene chert from Izbegli (Nachev, 1984), opaline chert from the East Rhodope Mountains (Goranov and Todorova, 1985), and chert beds and nodules from the Rhodope Mountains (Nachev and Kunchev, 1986). Three previous summaries of siliceous rocks in Bulgaria represented progress reports (Goranov et al., 1971; Nachev et al., 1981; Nachev and Nachev, 1986a).

The origin and various aspects of siliceous rocks have been increasingly addressed in the geological, mainly sedimentological, literature (Bushinskii,

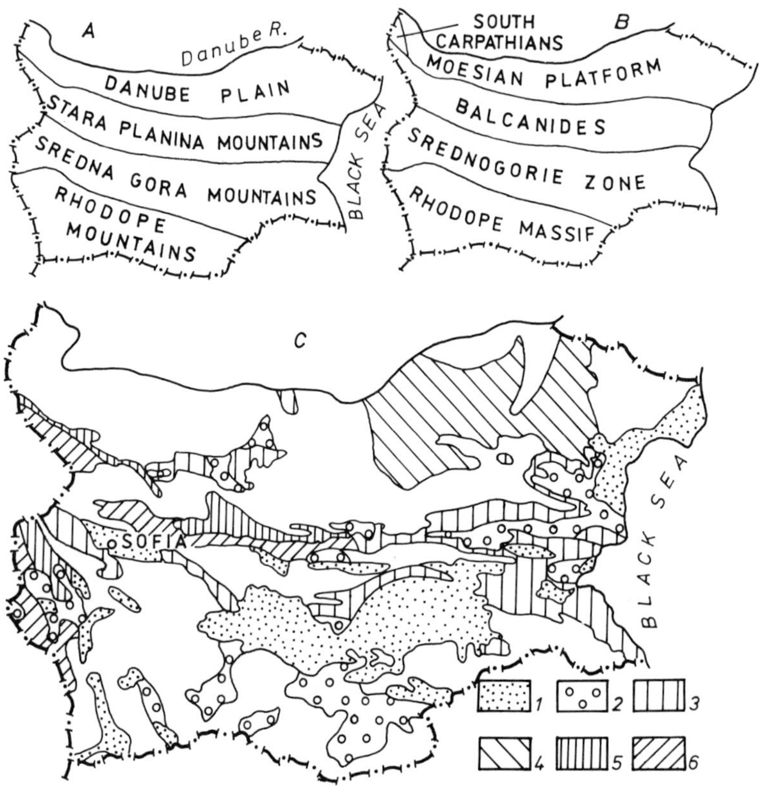

FIGURE 6.1. Distribution of chert-bearing sequences in Bulgaria. (A) Geographic sketch. (B) Morphotectonic sketch. (C) Geological sketch: 1, Neogene; 2, Paleogene; 3, Upper Cretaceous; 4, Lower Cretaceous (only Aptian); 5, Jurassic; 6, Cambrian, Ordovician, Silurian, Devonian, and Carboniferous. Sequences without associated chert are not shown.

1958; Atlas, 1973; Garrison, 1974; Khvorova, 1983; Aston, 1983; Iijima et al., 1983; Hein and Karl, 1983; Iijima and Utada, 1983; Tan, 1983; Moore, 1983; Murchey et al., 1983; Lisitzin, 1985; Barron, 1987; Hein, 1987).

Distribution

Geographic Distribution

Siliceous rocks (Fig. 6.1) occur in most geographic areas of Bulgaria—the Danube plain (north Bulgaria), the Balkan Mountains, the Sredna Gora Mountains, and the Rhodope Mountains (Fig. 6.1A). They also occur in most morphotectonic (geotectonic) zones of Bulgaria—the Moesian platform, the Balkanides, the Sredna Gora zone, and the Rhodope massif (Fig. 6.1B). Numerous occurrences are located near the cities of Vratza, Pleven, Russe, Razgrad, Shumen, Sofia, and Kardzhaly (Figs. 6.1C, 6.2). The siliceous rocks in northeast Bulgaria (Russe and Razgrad) have been called Dobrudza nodular chert (flint in Bulgarian), which is equivalent to the Rhodope bedded chert (also called flint) in the Rhodope Mountains (Plovdiv and Kardzhaly).

Stratigraphic Distribution

Siliceous rocks in Bulgaria have been found in 25 stratigraphic levels of the Phanerozoic (Figs. 6.1C, 6.2). The principal occurrences of siliceous rocks are associated with sedimentary rocks of dominantly Silurian, Devonian, Bathonian, Late Jurassic (Oxfordian and Tithonian), Aptian, Late Cretaceous (Coniacian-Maastrichtian), late Eocene, Oligocene, Pliocene, and Quaternary ages (Nachev et al., 1981; Nachev and Nachev, 1986a). The occurrences are few in number in the Triassic, Early Cretaceous (excluding Aptian), Paleocene, and Eocene sections. Siliceous rocks are not found in the sedimentary sections of the Ordovician, Permian, and Early and Middle Jurassic (excluding Bathonian) ages.

The stratigraphic distribution of silicious rocks in Bulgaria (Fig. 6.3) is similar to that in Japan (Iijima and Utada, 1983), southeast Asia (Tan, 1983), New Zealand (Moore, 1983), western North America (Murchey et al., 1983), and other areas. However,

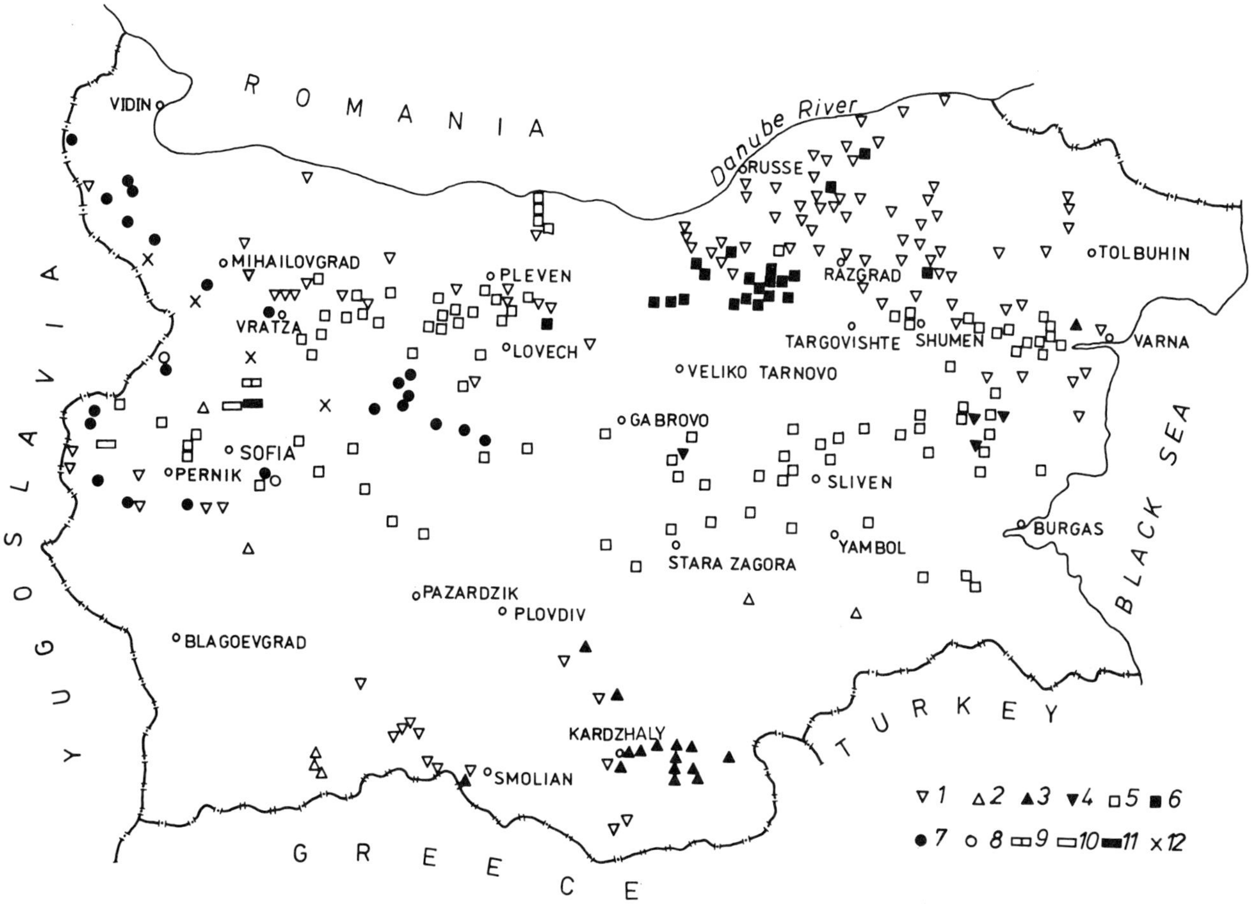

FIGURE 6.2. Geographic distribution of siliceous rocks in Bulgaria. (1) Quaternary: chert pebble. (2) Miocene and Pliocene: diatomite. (3) Upper Eocene and Oligocene: silicified limestone, bedded chert, agate, and diatomite. (4) Paleocene: spongolite. (5) Upper Cretaceous: nodular chert, spongolite, radiolarite, jasper, and silicified limestone. (6) Lower Cretaceous: nodular chert and calcareous chert. (7) Jurassic: nodular chert and radiolarite. (8) Triassic: nodular chert. (9) Upper Carboniferous: siliceous shale. (10) Lower Carboniferous and Devonian: bedded chert and siliceous shale. (11) Devonian and Silurian: bedded chert and siliceous shale. (12) Cambrian: metachert.

more Paleozoic siliceous rocks exist (Nachev and Nachev, 1986a) in Bulgaria in comparison to the worldwide distribution of siliceous rocks (Hein and Parrish, 1987).

Cambrian metachert associated with ophiolites and phillite may occur, but data are not complete.

Silurian bedded cherts and siliceous shale are composed of quartz, clay minerals, Radiolaria, and organic matter. They occur in the form of beds in sections up to 50 m thick. They are associated with graptolitic black shales and have been found north of Sofia.

Devonian and carboniferous bedded chert and siliceous shale are composed of quartz and abundant Radiolaria. They occur in the form of beds in sections up to 70 m thick. They are associated with limestone and shale in the Lower and Middle Devonian, with flysch including graywacke, siltstone, and shale in the Upper Devonian and Lower Carboniferous sections, and with shale, acid tuff, and volcanic rocks in the Upper Carboniferous sections. Most deposits occur near the cities of Sofia and Pernik (Nachev et al., 1981; Nachev and Nachev, 1986a).

Triassic nodular cherts have a chalcedonic quartz composition, are associated with limestone, and occur only very rarely.

Jurassic nodular cherts are also of chalcedonic quartz composition and are associated with sandy crinoidal limestone (Bathonian) or micritic limestone (Late Jurassic). The radiolarites are composed

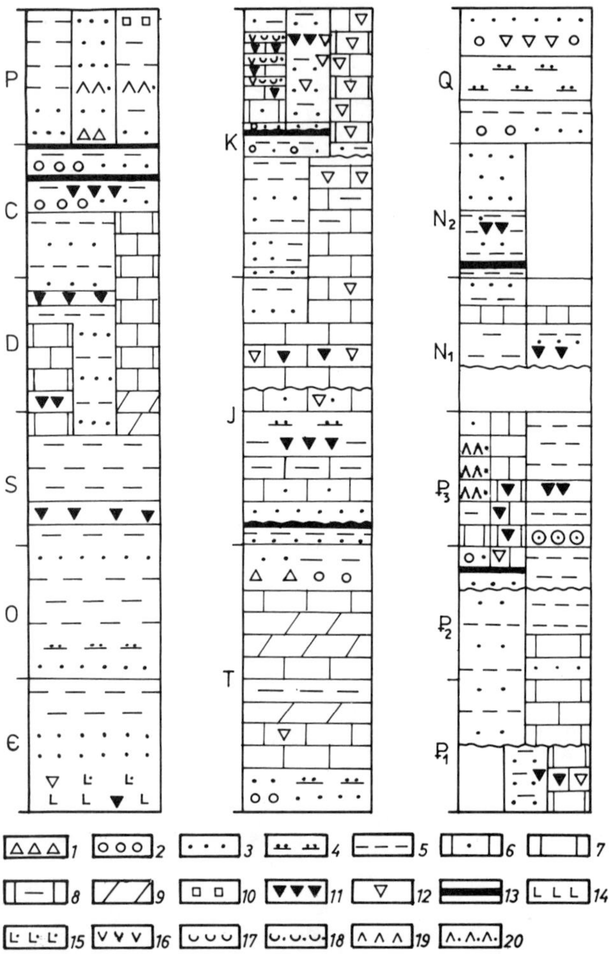

FIGURE 6.3. Stratigraphic distribution of the Phanerozoic siliceous rocks in Bulgaria: (1) breccia; (2) conglomerate; (3) sandstone, sand; (4) siltstone, silt, loess; (5) shale, clay; (6) sandy limestone; (7) limestone; (8) argillaceous limestone; (9) dolomite; (10) evaporite; (11) bedded chert; (12) nodular chert; (13) coal; (14) ophiolite, diabase; (15) diabasic tuff; (16) basalt, andesite, trachybasalt; (17) intermediate volcanic tuff; (18) tuffaceous rocks (tephraturbidite flysch); (19) rhyolite, dacite; (20) acid volcanic tuff.

of radiolarian skeletons, chalcedony, clay minerals, and calcite. They are associated with black shale (Bathonian) near Pernik or with shale, marl, and limestone (Oxfordian) near Sofia and Pernik (Nachev, 1976).

Aptian chert nodules and lenses and bedded calcareous chert with sections up to 20 m thick are mainly composed of chalcedonic quartz but contain abundant sponge spicules. They are associated with limestone. Numerous occurrences are located near the cities of Razgrad, Russe, and Targovishte, where they form the Opaka deposit (Goranov, 1965).

Upper Cretaceous nodular cherts have chalcedonic quartz composition and are about 20 cm in size. They are associated with chalk (Coniacian-Campanian) or fine-grained limestone (Maastrichtian). The chert occurs in the Moesian Platform and in part of the Sredna Gora zone and the East Balkan Mountains (Nachev and Slavova-Nacheva, 1961; Nachev and Yanev, 1980).

Rhythmically bedded cherts consist of chalcedony and are rich in sponge spicules or Radiolaria. They grade in composition into spongolite and radiolarite. These cherts occur in the form of beds of up to 0.4 m thick. They are associated with Turonian-Paleocene flysch composed of clastic limestone, micritic limestone, and hemipelagic argillaceous limestone in the East Balkan Mountains (Emine trough) (Nachev and Yanev, 1980).

Radiolarite, jasper, rhythmically bedded chert, and silicified limestone have mainly chalcedonic quartz compositions and contain Radiolaria, hematite, pyrolusite, chlorite, calcite, clay minerals, and volcaniclastic debris. They occur in the form of nodules, lenses, beds, and bodies of up to 20 m in size. They are associated with hemipelagic limestone, shale, manganese ore, and tephroturbidites or flysch. Vein agate is composed of chalcedony and quartz and is associated with volcanic tuff. Numerous occurrences are located in the Sredna Gora zone (Sredna Gora trough) and are of Coniacian, Santonian, and Campanian ages (Nachev and Nachev, 1986b,c).

Paleocene spongolites occur as beds up to 0.30 m thick and are associated with argillaceous limestone near Pleven (Sultanov, 1980). Oligocene spongolite and diatomite are associated with clay, marl, and manganese ore. Section thicknesses vary up to 12 m in the Varna region.

Silicified limestone and jasper consist of opal-CT, chalcedony, and quartz but also contain pyrolusite, hematite, and calcite. They occur in the form of lenses, beds, and irregular bodies with sections measuring from 1 to 40 m, rarely up to 100 m in thickness. They are associated with algal or numulitic limestone, bentonite, and acid volcaniclastic rocks. Geode and vein agates consist of chalcedony and quartz and are associated with rhyolitic tuff or rhyolite. They occur in the Rhodope area and are of late Eocene and Oligocene age.

Pliocene diatomite consists of diatoms, opal-A, and clay minerals (kaolinite). They occur as beds with sections measuring up to 40 m thick. They are associated with sand, clay, and coal. Many deposits occur in Pliocene sections near the cities of Sofia, Blagoevgrad, Stara Zagora, and Yambol.

Quaternary chert pebbles with chalcedony composition are well rounded with sections up to 10 m thick. They are associated with fluviatile sand and gravel. They originated from erosion and redeposition of nodular chert from Maastrichtian limestone in north Bulgaria (Chiren and Kameno pole deposits near Vratza and Borima deposit near Lovech) or from Aptian limestone in northeast Bulgaria (Kriva Reka deposits near Shumen and Driankovetz and Tetovo deposits near Razgrad). Quaternary karst chalcedony sands (Marshalite) originated as a result of weathering and leaching of the carbonate (calcite) component from Aptian calcareous chert and redeposition of chalcedony in karst caves. They are associated with Aptian calcareous chert, nodular chert, and limestone. They occur in the Opaka deposits near the town of Targovishte in the valley of the Cherni Lom River.

Evolution

Mineral Composition

The mineral composition of siliceous rocks in Bulgaria is variable (Fig. 6.4). Siliceous rocks of different ages have different mineral compositions. The main mineral phases are quartz in the Paleozoic deposits; quartz and chalcedony in Mesozoic and Paleocene deposits; quartz, chalcedony, cristobalite, and opal-CT in upper Eocene and Oligocene deposits; opal-CT and opal-A in Miocene and Pliocene sections; detrital chalcedony in Quaternary deposits (Atanasov, 1954; Nachev et al., 1981; Goranov and Todorova, 1985; Nachev and Nachev, 1986a).

Diagenesis

The siliceous rocks are dominantly sedimentary. Biogenic silica, $SiO_2 \bullet nH_2O$ or opal-A, was transformed diagenetically through solution and redeposition into opal-CT (poorly ordered mixed cristobalite and tridymite), which was in turn transformed into chalcedony or quartz, again through solution and reprecipitation. Some quartz forms through catagenic processes.

Types of Siliceous Rocks

The siliceous rock types in Bulgaria are highly variable (Fig. 6.4). The following types have been described: nodular chert (Fig. 6.5A,B) (Atanasov, 1954; Nachev and Slavova-Nacheva, 1961; Nachev, 1976; Sultanov, 1982); radiolarite (Nachev, 1976); spongolite (Fig. 6.5C) (Sultanov, 1980); diatomite; siliceous shale (phtanites or lidites) (Figs. 6.5D, 6.6A); jasper; rhythmically bedded chert; silicified limestone (Figs. 6.6B,C); calcareous chert; chalcedony sand (marshalite); and chert pebbles (Nachev et al., 1981; Nachev and Nachev, 1986a). Geode agate is known from Triassic deposits in the Lozen Mountains, Upper Cretaceous deposits in the Shumen District, and Paleogene deposits in the Rhodope Mountains. Vein agate occurs in Upper Cretaceous deposits of the Sredna Gora Mountains and in Paleogene deposits of the Rhodope Mountains (Fig. 6.6D).

Chemical Composition

SiO_2 (Table 6.1) is of course the main component of the siliceous rocks (52.34–97.45%). The second-order components include Al_2O_3 (0.09–12.92%), attributed mainly to illite and smectites; Fe_2O_3 (0.11–7.07%), chiefly in hematite; CaO (0.08–24.22%), chiefly in calcite; K_2O (0.06–2.55%), mainly in illite; H_2O^+ (0.54–2.55%, rarely up to 6%), in opal-A and opal-CT. Other mineral phases reflected in the chemistry include detrital quartz, feldspar, kaolinite, and others.

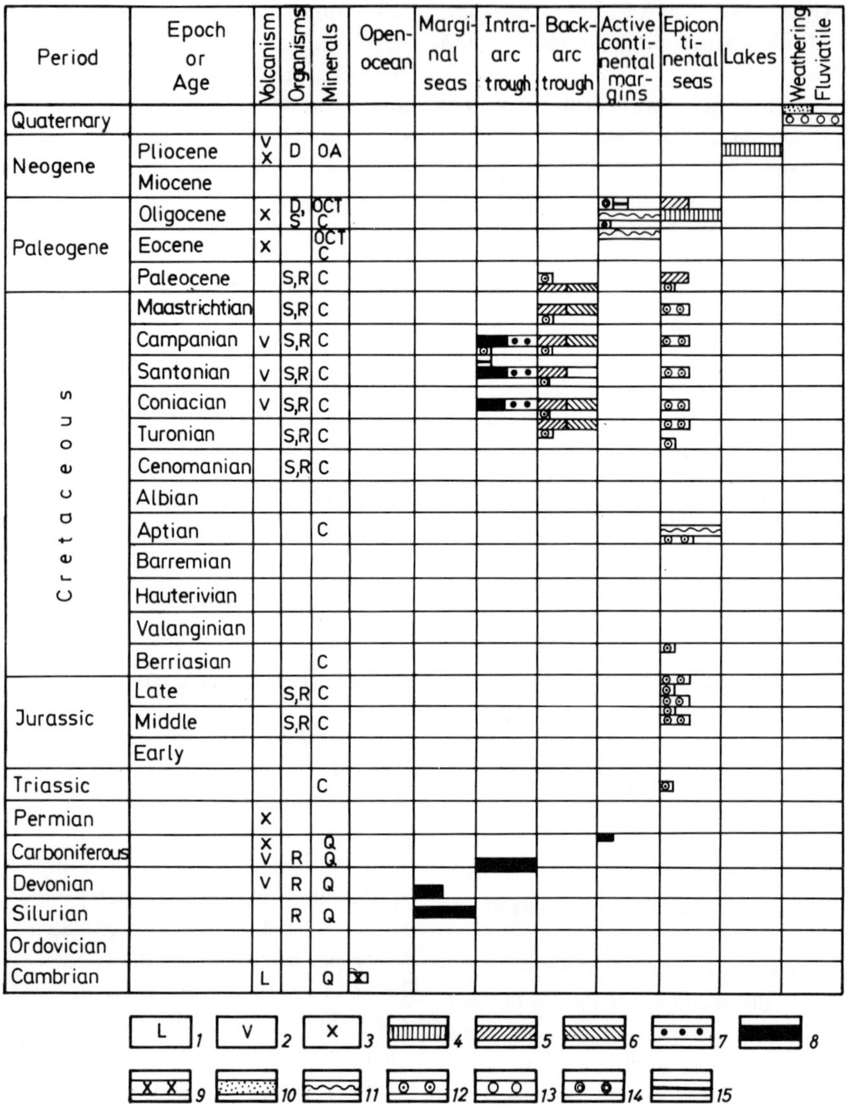

FIGURE 6.4. Evolution of the siliceous rocks in Bulgaria: (1) ophiolite, tholeiitic basalt, and diabase; (2) andesite, basalt, and trachybasalt; (3) andesite and rhyolite; (4) diatomite; (5) spongolite; (6) radiolarite; (7) jasper; (8) bedded chert and siliceous shale; (9) metachert; (10) chalcedony sand; (11) calcareous chert and silicified limestone; (12) nodular chert; (13) chert pebbles in Quaternary deposits; (14) agate geodes; (15) agate veins. Symbols: D, diatoms; S, sponge spicules; R, radiolarians; OA, opal-A; OCT, opal-CT; C, chalcedony; Q, quartz.

Origin and Correlation of Siliceous Rocks

The different types of siliceous rocks are mainly of sedimentary origin. The diatomite, spongolite (Fig. 6.5C), and radiolarite are biogenic sedimentary rocks. Nodular chert (Fig. 6.5A,B) is a replacement product of limestone and is therefore a diagenetic rock. Chalcedony sand results from weathering and leaching of the carbonate component of Aptian calcareous cherts and is a secondary chert. Chert pebbles in Quaternary deposits were derived from erosion of nodular cherts in Lower and Upper Cretaceous limestones. The chert pebbles form a secondary or weathering deposit. Some of the bedded siliceous rocks, including jasper (Figs. 6.5D, 6.6A) and silicified limestone (Fig. 6.6B,C) in the Sredna Gora Mountains (Upper Cretaceous age) and in the Rhodope Mountains (Paleogene age) are related to

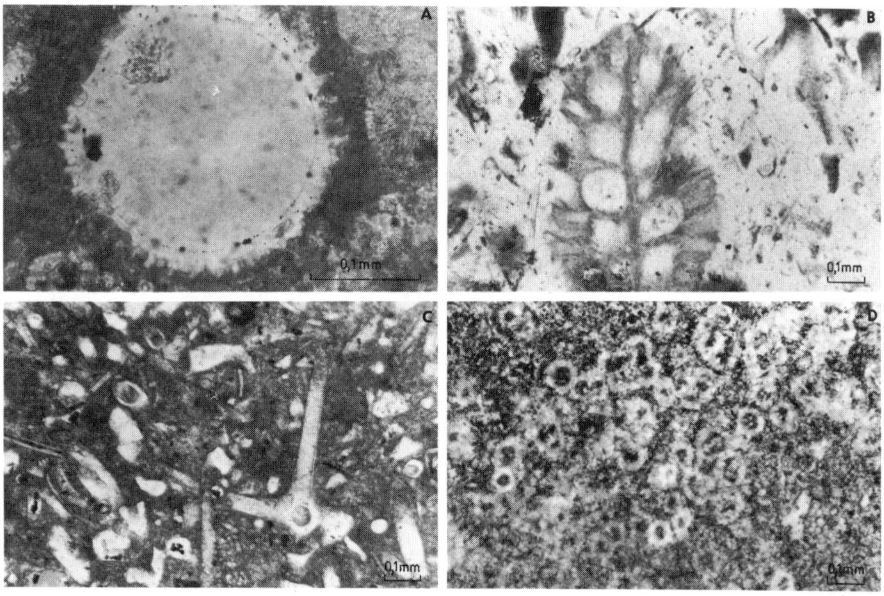

FIGURE 6.5. Nodular chert, spongolite, and jasper in Bulgaria. (A) Radiolaria in Turonian nodular chert from a section at Kaspichan (Shumen), sample 47, polarized light. The circular object is the remaining part of Radiolaria, now filled with fibrous chalcedony. The matrix is of dark cryptocrystalline and microcrystalline calcite. (B) Bryozoa in Santonian nodular chert from a section at Shuman, sample 44, polarized light. The remains of the bryozoa have preserved cell structure but are replaced by chalcedony. The nodule is of cryptocrystalline and microcrystalline chalcedony with some chalcedony sponge spicules. (C) Sponge spicules in Campanian spongolite from a section at Varna, sample 64, polarized light. Numerous remains of the sponge spicules are chalcedony. Some detrital quartz, calcite, and pyrite occur. The matrix is of dark microcrystalline and cryptocrystalline chalcedony. (D) Chalcedony spherules in Oligocene jasper from a section at Kochash (Kardzhaly), sample 342, polarized light. The spherules are fibrous chalcedony with pyrolusite at the centers. The matrix is cryptocrystalline and microcrystalline chalcedony with pyrolusite.

devitrification of volcanic glass or ash and to catagenic replacement processes of limestone by silica. Part of the agate (Fig. 6.6D) is thought to be of hydrothermal origin. Silica may have been partly derived from volcanic sources.

The biogenic origin of most of the siliceous rocks in Bulgaria is related to the evolution of the main rock-forming siliceous organisms during the Phanerozoic (Fig. 6.4)—i.e., radiolarians (Fig. 6.5B) in the Proterozoic to Paleocene, sponges (Fig. 6.5C) in the Jurassic to Paleogene, and diatoms in the Eocene to Pliocene (Nachev et al., 1981; Nachev and Nachev, 1986a).

The origin of some siliceous rocks in Bulgaria is related to volcanic activity. The devitrification of volcanic ash and glass released SiO_2 which replaced calcite in bedded limestone. The silicification resulted in formation of layered siliceous rocks, which in the field appear as chert and jasper lenses and layers in limestone. The presence of some siliceous rocks correlates well with volcanic activity (Fig. 6.4): metachert in the Proterozoic and Cambrian were coeval with spreading-center basaltic volcanism (diabases); siliceous shale and bedded chert in Silurian, Devonian, and Carboniferous sections occurred with island-arc basaltic volcanism (diabases); bedded jasper, bedded chert, nodular chert, and agate in Upper Cretaceous sections from the Sredna Gora Mountains also contain island-arc submarine andesite, basalt, and trachybasalt volcanic rocks; bedded jasper, bedded chert, and agate in upper Eocene and Oligocene sections from the Rhodope Mountains occur with mainly subaerial acid volcanic rocks (Nachev and Nachev, 1986a).

Jasper and silicified limestone—hemipelagic micritic limestone with calcareous foraminifera (Globotruncana)—of Coniacian, Santonian, and Campanian ages from the Sredna Gora zone (Nachev and Nachev, 1986b,c) and shallow algal and numulitic limestone of late Eocene and Oligocene ages

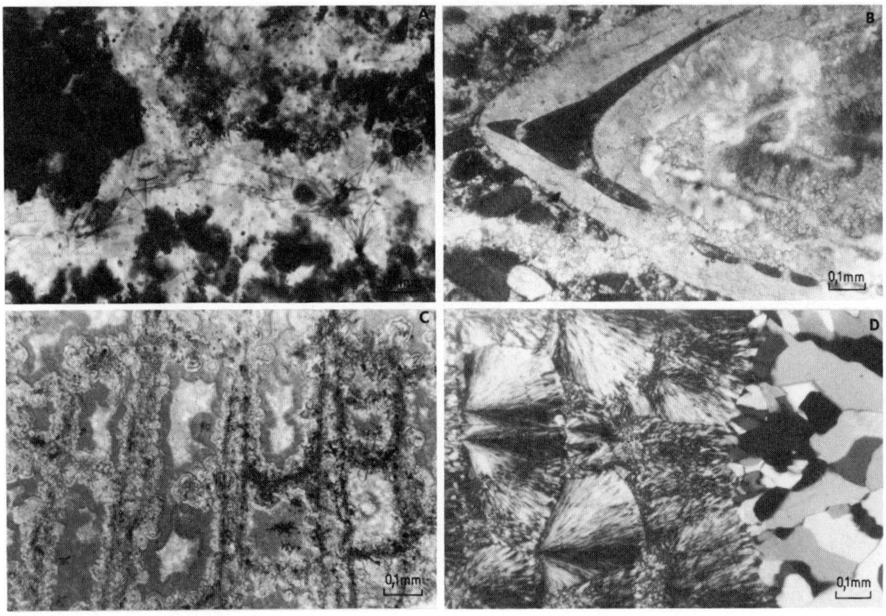

FIGURE 6.6. Jasper, silicified limestone, and agate in Bulgaria. (A) Pyrolusite and chalcedony in Oligocene jasper from a section at Kochash (Kardzhaly), sample 340–2, polarized light. The jasper consists of light fibrous and microcrystalline chalcedony. The pyrolusite (black) forms aggregates or needle-shaped crystals. The jasper is associated with manganese ore. (B) Partly silicified Oligocene limestone from a section at Perperek (Kardzhaly), sample 336-1, polarized light. Numerous algae (black) and numulite foraminifera (gray) in cryptocrystalline and microcrystalline calcite. The chalcedony (light) is microcrystalline or fibrous or forms spherules. Thus, shallow-water limestones are completely silicified in some sections. (C) Silicified coral limestone from Oligocene section at Skalishte (Kardzhaly), sample 328-3, polarized light. It is a cross section of silicified coral. The chalcedony is cryptocrystalline (dark), microcrystalline, and fibrous (light). The quartz (white) is microcrystalline and is located in centers of coral. The pyrolusite (black) forms aggregates and is formed after the chalcedony. (D) Geode agate from Oligocene section at Gledka (Kardzhaly), sample 355, crossed polars. The chalcedony is mainly fibrous, occasionally spherical. The quartz (white and black) is microcrystalline and coarsely crystalline in the central part of the geode (to the right).

from the Rhodope massif are especially likely to have had a volcanic source of the silica. Jasper and silicified limestone are volcanosedimentary bedded cherts.

Alternatively, the jasper and silicified limestone can be considered as direct or indirect products of submarine volcanism. The volcanic-sedimentary hypothesis (see Garrison, 1974) has several variants, as follows: (1) Silica, supplied either directly from submarine eruptions or leaching of congealed lava on the sea floor, precipitated silica inorganically. (2) Silica released during submarine volcanism or during submarine weathering of volcanic ash enhances the productivity and/or preservation of siliceous microplankton. (3) The CCD is elevated to relatively shallow depths during periods of intense volcanic activity. For most siliceous deposits, the composition of the pelagic rocks (e.g., bedded chert), their relationships to igneous rocks, and data from modern ocean basins all argue against a direct connection between volcanism and formation of the siliceous sediments, such as volcanically induced chemical precipitation or stimulation of plankton blooms (Garrison, 1974).

Paleogene jasper and silicified limestone in the Rhodope Mountains are associated with shallow marine limestone, pyrolusite, hematite, bentonite, zeolite minerals, volcaniclastic rocks, and lavas, but without siliceous plankton. The silica in these rocks was probably released during subaqueous volcanic hydrothermal activity or during submarine weathering of volcanic ash.

Upper Cretaceous (Campanian) jasper and silicified limestone in the Sredna Gora zone are associated with hemipelagic micritic limestone, marls, pyrolusite, hematite, bentonite, zeolite, tephraturbidite, and pillow lavas. Probably part of the silica in these rocks was derived from submarine weathering

TABLE 6.1. Chemical composition in weight percent of siliceous deposits.

	1	2	3	4	5	6	7	8	9	10	11	12	13	14	15	16
SiO_2	97.45	96.23	69.09	96.51	95.00	87.86	91.57	96.26	52.34	87.95	79.98	89.20	60.56	86.36	91.08	78.10
TiO_2	0.01	0.11	0.83	0.01	0.02	0.02	0.02	0.46	0.03	0.16	0.01	0.01	0.32	0.05	0.13	0.52
Al_2O_3	0.30	0.60	12.92	0.09	1.32	0.36	1.80	0.25	0.98	4.08	0.95	0.10	6.25	1.50	2.42	9.20
Fe_2O_3	0.23	0.43	4.47	0.11	0.16	0.43	0.51	0.12	0.27	1.49	3.81	7.07	0.75	0.21	0.73	1.43
FeO	–	–	–	–	0.18	–	–	–	–	0.60	–	–	1.40	0.32	–	–
MnO	0.004	0.01	–	0.02	0.01	–	0.03	0.01	0.02	0.02	9.05	0.04	0.06	0.03	0.01	0.11
MgO	0.05	0.20	0.96	0.05	0.03	0.90	0.13	0.01	0.35	0.69	0.01	0.90	0.75	0.10	0.10	1.00
CaO	0.08	0.59	1.56	0.28	0.95	5.73	0.81	0.08	24.22	0.42	0.96	1.10	14.53	5.54	0.17	1.00
Na_2O	–	0.39	0.40	–	0.14	–	–	–	0.03	0.48	–	0.03	0.36	0.18	0.16	0.60
K_2O	0.06	0.42	1.50	0.06	0.15	0.06	0.08	0.05	0.03	1.62	–	0.03	1.27	0.33	0.60	2.55
P_2O_5	–	0.09	–	–	0.02	–	–	–	–	0.03	0.01	–	0.06	0.02	0.06	0.05
CO_2	–	0.46	–	–	0.88	–	–	–	–	0.02	2.24	–	11.00	4.49	0.09	–
H_2O^+	–	0.64	–	–	0.87	–	–	–	–	1.94	–	–	2.46	0.54	–	–
H_2O^-	–	0.20	–	–	0.08	–	–	–	–	0.45	–	–	0.22	–	0.24	–
LOI	1.53	–	8.21	0.91	1.71	5.35	5.09	1.05	22.07	1.88	2.54	2.45	13.14	4.98	4.32	4.95
Total	99.71	100.30	99.94	98.04	98.80	100.71	97.27	99.29	100.31	99.96	97.34	100.89	100.08	100.67	100.10	99.51

Key: (1) Chert pebble, Quaternary, Razgrad. (2) Chalcedony sand, Quaternary, Opaka, Targovishte. (3) Diatomite, Neogene, Dragovishtitza, Sofia. (4) Nodular chert, Paleocene, Chudny skaly, Varna. (5) Nodular chert, Oxfordian, Oreshetz, Vidin. (6) Nodular chert, Triassic, Vardalovtzi, Sofia. (7) Bedded chert, Oligocene, Komouniga, Plovdiv. (8) Bedded silicified limestone, Oligocene, Kremene, Smolian. (9) Bedded calcareous chert, Aptian, Krepcha, Targovishte. (10) Green jasper, Campanian, Toplika, Pazardzik. (11) Black jasper, Campanian, Toplika, Pazardzik. (12) Red jasper, Oligocene, Kochash, Kardzhaly. (13) Bedded calcareous chert, Santonian, Mirkovo, Sofia. (14) Calcareous bedded radiolarian chert, Oxfordian, Losen, Sofia. (15) Bedded radiolarian chert, Upper Devonian, Sevil, Sofia. (16) Siliceous shale, Silurian, Vlado Trichov, Sofia. 2,5: Classical wet chemical silicate technique. 1, 3, 4, 6, 7, 8: By X-ray fluorescence. 9, 10, 11, 13, 14, 15: Determined by classical wet chemical silicate technique. 12, 16: By X-ray fluorescence. –, Not determined. LOI, loss on ignition.

of volcanic ash. Thin-bedded radiolarian chert in the Sredna Gora zone are biogenic deposits related to fluctuations of the calcite compensation depth (Garrison, 1974). More recent studies have linked the deposition of radiolarian chert to regions of upwelling and high plankton productivity (Garrison, 1974; Hein and Parrish, 1987), but this mechanism cannot be applied to radiolarites in the Bulgarian epicontinental seas and in the interarc Sredna Gora trough (Nachev and Nachev, 1986b,c).

The siliceous rocks in Bulgaria were formed in different depositional environments (Fig. 6.4). Indicators such as mineralogy, lithologic associations, geochemical compositions, sedimentary structures, thickness, stratigraphic sequences, and lateral facies changes are used for interpretation of source, type of sedimentation, and depositional environments (Nachev, 1976; Nachev and Yanev, 1980).

An open-ocean environment (Proterozoic and Cambrian) is suggested by the lithologic association of chert, shale, and ophiolitic rocks and is probably related to a sea-floor spreading environment. A marginal sea environment (Silurian and Devonian) is identified by oceanic sedimentation consisting of shale, black shale, bedded chert, and siliceous shale. The Kuchay-Cerna Gora intraarc trough (Devonian and Carboniferous) is reconstructed on the basis of the association of explosive basalt (diabase, tuffs) volcanism with graywacke-siltstone-shale flysch, siliceous shale, and bedded radiolarite. An active continental margin of the Andian type (Late Carboniferous) is suggested for this depression based on the occurrence of continental sedimentation with coal, subaerial volcanism, siliceous shale, and bedded chert. The Sredna Gora intraarc trough (Coniacian, Santonian, and Campanian ages) is marked by hemipelagic sedimentation (Globotruncana micrite limestone), turbidites, tephraturbidites (tuffaceous flysch), tuff, andesite, basalt, and trachybasalt volcanic rocks of island-arc submarine volcanic origin. Radiolarian bedded chert, siliceous shale, jasper associated with manganese ore, silicified limestone, nodular chert, and vein agate formed in separate areas within these major depositional basins. Reconstruction of the Emine back-arc trough (Turonian-Paleocene) suggests hemipelagic and turbidite sedimentation (flysch) with episodic deposition of bedded radiolarite, spongolite and nodular chert with a transition to shallow-water limestone (carbonate platform). The Rhodope active continental margin is also of the Andian type (late Eocene and Oligocene) as indicated by freshwater to shallow marine sedimentation (algal, nummulitic, or coral limestone), tuff, tuffite, and rhyolite (subaerial acid volcanism) with silicified limestone, bedded chert, jasper, nodular chert, and agate.

Mesozoic and Cenozoic epicontinental seas are identified on the basis of shallow-water sedimentation, mainly carbonate and argillaceous clastic sedimentation with diagenetic nodular chert or primary diatomite and spongolite. The following deposits are interpreted to have formed in epicontinental seas:

Triassic period – dolomite and limestone with nodular chert;

Bathonian age – sandy crinoidal limestone with nodular chert and bedded radiolarite;

Callovian, Oxfordian, Kimmeridgian, and Tithonian ages – micrite limestone with nodular chert and Oxfordian bedded radiolarite;

Aptian age – limestone with nodular chert and calcareous chert;

Cenomanian and Turonian ages – limestone with nodular chert;

Coniacian, Santonian, Campanian, and Maastrichtian ages – chalk with nodular chert;

Maastrichtian and Paleocene age – limestone with nodular chert, partly (Paleocene) with bedded spongolite;

Eocene, Oligocene, and Miocene age – bedded spongolite and diatomite.

Freshwater lake deposits (Miocene and Pliocene) in south Bulgaria are identified by clastic and argillaceous sedimentation and bedded freshwater diatomite. Deposition by fluviatile processes (Quaternary) in north Bulgaria originated through erosion and redeposition of chert pebbles. Weathering of the karst terrain (Quaternary) caused carbonate leaching of Aptian calcareous chert and produced chalcedony sand in karst caves (Opaka deposit near the town of Targovishte).

Conclusions

The compositions and types of siliceous rocks in Bulgaria correlate well with the evolution of siliceous organisms and changes in tectonic and geologic environments. The main rock-forming siliceous organisms were Radiolaria (Cambrian-Paleocene), silica sponges (Bathonian-Oligocene), and diatoms (Oligocene-Pliocene). The type of associated volcanism is basalt, spreading center (Cambrian), island-arc submarine basalt (Silurian–Lower Carboniferous; Coniacian-Campanian) or subaerial acid volcanism (upper Eocene and Oligocene). Sedimentary depositional environments were open-ocean, marginal sea, intraarc trough, back-arc trough, active continental margin of the Andian type, epicontinental sea, freshwater lake, fluviatile, and weathering.

Acknowledgments. We are grateful to the Bulgarian IGCP Project 187 Working Group for their cooperation, to James R. Hein for his critical comments on the manuscript, and to Maria Slaveikova for the translation and typing.

References

Aston SR (1983) Silicon Geochemistry and Biogeochemistry. Academic Press, London, 248 pp.

Atanasov GD (1954) Flint in Oxfordian Limestones, annuaire de L'Universite of Sofia, 48, 2 Géologie: 61–84 (in Bulgarian).

Atlas of Structure and Texture in Sedimentary Rocks (1973) Nedra, Moscow, 3, 395 pp (in Russian).

Barron JA (1987) Diatomite: Environmental and geologic factors affecting its distribution. In: Hein JR (ed) Siliceous Sedimentary Rock-Hosted Ores and Petroleum. Van Nostrand Reinhold, New York, pp 164–178.

Bushinskii GI (1958) Siliceous rocks. In: Hand-book by Petrography on Sedimentary Rocks. "Gosnauktehisdat," Leningrad, 2, pp 255–268 (in Russian).

Garrison RE (1974) Radiolarian cherts, pelagic limestones, and igneous rocks in eugeosynclinal assemblages. In: Hsü KJ, Jenkyns HC (eds) Pelagic Sediments: On Land and Under the Sea. Special Publication No. 1 of the International Associations of Sedimentologists, Oxford, U.K., pp 367–399.

Goranov A (1965) Silicites (Chalcedonites) from Aptian Deposits in Popovo District (NE Bulgaria). Carpato-Balkan Geological Association. 7th Congress, Sofia, BAS, Report 2, 1, pp 329–332 (In Russian).

Goranov A, Nachev IK, Todorova T, Trashliev S, Yanev S (1971) Association features and distribution of sedimentary mineral resources in Bulgaria. Bulletin of Geological Institute, Series Stratigraphy and Lithology 20:87–109 (in Bulgarian).

Goranov A, Todorova T (1985) Opal siliceous rocks from the eastern Rhodopes. Comptes Rendus de l'Academie Bulgare des Sciences, 38(3):353–355.

Hein JR (1987) Siliceous Sedimentary Rock-Hosted Ores and Petroleum. Van Nostrand Reinhold, New York, 304 pp.

Hein JR, Karl SM (1983) Comparisons between open-ocean and continental margin chert sequences. In: Iijima A, Hein JR, Siever R (eds) Siliceous Deposits in the Pacific Region. Elsevier, Amsterdam, pp 25–43.

Hein JR, Parrish JT (1987) Distribution of siliceous deposits in space and time. In: Hein JR (ed), Siliceous Sedimentary Rock-Hosted Ores and Petroleum. Van Nostrand Reinhold, New York, pp 10–57.

Iijima A, Hein JR, Siever R, eds (1983) Siliceous Deposits in the Pacific Region. Elsevier, Amsterdam, 472 pp.

Iijima A, Utada M (1983) Recent developments in the sedimentology of siliceous deposits in Japan. In: Iijima A, Hein JR, Siever R (eds) Siliceous Deposits in the Pacific Region. Elsevier, Amsterdam, pp 45–64.

Khvorova IV (1983) Siliceous rocks. In: Hand-book by Sedimentology. Nedra, Moscow, pp 163–176 (in Russian).

Lisitzin AP (1985) The silica cycle during the last ice age. Palaeogeography, Palaeoclimatology, Palaeoecology 50:241–270.

Moore PR (1983) Chert-bearing formations of New Zealand. In: Iijima A, Hein JR, Siever R (eds) Siliceous Deposits in the Pacific Region. Elsevier, Amsterdam, pp 93–108.

Murchey B, Jones DL, Holdsworth BK (1983) Distribution, age, and depositional environments of radiolarian chert in western North America. In: Iijima A, Hein JR, Siever R (eds). Siliceous Deposits of the Pacific Region. Elsevier, Amsterdam, pp 109–125.

Nachev CI, Nachev IK (1986a) Distribution and evolution of the siliceous rocks in Bulgaria. Comptes Rendus de l'Academie Bulgare des Sciences, 39(8):81–83.

Nachev IK (1962) An attempt at formational analyses of the Jurassic system in Bulgaria. Travaux sur la Geologie de Bulgarie Serie Stratigraphie et Tectonique, 4:61–93 (in Bulgarian).

Nachev IK (1976) Sedimentology of Jurassic Sediments in Bulgaria. Bulgarian Academy of Sciences, Sofia, 160 pp (in Bulgarian).

Nachev IK (1984) Composition and origin of the siliceous rocks from Izbegly Deposit (Plovdiv District). In: Third Seminar on Petroarchaeology, University of Plovdiv, Plovdiv, Bulgaria, pp 45–54.

Nachev IK, Kunchev K (1984) Aptian and Quaternary flint in north-east Bulgaria. In: Third Seminar on Petroarchaeology. University of Plovdiv, Plovdiv, Bulgaria, pp 65–82.

Nachev IK, Kunchev K (1986) Paleogene siliceous rocks in the Rhodopes (Bulgaria). International Conference for Prehistory on Flint Mining and Lithic Raw Material Identification, Carpathian Basin, Budapest-Sumeg, pp 201–206.

Nachev IK, Nachev CI (1986b) Sedimentology of the upper Cretaceous in the Panagjuriste District. I. Results of the studies. Paleontology, Stratigraphy and Lithology, 22:35–70.

Nachev IK, Nachev CI (1986c) Sedimentology of the upper Cretaceous in the Panagjuriste District. II. Interpretation of the results. Paleontology, Stratigraphy and Lithology, 23:65–77.

Nachev IK, Slavova-Nacheva L (1961) Upper Cretaceous sedimentology of NE Bulgaria. Travaux sur la Geologie de Bulgarie Serie Stratigraphie et Tectonique, 3:171–233 (in Bulgarian).

Nachev IK, Yanev SN (1980) Sedimentary Geocomplexes in Bulgaria. Nauka and Iskustvo, Sofia, 204 pp (in Bulgarian).

Nachev IK, Kovnurko GM, Kanchev K (1980) Siliceous rocks in Bulgaria and their exploitation. In: Second International Seminar on Petroarchaeology. University of Wrotzlaw, Wrotzlaw, Poland pp 98–100.

Nachev IK, Kovnourko GM, Kanchev K (1981) Siliceous rocks in Bulgaria and their exploitation. In: Intradisciplinary Research, 7–8. Sofia, Archaeological Institute, Bulgarian Academy of Sciences, pp 41–58 (in Bulgarian).

Sultanov AT (1980) Spongolites in the Paleocene near the village of Komarevo, Pleven District. Paleontology, Stratigraphy and Lithology 13:45–49 (in Bulgarian).

Sultanov AT (1982) Upper Cretaceous flint concretions in north Bulgaria. Paleontology, Stratigraphy and Lithology, 17:51–60 (in Bulgarian).

Tan DNK (1983) Cherts of southeast Asia. In: Iijima A, Hein JR, Siever R (eds) Siliceous Deposits in the Pacific Region. Elsevier, Amsterdam, pp 79–91.

7
Early and Late Diagenetic Chert in Carbonate Turbidites of the Senonian Flysch, Northeast Bilbao, Spain

J.J. Elorza and M.A. Bustillo

Overview

Sedimentological and petrographical analysis of the lower Senonian flysch (Plencia Fm.), near Barrika village, and a part of the upper Senonian flysch (Eibar Fm.) show three types of diagenetic silicification: Bedded and nodular chert, chert in poorly developed laminites, and fracture-related chert.

The beds and nodules of early formed chert and chert in laminites occur in the Tb, Tc, and Td Bouma intervals in the carbonate turbidites. The original sedimentary structures are preserved in the cherts. The fracture-related chert has a later diagenetic origin, and cross-cuts the sedimentary structures and the early-formed chert.

Silica minerals include microquartz and length-fast chalcedony in the bedded, nodular, and fracture-related chert. Microquartz and length-slow chalcedony (quartzine) compose laminite-hosted chert.

No differences exist in the abundance of minor elements among all the types of chert. The contents in Zr, Pb, Rb, and Zn are higher than the amounts commonly found in these types of cherts from other areas, and these elements may be associated with the iron oxides.

The silica of the bedded and nodular chert was derived from the early dissolution and calcitization of sponge spicules and radiolarians mixed in with the carbonate turbidites. The late stage fracture-related chert apparently came from a later-stage calcitization of siliceous tests and perhaps also from a partial silica remobilization of the early-formed chert.

Introduction

Silicification is one the most important diagenetic characteristics of the calcareous flysch (early Senonian age) in the Biscay Sinclinorium (Fig. 7.1A,B). This area belongs to the so-called Basque folded arc (Rat, 1962), which is a part of the passive continental margin of northern Spain (Boillot and Capdevila, 1977).

Siliceous deposits occur chiefly as beds and nodules, as noted in papers on regional geology (Lamare, 1936; Rat, 1959; Jerez Mir et al., 1971; Campos, 1979; Mathey, 1982, 1986), although the siliceous deposits have not been specifically studied, except that Elorza et al. (1985) determined the quartz crystallinity index (Murata and Norman, 1976; Gregg et al., 1977) of the bedded chert from Barrika, northeast Bilbao, Spain.

We studied the silicification of the calcareous flysch along an excellent coastal exposure near Barrika village (Fig. 7.1C). We attempt to establish a silicification model and to compare the Barrika section with silicified carbonate turbidites to the northeast in Saint Jean de Luz, Haicabia, Bidache, and other areas. These various sections are considered to have been deposited in different subgrabens (Henry et al., 1971; Rosell et al., 1985; Mathey, 1986). We also describe the different chert types and the petrographical and mineralogical characteristics of the chert and the host carbonate rock. The biogenic source of the silica and the major and minor element chemistry of the chert and carbonate host rock are discussed.

Geological Setting

A sea-floor spreading stage during the Jurassic in the Bay of Biscay produced the continental margin of northern Spain, at the boundary of the Iberian plate and the Eurasian plate. Different hypotheses exist about the evolution of the continental margin of

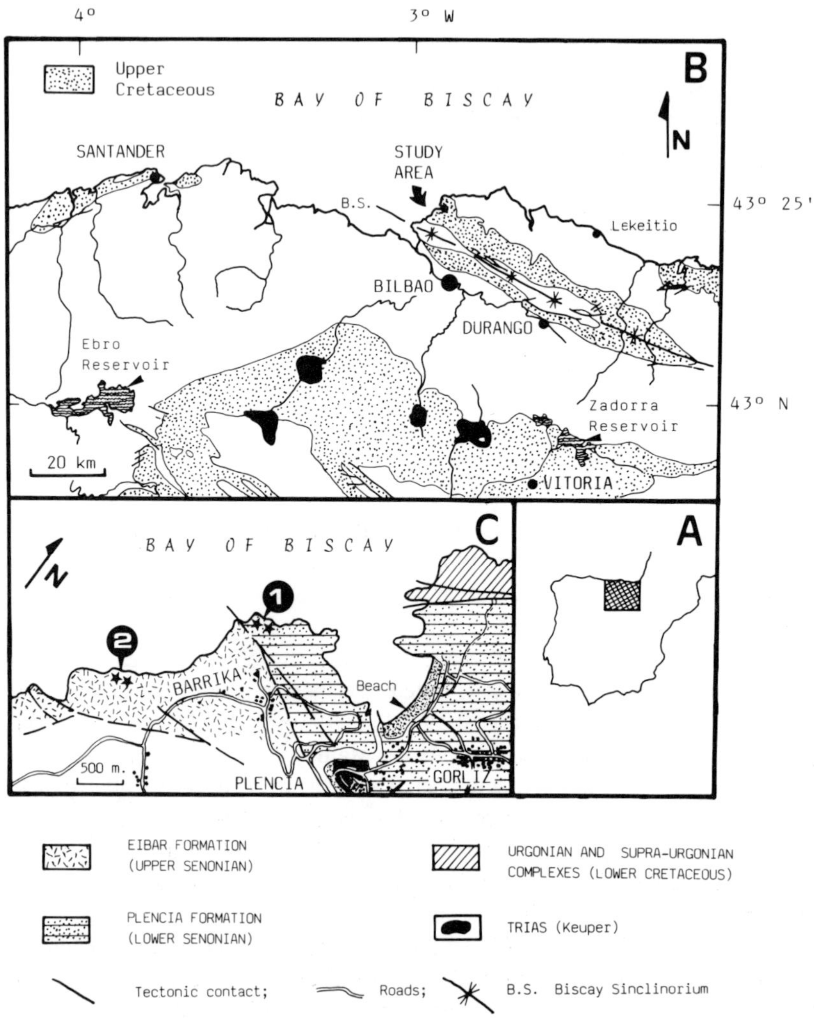

FIGURE 7.1. (A) Outline of the Iberian Peninsula showing the location (cross-hatched area) of a part of the Basque-Cantabric Basin, shown in (B). (B) A geological map of the Basque-Cantabric Basin showing Upper Cretaceous outcrops and the area studied. (C) Study area and localities mentioned in the text; the location of outcrop 1 (Plencia Formation) and outcrop 2 (Eibar Formation) contain bedded chert.

northern Spain (Le Pichon et al., 1971; Ries, 1979; Boillot et al., 1979; Vanderberg, 1980); however, it is possible to identify several tectonic phases that occurred during the extensional period (ocean basin formation): (1) A rifting phase occurred during the Late Jurassic, with the subsequent formation of "Weald basins." (2) A phase of distension flexure and rotation of crustal blocks along curved faults in the outer margin occurred during the Aptian and Albian, with the formation of the Urgonian carbonate platform. (3) The generalized collapse of crustal blocks and significant fracturing occurred during the late Albian to Eocene (Rat, 1982; Rat et al., 1982). During this latest phase there appeared a northeast-southwest trough, formed by development of transverse structures; formation of this trough was followed by formation of a distal platform (Navarro-Cantabric trough) and a proximal platform (north Castellain carbonate platform) (Fig. 7.2A).

During the Late Cretaceous (early Senonian), the trough was filled with carbonate turbidites (calcareous flysch) along two separate subgrabens (Plencia and Saint Jean de Luz). Later, during the late Senonian, a single graben (Orio Basin) was filled with siliciclastic turbidites (sandy flysch). At the end of the Late Cretaceous (middle-late Maastrichtian), limestone/marl alternations were deposited (Mathey, 1983, 1986; Mount and Ward, 1986).

The chert studied is associated with carbonate turbidites outcropping near Barrika (northeast Bilbao) and belongs to the Plencia Formation (Mathey, 1982), which was deposited in the Plencia Basin. It is a deep marine deposit regarded to have been deposited on the middle or outer part of a fan (Mathey, 1986).

The Fruniz Formation underlies the Plencia Formation and is composed of alternating marlstone and limestone beds together with submarine vol-

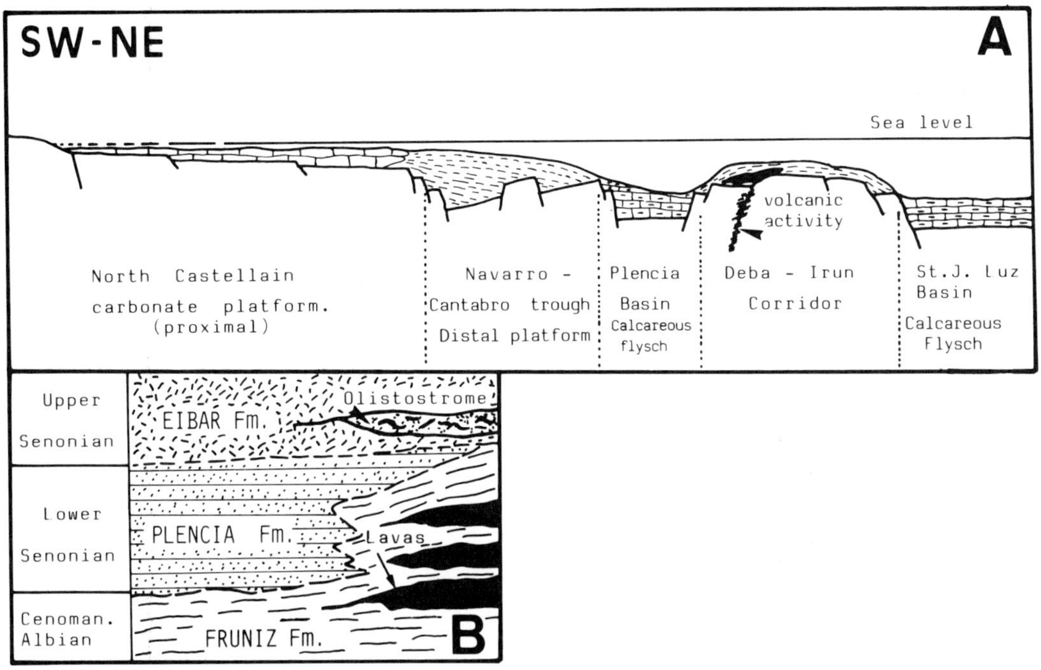

FIGURE 7.2. (A) Schematic portrayal of paleogeography and major structures for the early Coniacian time in the passive continental margin of northern Spain (not to scale). (Based on Rat et al., 1982, Mathey, 1986.) (B) Composite stratigraphic section of the Fruniz, Plencia, and Eibar Formations. (Based on Mathey, 1982.)

canic rocks (lava flows and breccias). This formation is of late Albian to early Coniacian age and comprises the lateral facies of the Plencia Basin (Fig. 7.2B). The Eibar Formation (late Senonian), concordantly and gradationally overlies the Plencia Formation and is composed of sandstone or micaceous argillite alternating with calcareous siltstone or grayish marlstone beds, making up siliciclastic turbidites (sandy flysch). A large olistostrome (≅50 m thickness) composed of volcanic rock fragments (Upper Cretaceous), conglomerate and sandstone blocks (black flysch; late Albian–early Cenomanian), carbonate turbidite beds including bedded and nodular chert (Plencia Formation) etc., and a matrix basically of carbonate debris occurs and marks an important depositional event within the sandy flysch section.

Occurrence of Chert

In the Plencia Formation (outcrop 1, Fig. 7.3A) and in the olistostrome deposit of the Eibar Formation (outcrop 2, Fig. 7.3B), three types of chert occur: (1) bedded and nodular chert; (2) chert in poorly developed laminites; (3) fracture-related chert.

Bedded and Nodular Chert

The beds and nodules of chert are located in the Tb, Tc, and Td intervals of Bouma's turbidite sequence (Fig. 7.4). The chert nodules are 2 to 20 cm thick, and the longer dimension is commonly parallel to the bedding. The bedded chert contains patches of carbonate representing relict host rock (Fig. 7.5A). Bedded chert may have parallel or convolute lamination of the Tb, Td, or Tc turbidite intervals inherited from the carbonate turbidite the silica replaced (Fig. 7.5B).

The morphology of chert nodules is also controlled by primary sedimentary structures of the carbonate turbidite host rocks. Straight boundaries were produced if the host rock had parallel lamination (Tb, Td), and irregular boundaries were produced if the host rock had convolute lamination (Fig. 7.5C). Finally, the internal structure of the host rock may have determined the location of the nodular chert; nodules preferentially formed in the anticlinal cores of undulate lamination (Tc interval) (Fig. 7.5D).

At outcrop 2 (Eibar Formation), large blocks of carbonate turbidite beds belonging to the Plencia Formation form part of the large debris flow deposit.

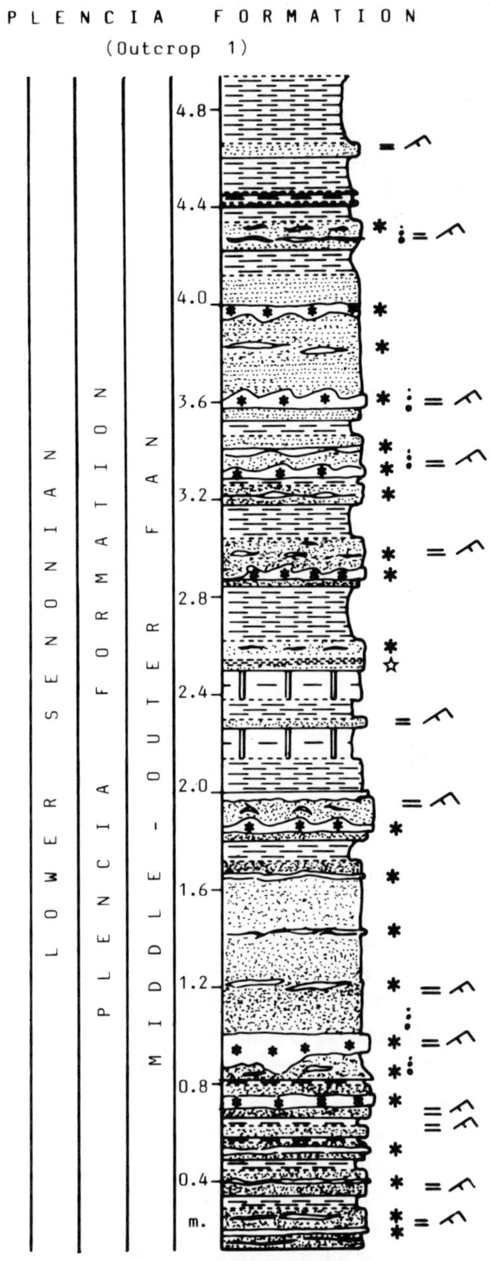

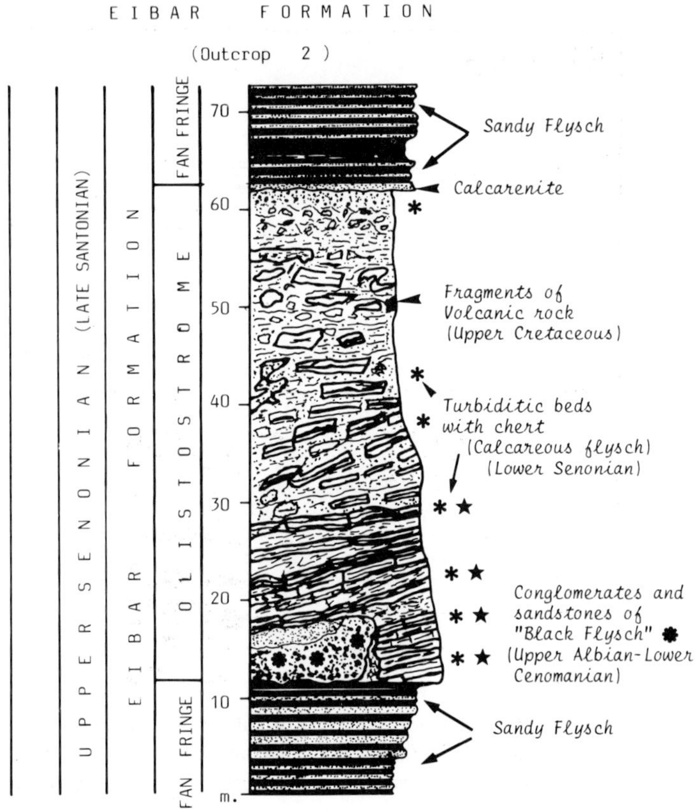

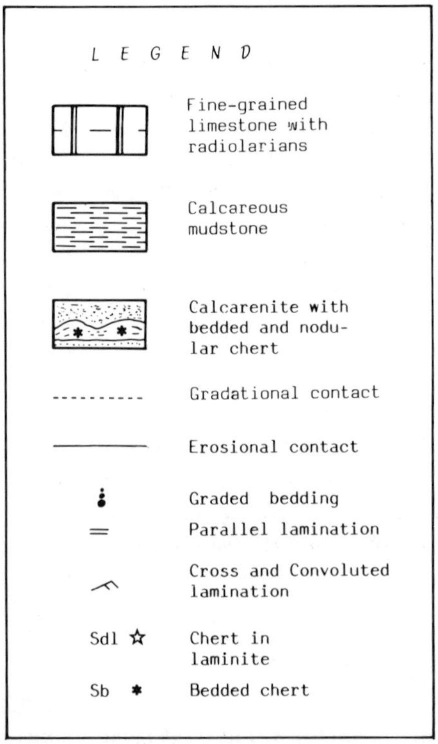

FIGURE 7.3. (A) Lithostratigraphic sketch of the Plencia Formation (lower Senonian), in Barrika (northeast Bilbao). (B) Lithostratigraphic sketch of the Eibar Formation (upper Senonian), with the olistsrostrome studied in Barrika (northeast Bilbao).

The bedded and nodular cherts included in the blocks were fractured and brecciated. The fragments are separated by several millimeters to centimeters and were rotated in places, whereas the carbonate host rock apparently did not undergo these modifications. The chert must have formed before the complete lithification of the carbonate host sediment (Fig. 7.6A).

Chert in Poorly Developed Laminites

Chert in poorly developed laminites consists of thin and poorly defined alternating chert and limestone (calcarenite) laminae produced by the preferential silicification of alternate laminae of the Tb and Td intervals (Fig. 7.6B). The silicified laminae range in thickness from several tens of microns to 1 to 2 mm.

Fracture-Related Chert

Fracture-related chert has irregular, rectilinear, or curved shapes and randomly crosscuts the carbonate turbidite beds. This chert is recognized at the bottom of the olistostrome in the Eibar Formation. The fracture-related chert is never found in the hemipelagic mudstone. The irregular contact with the host rock shows that silicification is not a simple infilling of an open fracture but is mainly a replacement of the adjacent carbonate rock. These fractures cross the bedded chert, nodular chert, and laminite-hosted chert. The width of the fracture-related chert increases when the fracture passes from the bedded chert to the carbonate host rock (Fig. 7.6B,C). In places, a complex fracture network occurs between the two lithologies (Fig. 7.6D).

The Plencia Formation contains millimeter-size fractures filled with chert. These fractures are local and limited in thickness, and silica did not replace the adjacent host rock.

Petrography and Mineralogy

The chert host rock is a calcarenite to coarse-grained calcisiltite that can be classified as a biopelsparite (grainstone). The most abundant components are siliceous sponge spicules and carbonate pellets (60%); radiolarians, fragments of foraminifers, and detrital quartz are present in minor amounts (<20%). Feldspar, mica, tourmaline, zircon, and apatite occur locally. Clay minerals comprising less than 10% of

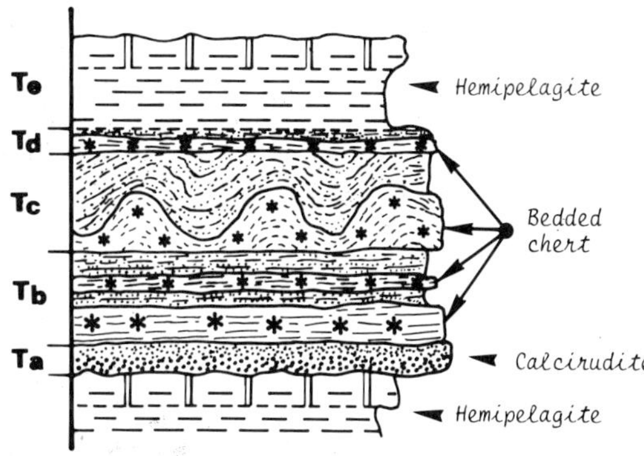

FIGURE 7.4. Schematic turbidite bed, belonging to outcrop 2, with bedded chert in Tb, Tc, and Td of Bouma's sequence; chert never occurs in Ta or Te intervals. (Not to scale.)

the rock were detected and determined by Mathey (1986).

Siliceous sponge spicules occur in all the carbonate turbidite intervals and define the laminations of the Tb, Tc, and Td intervals, with a strong decrease in size toward the top of the beds. The cement for the turbidites is chiefly ferroan calcite spar, although microsparite and micrite appear locally (Fig. 7.7A). In thicker calcarenite beds (Ta interval) larger bioclasts of a different nature (orbitoline, gastropods, coralline algae, echinoid spines, etc.) occur together with weathered volcanic rock fragments.

The Te interval (hemipelagite) does not contain chert and is composed of ferroan micrite (mudstone) containing planktonic foraminifers, radiolarians, and sponge spicules, which are small and always calcified (Fig. 7.7B). The allochems are greater than 10% in places, and the rock is a wackestone, with a larger proportion of radiolarians than sponge spicules. The clay minerals (illite, smectite, kaolinite, and chlorite) range from about 20% to 60% in these intervals (Mathey, 1986).

The chert is present only within the calcarenite beds. Most of the silica is in the form of microquartz and cryptocrystalline quartz, with grain sizes <20 μm (Fig. 7.7C). A fibrous variety of quartz is a minor component. Length-fast chalcedony (chalcedonite) appears to have (1) replaced the walls of sponge spicules; (2) infilled the axial canals of sponge spicules, radiolarian chambers, and other original sediment voids; and (3) replaced some carbonate fossils (foraminifers, bivalves, etc.). Locally within the sponge spicules there occurs a brownish-yellow semiisotropic material considered as a relict

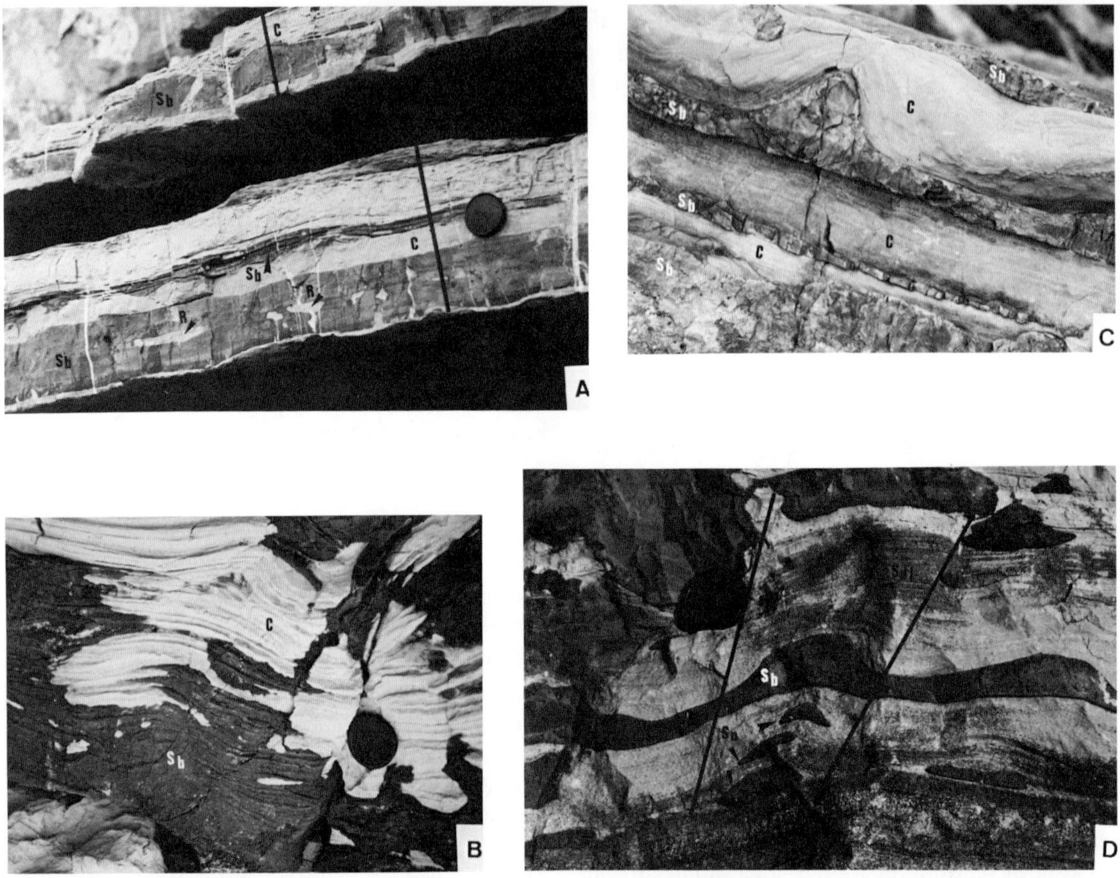

FIGURE 7.5. (A) Diagenetic silicification forming bedded chert (Sb) in calcarenite turbidite (C); primary sedimentary structures are preserved in the chert. Remains of the host rock (R) that were not silicified occur in the bedded chert. (B) Detail of silicification in calcarenite (C) with convolute laminations. When the silica supply was low, silicification was patchy. (C) Bedded chert (Sb) replacing and preserving the primary sedimentary structures, convolute and planar lamination (C). (D) Bedded and nodular cherts (Sb) replacing primary sedimentary structures. There were a migration and concentration of silica in the anticlinal cores of undulate laminations. Lens cap for scale is 5 cm in diameter.

opaline precursor phase. Fine-grained iron oxides are also distributed irregularly in the rock.

Relict carbonate is commonly included within the chert; in places it is the product of neoformation. The neoformed carbonates are frequently euhedral rhomb-shaped crystals of ferroan calcite: (1) scattered in the microquartz; (2) replacing a part of chalcedonic spicules; and (3) infilling later-stage fractures which cut the early-formed chert. A small amount as replacement of massive microquartz occurs along the fractures. Subrounded to subangular fine-grained detrital quartz occurs in the chert and shows no signs of alteration.

The chert in poorly developed laminites is made up dominantly of small spherulites of length slow chalcedonite (quartzine) and lesser amounts of microquartz arranged in laminae. These laminae of chert alternate with carbonate laminae of similar thickness (Fig. 7.7E,F). The quartzine tends to form aggregates of spherules in all laminae containing carbonate and iron oxide particles.

There are no textural differences in quartz in bedded, nodular, and fracture-related chert. All consist mainly of microquartz and chalcedonite. Fracture-related chert is fundamentally a replacement of the adjacent rock and does not correspond solely to a fracture infilling. For this reason, the silicification textures are similar to those in bedded and nodular cherts; only the timing of silicification is different. In places, barite occurs in the fracture-related chert.

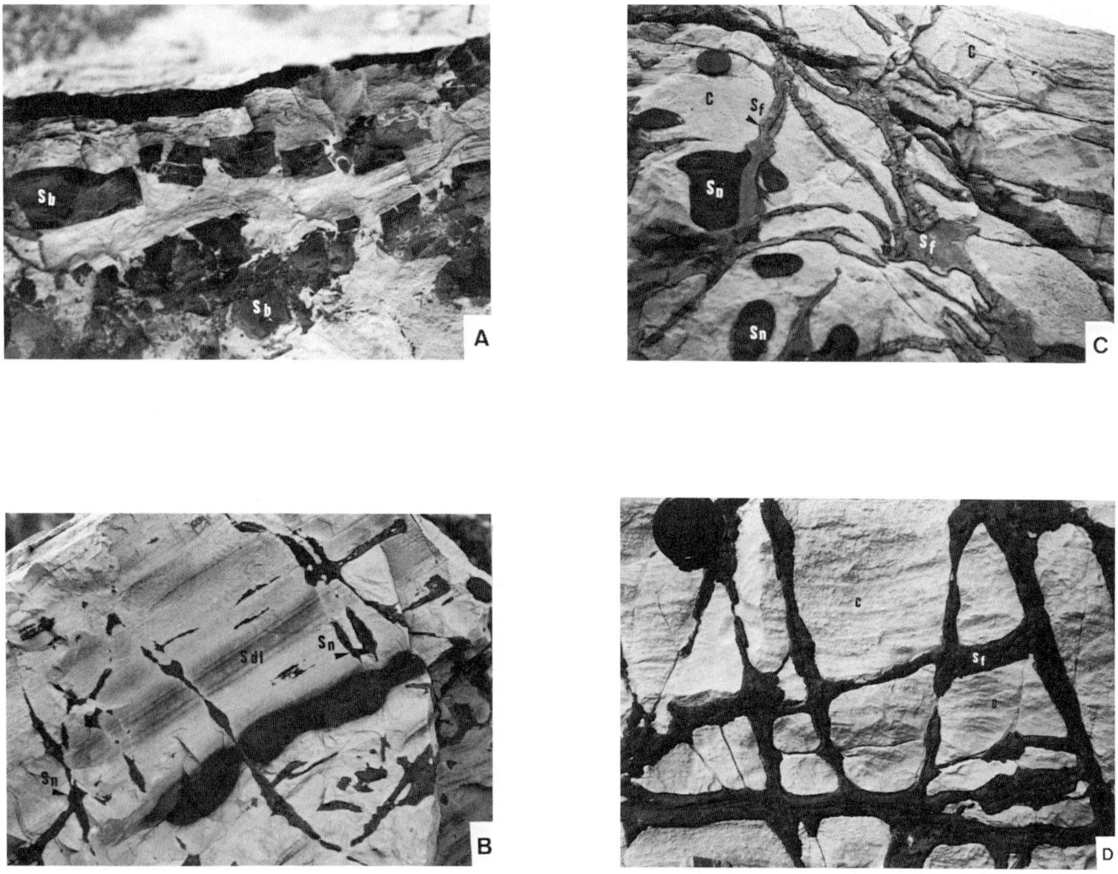

FIGURE 7.6. (A) Fractured and brecciated bedded chert (Sb). (B) Laminite (Sdl) cut by fractures. These fractures were silicified at a later stage than the laminite. Note the thickening of fracture-related chert as it crosses the carbonate host rock. (C) Fractures crosscutting nodular chert (Sn) and carbonate host rock (C). Silicification of the host rock along the fractures was a late-stage process. (D) Complex network of late-stage chert (Sf) replacing the host rock along fractures. Lens cap for scale is 5 cm in diameter.

Geochemistry

Trace and major element compositions of cherts and trace element composition of biopelsparites (host rock) were determined by X-ray fluorescence spectroscopy. Eight major elements Si, Al, Fe, Mn, Ca, Na, K, and Ti were determined for cherts using a Philips PW 1410 XRF spectrometer, with a chromium tube and glass disc (Table 7.1).

Twelve minor elements were determined for a variety of rock types using tungsten and chromium tubes and pressed powder samples (Table 7.2). Carbonate was removed with HCl from the cherts before analyses were made, because the carbonates included in the cherts are remains of the host rock. In two rocks, the bulk chert samples and the insoluble residues after HCl treatment were analyzed. The silica content of the untreated chert ranges from 84% to 98% (Table 7.1). Untreated cherts contains more Fe and Na than their HCl-treated counterpart. Fe and Na oxides are mainly contained in the carbonate minerals. The other oxides, except silica, either are minor or do not show consistent behavior.

Trace element compositions (Table 7.2) are compared with published data for bedded chert (Audley-Charles, 1965; Barrett, 1981; Matsumoto and Iijima, 1983; Adachi et al., 1986). The contents of Zr, Pb, Rb, and Zn are relatively high in the Barrika cherts. These elements are below the limits of detection in the host rocks. In untreated chert, Rb, Pb, Th, Zn, Ga?, and Zr are less abundant than in HCl-treated chert, suggesting that these trace elements are not associated with the calcite.

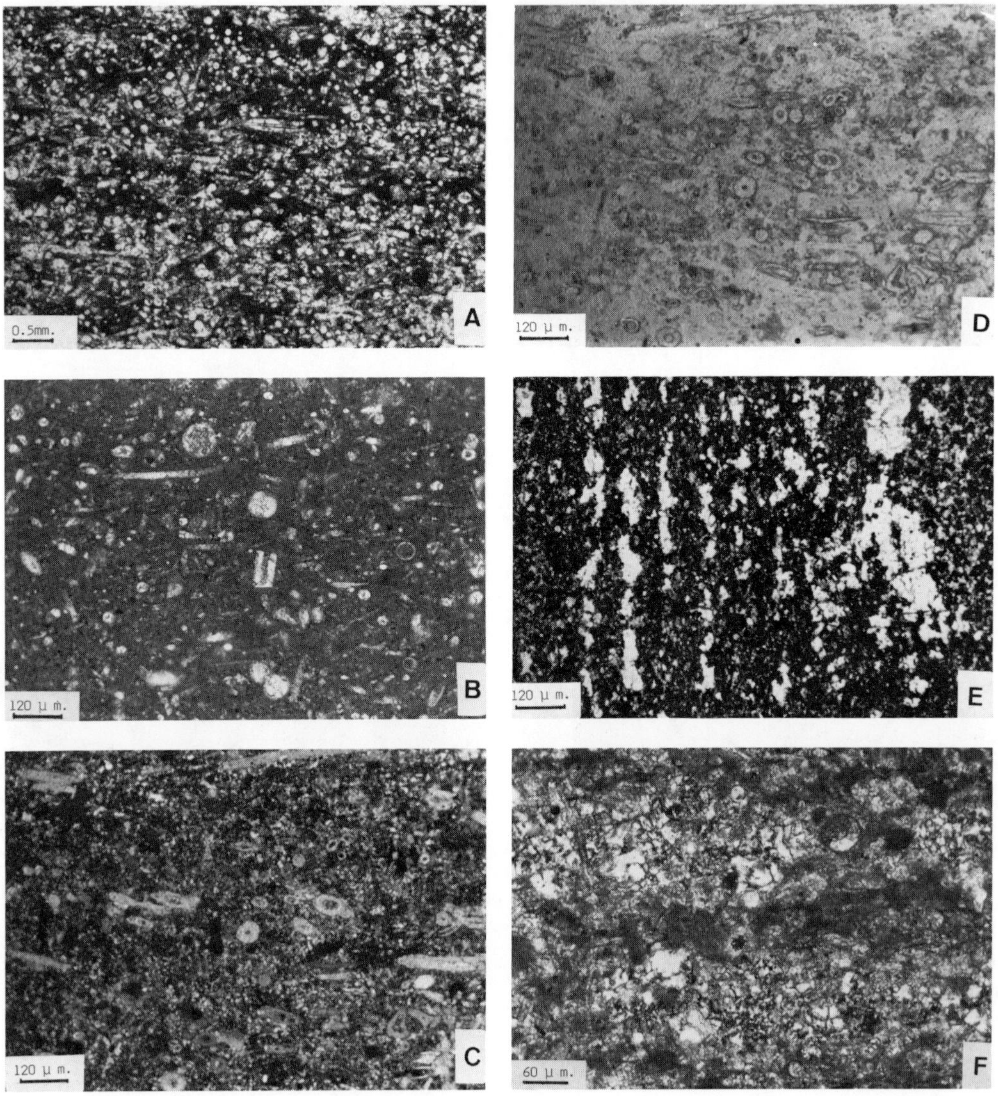

FIGURE 7.7. (A) Photomicrograph of carbonate host rock (biopelsparite) with sections of sponge spicules. Crossed nicols. (B) Photomicrograph of fine-grained limestone (wackestone) with radiolarians and sponge spicule fragments, a hemipelagite bed. Crossed nicols. (C) Photomicrograph of microquartz with sections of calcified sponge spicules. Plane polarized light. (D) Same as (C) with crossed nicols. (E) Photomicrograph of indistinct quartzine laminae in laminated carbonate host rock (biopelsparite). Crossed nicols. (F) Photomicrograph of quartzine laminae composed of spherulites.

Our results are different from those determined for carbonate-hosted cherts from DSDP legs 62 and 69 (Hein et al., 1981, 1983). In the DSDP samples, SiO_2 and B increase and all other elements decrease with increased silicification of the chalk.

No difference exists between the content in minor elements in bedded, nodular, and fracture-related chert, because all these cherts formed by the replacement of the host carbonate. Sr and Zr are the most abundant trace elements; Sr is also associated with the carbonate host rock, but Zr is not. Zr may be incorporated in iron minerals (Goldberg and Arrhenius, 1958; Barrett, 1981). Barrett (1981) thinks that Pb, Zn, and a lesser extend Rb may be included in the hematite. Most of the minor elements detected may be associated with the hematite, which is common in many cherts.

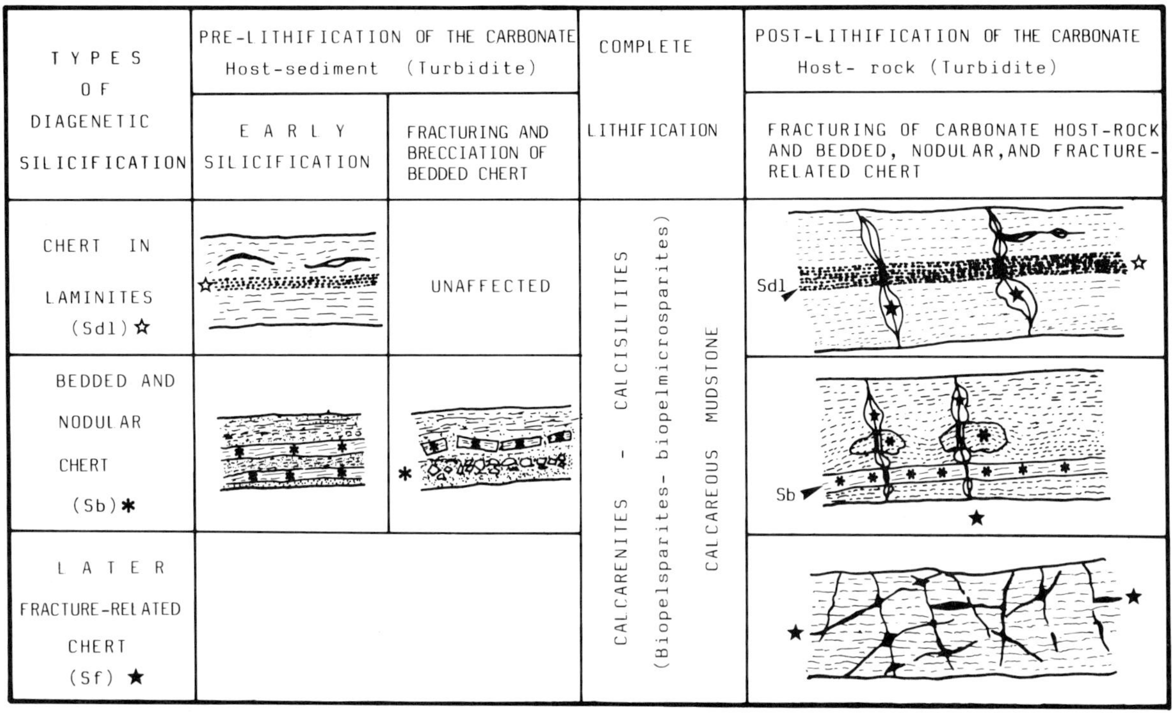

FIGURE 7.8. Schematic diagram of the different types of diagenetic silicification generated during pre- and postlithification of the carbonate turbidite.

Discussion

The formation of bedded and nodular chert was an early diagenetic process, because when the olistostrome was deposited by sediment gravity-flow processes in the Eibar Formation, the bedded and nodular cherts (included within the carbonate turbidite blocks belonging to the Plencia Formation) were fractured and brecciated before the carbonate host sediment was completely lithified (Fig. 7.6A). Therefore, the initial silicification must have been completed before deposition of the olistostrome, dated as late Santonian (Orue-Etxebarria, personal communication).

The formation of fracture-related chert (outcrop 2) was a late-stage diagenetic process that occurred after lithification and fracturing of the carbonate turbidite blocks inside the olistostrome. The fracturing may have been generated after deposition of the olistostrome as a consequence of readjustment of the car-

TABLE 7.1. Chemical composition in weight percent of cherts, by X-ray fluorescence spectroscopy.

	N-1	N-3	N-4	N-4*	N-5	N-12	N-12*	NF-1	NF-5
SiO_2	97.58	98.00	97.30	83.98	97.51	97.65	88.59	98.29	98.15
Al_2O_3	0.16	0.07	0.74	0.37	0.45	0.31	0.30	0.15	0.25
Fe_2O_3	0.26	0.13	0.13	0.48	0.16	0.13	0.59	0.06	0.05
MnO	0.01	0.01	0.01	0.01	0.01	0.01	0.01	0.01	0.01
CaO	1.02	0.80	0.80	7.91	0.80	0.80	5.44	0.81	0.80
Na_2O	0.01	–	–	0.21	–	0.01	0.13	–	–
K_2O	–	–	–	–	–	–	–	–	–
TiO_2	0.02	0.02	0.03	0.02	0.04	0.03	0.02	0.02	0.02
Loss on fusion at 1050°C	0.88	0.71	0.80	6.72	0.81	0.77	4.84	0.81	0.85
Total	99.94	99.74	99.81	99.70	99.78	99.71	99.92	100.15	100.13

Key: N, HCl-insoluble residues of nodular and bedded chert. N*, bulk untreated chert. NF, chert related to fractures (total iron as Fe_2O_3).

TABLE 7.2. Trace elements composition of rocks from northeast Bilbao, in parts per million.

	Rb	Ba	Pb	Sr	Ce	Y	Th	Zn	Cu	Ni	Ga	Zr
Bedded and nodular chert, HCl-treated												
N–1	64	7	38	180	10	23	25	34	11	7	24	165
N–2	69	–	35	177	9	26	28	51	10	7	25	185
N–3	72	8	39	182	9	26	27	45	11	7	26	181
N–4	69	23	37	185	9	25	28	61	12	8	26	180
N–4*	51	–	28	291	10	20	21	23	9	7	21	131
N–5	70	18	38	181	9	25	28	38	10	6	26	181
N–6	73	12	37	182	9	26	28	38	11	8	25	182
N–7	70	–	36	179	10	26	29	40	10	9	25	182
N–8	70	–	38	182	9	26	28	39	10	7	24	181
N–9	71	7	37	181	9	26	28	47	10	7	26	184
N–10	69	16	35	183	9	26	30	47	11	7	24	185
N–11	71	–	39	181	9	25	26	44	10	7	25	182
N–12	69	–	39	181	9	22	27	41	10	7	24	184
N–12*	55	10	33	251	8	22	22	27	12	9	22	145
Fracture-related chert, HCl-treated												
NF–1	69	–	37	186	9	25	27	54	12	7	25	178
NF–2	69	–	37	192	8	26	28	48	10	6	25	181
NF–3	69	–	38	90	8	26	27	45	9	7	25	184
NF–4	70	–	39	184	8	26	28	34	9	7	24	178
NF–5	70	–	35	188	9	26	28	39	11	6	24	186
Calcarenite												
LF–29	17	–	6	410	8	7	–	–	8	–	13	26
LF–30	–	–	6	390	8	–	–	–	8	–	11	–
LF–32	–	–	–	379	8	–	–	–	–	–	11	–
LF–36	–	–	–	361	8	–	–	–	8	–	11	–
LF–48	–	–	7	326	8	–	–	–	7	–	12	19
LF–61	–	–	–	438	8	–	–	–	7	–	13	12

Key: –, less than 5 ppm. *Untreated sample.

bonate turbidite blocks. Along these irregular fractures, the silica-rich fluids migrated and replaced the carbonate host rock which had been already lithified.

No evidence was found to determine the time of silicification of the laminites. However, we speculate that silicification occurred near the time of formation of bedded and nodular cherts. Silicification was before the formation of fracture-related chert, because the fractures crosscut the laminites and nodular chert (Figs. 7.6B,C; 7.8).

The source of silica for the formation of bedded, nodular, and laminite-hosted cherts is from the siliceous biogenic debris deposited with the carbonate turbidite beds. These beds have a high abundance of siliceous sponge spicules and their traces (≅40% bulk rock). The calcitization of some and direct dissolution of other biogenic opal-A tests and spicules supplied enough silica to form the bedded and nodular cherts.

The silica-rich pore fluids remained localized or migrated laterally, partially or completely replacing the high-porosity carbonate sediment in favorable places. Solubility values for biogenic opal are 100 to 140 ppm, cristobalite 40 to 70 ppm, and quartz 6 ppm at 25°C at a pH<9 (Morey et al., 1962; Jones and Segnit, 1971; Krauskopf, 1979). The lateral migration of silica to more permeable zones produced chert nodules and bands. Primary sedimentary structures (convolute, parallel lamination, etc.) are preserved in the chert and, in fact, controlled the process of silicification by providing zones of variable porosity and permeability.

The number of chert beds within each turbidite bed is exclusively due to the original amount of biogenic silica. When there was not enough silica for complete replacement, remains of the host carbonate appear inside the chert, with the original sedimentary structure preserved.

The source of silica for the late-stage fracture-related chert may have been from later-stage calcitization of siliceous tests, or perhaps from partial silica remobilization of the early-formed chert.

The conditions of silicification for the bedded, nodular, and fracture-related chert are typical of deep marine, but not necessarily open-ocean diagenetic environment, with formation of microquartz, cryptocrystalline quartz, and fibrous chalcedonite. Pore waters were neutral to acid with variable concentrations of silica (Folk and Pittman, 1971). In chert associated with laminites, the fibrous texture of quartzine was probably formed in an alkaline or sulfate diagenetic microenvironment (Folk and Pittman, 1971); also, the precipitation of quartzine may indicate that pore fluids were rich in magnesium, as well as being saturated with respect to quartz (Jacka, 1974; Hatfield, 1975; Folk, 1975; Keene, 1983). We detected barite by X-ray diffraction only within the fracture-related chert; Mathey (1986) also notes authigenic barite in the carbonate host rock from the Plencia Formation. The presence of barite is consistent with the occurrence of quartzine, which in pelagic sediment is always associated with authigenic barite and in many places with dolomite (Keene, 1983).

Upwelling along the carbonate platform caused high productivity of siliceous organisms (sponges, radiolarians) that were later mixed with other allochems as carbonate-bearing, low-density turbidity currents that deposited the sediment above the CCD (Mathey, 1986).

Silicification in the marlstones and fine-grained limestone (hemipelagite) is not found, despite the presence of calcified sponge spicules and radiolarians (Fig. 7.7B). The most probable explanation is that the liberated silica migrated and was lost to the overlying seawater (Bustillo and Ruiz-Ortiz, 1984, 1987) or migrated and was concentrated in the calcarenite beds.

The silica-rich fluids were also somewhat enriched in minor elements (Zr, Pb, Zn, Pb) derived from the dissolution of biogenic silica, but primarily from the leaching of the hemipelagitc beds, mainly the clay minerals. During early diagenesis, silicification caused rejection of the minor elements from the chert and their concentration in the calcarenite beds, which had a greater effective porosity than the mudstone beds. Throughout the silicification process, the minor elements were incorporated in the silica-rich solutions from which the initial diagenetic opaline phase was precipitated. Later, when the diagenetic opaline phase transformed to quartz, the minor elements were probably incorporated into the iron oxides (hematite).

Conclusions

Bedded and nodular cherts originated by an early-stage silicification of carbonate turbidites (early Senonian). Formation of chert in laminites was apparently not synchronous with formation of bedded and nodular cherts, because the different fibrous silica textures (quartzine and chalcedonite, respectively) indicate silicification under different diagenetic conditions. However, both episodes of silicification occurred before formation of the fracture-related chert, because this late-stage chert crosscuts the bedded, nodular, and laminite-hosted cherts.

Fracture-related chert was formed during late-stage silicification of carbonate rock adjacent to fractures. The fractures may have formed in the calcarenite blocks after the olistostrome was deposited and the carbonate host rock was already completely lithified.

The source of silica for the formation of bedded and nodular cherts was from the dissolution and calcitization of siliceous organisms, mainly sponge spicules and secondarily radiolarians, which were included in the carbonate turbidite beds. Upwelling along the carbonate platform produced high productivity of siliceous sponges and radiolarians, which were deposited in the flysch trough with other allochems as carbonate turbidites. The radiolarians were more plentiful in the trough environment and are better preserved in the hemipelagic beds.

The source of silica for the fracture-related chert came from late-stage calcitization of siliceous organisms or from remobilization of silica in the bedded and nodular cherts.

The relatively high concentration of minor elements in all chert types resulted from leaching of clay minerals in hemipelagic beds. The initial diagenetic opaline phase formed from interstitial solutions rich in minor elements. During transformation of the initial diagenetic opal to quartz, the minor elements were released and subsequently incorporated into the authigenic iron oxides (hematite).

Acknowledgments. This work forms part of the CSIC and CACYT project 200, "Estudio Mineralógico, Petrológico y Sedimentológico de las Rocas

Silíceas (Silex, Opalos, etc.) y Sedimentos Asociados en el Ambito de Tethys (Cuenca Vasco-Cantábrica, Meseta Central y Cordilleras Béticas)." We thank James R. Hein, U.S. Geological Survey, for critical review of the manuscript and suggestions for its improvement.

References

Adachi M, Yamamoto K, Sugisaki R (1986) Hydrothermal chert and associated siliceous rocks from the northern Pacific: Their geological significance as indication of ocean ridge activity. Sedimentary Geology 47:125–148.

Audley-Charles MG (1965) Some aspects of the chemistry of Cretaceous siliceous sedimentary rocks from eastern Timor. Geochimica et Cosmochimica Acta 29:1175–1192.

Barrett TJ (1981) Chemistry and mineralogy of Jurassic bedded chert overlying ophiolites in the north Apennines Italy. Chemical Geology 34:289–317.

Boillot G, Capdevila R (1977) The Pyrenees: Subduction and collision? Earth Planetary Science Letters 35: 151–160.

Boillot G, Dupeuble PA, Malod J (1979) Subduction and tectonics on the continental margin of northern Spain. Marine Geology 32:53–70.

Bustillo MA, Ruiz-Ortiz P (1984) Chert formation in carbonate turbidite (upper Jurassic, Betic Mountains, southern Spain). Abstracts 27th International Geological Congress (Moscow) II:26.

Bustillo MA, Ruiz-Ortiz P (1987) Chert occurrences in carbonate turbidites: Examples from the upper Jurassic of the Betic Mountains. (Southern Spain), Sedimentology 34:611–621.

Campos J (1979) Estudio geológico del Pirineo vasco al W del rio Bidasoa. Munibe 1-2:3–139.

Elorza JJ, Arriortua MI, Amigo JM (1985) Indices de cristalinidad en los silex de carácter turbiditíco de Barrika (NE de Bilbao). Boletin Geológico y Minero XCVI(I):74–81.

Folk RL (1975) Third-part reply to Hatfield: Discussion of Jacka A.D. 1974. "Fossils by length-slow chalcedony and associated dolomitization." Jour Sedimentary Petrology 44:421–427.

Folk RL, Pittman S (1971) Length-slow chalcedony: A new testament for vanished evaporites. Journal of Sedimentary Petrology 41:1045–1058.

Goldberg ED, Arrhenius GOS (1958) Geochemistry of Pacific pelagic sediments. Geochimica et Cosmochimica Acta 13:153–212.

Gregg JM, Goldstein ST, Walters LJ (1977) Occurrence of strained quartz in the siliceous frustules of cultured freshwater diatoms. Journal of Sedimentary Petrology 47:1623–1629.

Hatfield CB (1975) Discussion: Replacement of fossils by length-slow chalcedony and associated dolomitization. By Alonso D. Jacka. Journal of Sedimentary Petrology 44:421–427; Journal of Sedimentary Petrology 45: 951–952.

Hein JR, Vallier TL, Allan MA (1981) Chert petrology and geochemistry, Mid-Pacific Mountains and Hess Rise, Deep Sea Drilling Project leg 62. In: Thiede J, Vallier TL, et al. (eds), Initial Reports DSDP 62. U.S. Government Printing Office, Washington, pp 711–748.

Hein JR, Sancetta C, Morgenson LA (1983) Petrology and geochemistry of silicified upper Miocene chalk, Costa Rica Rift, Deep Sea Drilling Project leg 69. In: Langseth MG, Cann J, et al. (eds), Initial Reports DSDP 69. U.S. Government Printing Office, Washington, pp 395–422.

Henry J, Lanusse R, Villanova M (1971) Evolution du domaine marine pyrénéen du Sénonien supérieur à l'Eocene inférieur. In: Histoire Structurales du Golfe de Gascogne. Technip Ed. I, p IV 7–1 à IV 7–7.

Jacka AD (1974) Replacement of fossils by length-slow chalcedony and associated dolomitization. Journal of Sedimentary Petrology 44:421–427.

Jerez Mir L, Esnaola JM, Rubio V (1971) Estudio geológico de la provincia de Guipúzcoa. Memorias IGME 79:130.

Jones JB, Segnit ER (1971) The nature of opal. I. Nomenclature and constituent phases. Journal of Geological Society Australia 18:57–68.

Keene JB (1983) Chalcedonic quartz and occurrence of quartzine (length-slow chalcedony) in pelagic sediments. Sedimentology 30:449–454.

Krauskopf KB (1979) Introduction to Geochemistry, 2d Ed. McGraw-Hill, New York, 617 pp.

Lamare P (1936) Recherches géologiques dans les Pyrénées basques d'Espagne. Mem Soc Geol France 27:465.

Le Pichon X, Bonnin J, Franchetau J, Sibuet JC (1971) Une hypothèse d'évolution tectonique du Golfe de Gascogne. In: Histoire Structurale du Golfe de Gascogne. Technip, p VI 11-1 VI II–44.

Mathey B (1982) El cretácico superior del Arco vasco. In: El Cretácico de España. Univ. Complutense, Madrid,pp 11–136.

Mathey B (1983) Le système du flyschs calcaires. In: Vue sur le Crétacé Basco-Cantabrique et Nord-Ibérique. Mem Geol Univ Dijon 9:83–86.

Mathey B (1986) Les flysch crétacé supérieur des Pyrénées basques. Age, anatomie, origine du material, milieu de depot et rélations avec l'ouverture du Golfe de Gascogne. Thèse, Université de Bourgogne, Bourgogne, France, 403 pp.

Matsumoto R, Iijima A (1983) Chemical sedimentology of some Permo-Jurassic and Tertiary bedded cherts in Central Honshu, Japan. In: Iijima A, Hein JR, and Siever R (eds) Siliceous Deposits in the Pacific Region, Elsevier, Amsterdam, pp 175–191.

Morey GW, Fournier RO, Rowe JJ (1962) The solubility of amorphous silica at 25°C. Journal Geophysical Research, 69:1995–2002.

Mount JF, Ward P (1986) Origin of limestone/marl alternations in the upper Maastrichtian of Zumaya, Spain. Journal of Sedimentary Petrology 56:228–236.

Murata KJ, Norman MB (1976) An index of crystallinity for quartz. American Journal of Science 276:1120–1130.

Rat P (1959) Les pays crétacés basco cantabriques. Publ University Dijon 18:525.

Rat P (1962) Présentation géologique du versant cantabrique entre Pyrénées et Asturies. Inst Est Pyrenaicos CSIC Zaragoza 9-26.

Rat P (1982) Subsidence et évolution des environnements sédimentaires sur la marge cantabrique (Espagne) au Crétacé. Neues Jb Geol Palaont Abh 165:32–45.

Rat P, Amiot M, Feuillee P, Floquet M, Mathey B, Pascal A, Salomon J (1982) Etapes et style de l'evolution de la marge cantabrique et de son arriére-pays (Espagne). Acad Sci Paris 295:247–250.

Ries AC (1979) The opening of the Bay of Biscay. A review. Earth Science Review 14:35–65.

Rosell J, Remacha R, Zamorano M, Galbaldon V (1985) Serie turbiditica del Cretácico superior del País-Vasco. Boletin Geológico y Minero XCVI(IV):361–366.

Vandenberg J (1980) New paleomagnetic data from the Iberian Peninsula. Geologie en Mijnbouw 59:49–60.

8
Radiolarite Sequences of the Subbetic, Betic Cordillera, Southern Spain

P.A. Ruiz-Ortiz, M.A. Bustillo, and J.M. Molina

Overview

The radiolarite sequence of the middle Subbetic (Betic cordillera, southern Spain) is composed of three main lithologies: radiolarite, siliceous mudstone and marl, and pelagic limestone with radiolarians. Virtually all transitional lithologies among the three listed exist. Clay minerals, calcite, and hematite make up the radiolarite impurities. Smectite, illite, and locally kaolinite are the clay minerals present, smectite always being most abundant. No radiolarite bed with more than 85% silica has been found; the total SiO_2 content ranges from 61 to 85%. Frequently the compact appearance of the radiolarite results from a high carbonate content. The ratios Si/Si + Al + Fe + Ca, Si/Si + Al + Fe, and Al/Al + Fe + Mn are calculated and used to compare the beds of couplets in rhythmic sequences and to obtain the relative distance of sections from their terrigenous source. These three chemical indices do not give consistent results. Fe enrichment of radiolarite relative to siliceous mudstone is probably related to the low mean sedimentation rate of these sequences from the Subbetic (3.5 mm/1,000 y). Pelagic settling in a basin with irregular bottom topography and with areas topographically protected from sediment input is proposed for deposition of these deposits. Productivity cycles and terrigenous dilution gave rise locally to rhythmically bedded sequences, usually composed of radiolarite-mudstone alternations. Locally, contour and distal dilute turbidity currents reworked and deposited the sediment, and gave rise to thin sequences of layered sedimentary rocks and to laminated beds. Deposition above a relatively shallow CCD is deduced, and a water depth shallower than previously thought is suggested for these deposits, which are considered as the deepest water deposits of the Betic External Zones.

Introduction

In the Alpine Mediterranean realm, radiolarian-rich sedimentary rocks, commonly referred to as radiolarites, outcrop frequently in the Middle and Upper Jurassic sections. A wide variety of siliceous rocks deposited in different sedimentary environments and tectonic settings occur. The sedimentology of these rocks remains poorly known, as stated by Jones and Murchey (1986): "The field of radiolarian chert sedimentology is in its infancy." Commonly, few significant sedimentological data can be obtained from radiolarites, partly because complex diagenesis masks many sedimentary features. Geochemical and mineralogical studies may be alternative tools used to understand the origin and significance of these siliceous deposits.

In the Subbetic (Betic Cordillera), siliceous radiolarian-rich sequences of Middle and Late Jurassic age have been referred to as radiolarites in regional tectonic and stratigraphic papers (e.g., Vera, 1966; Garcia-Dueñas, 1967; Azema et al., 1979); no specific studies of these radiolarite sequences had been done before our work (Bustillo et al., 1986). Subbetic radiolarites have a comparatively low (85%) silica content, which creates a problem because of the lack of a clear and widely accepted classification of siliceous rocks and transitional carbonate and claystone lithologies. Also it is difficult to assign a specific rock name to the lithologies in the field.

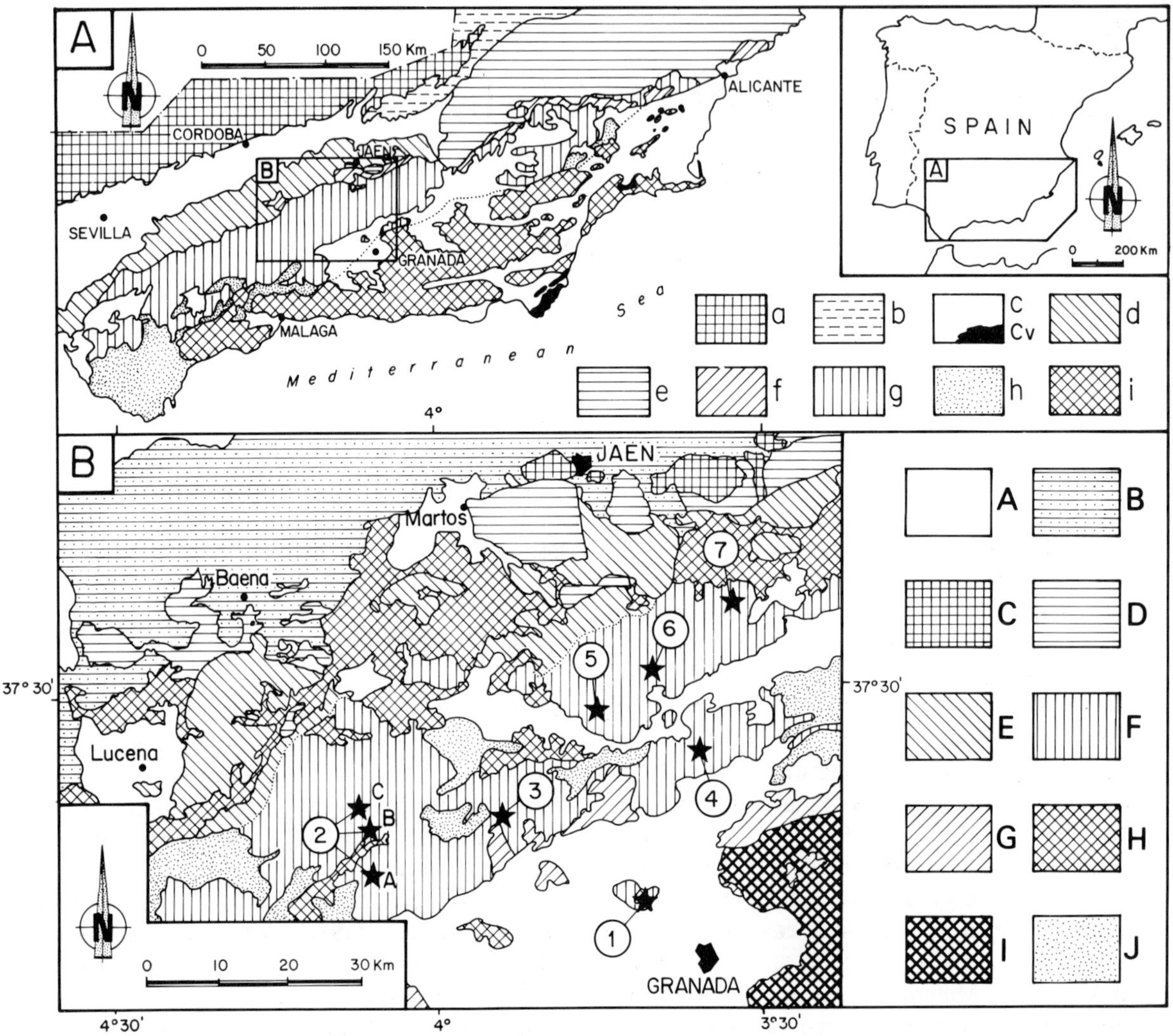

FIGURE 8.1. Geologic and location maps and position of the outcrops studied. (**A**) Key: (a) Iberian Hercinian fold-belts (Spanish Meseta); (b) horizontal cover of the Spanish Meseta (Triassic and Jurassic); (c) Neogene-Quaternary of the Guadalquivir and other basins; in black (Cv) volcanic rocks; (d) Guadalquivir basin areas with Subbetic nappes and olistostromes (Guadalquivir units and Carmona nappe); (e) Prebetic zone; (f) intermediate units between the Prebetic and the Subbetic; (g) Subbetic zone; (h) Campo de Gibraltar units (Numidian flysch and other turbidite formations making up several nappes); (i) internal zones. Dotted line is boundary between the internal and external zones of the Betic cordillera. (**B**) Key: (A) Neogene and Quaternary; (B) olistostromes of the Guadalquivir basin; (C) Prebetic zone; (D) intermediate units between the Prebetic and the Subbetic; (E) external Subbetic; (F) middle Subbetic; (G) internal Subbetic; (H) Triassic (mainly Keuper) rocks; (I) internal zones; (J) Paleogene and/or Campo de Gibraltar units. Dotted line is boundary between the external and middle Subbetic. **Studied localities** in circles: (1) Sierra Elvira (SE); (2) Algarinejo (AG); (3) Sierra Pelada (SP); (4) Zegrí (ZG); (5) Jarandilla-Hachazo (JH); (6) Navalcan (NAV); (7) Arbuniel (AB).

In this brief report, we present the results of a comprehensive study of the radiolarite sequences of the Subbetic. Sedimentological, petrological, and geochemical data are provided and integrated into the paleogeography of the external zones of the Betic Cordillera.

Middle Subbetic Zone: Geologic Setting

The Subbetic zone makes up one of the major paleogeographic realms of the Betic Cordillera (Fig. 8.1A). Sedimentation began in the Jurassic with shallow marine carbonate deposits. From the Carixian, the carbonate shelf disintegrated (Azema et al., 1979; Garcia-Hernandez et al., 1980), which marked the beginning of pelagic sedimentation, similar to what occurred in other Alpine Mediterranean areas (Bernoulli and Jenkyns, 1974). This breakup of the shelf has been interpreted as intracontinental rifting (Garcia-Hernandez and Vera, 1986). During the rest of the Jurassic, the Subbetic was a basin receiving mainly pelagic and hemipelagic deposits and exhibited a sharp differential subsidence that gave rise to troughs and swells. Two geologic and topographic environments identified as northern and southern topographic swells (external and internal Subbetic, respectively) were separated by an elongate basin (middle Subbetic) (Fig. 8.9). Pelagic, hemipelagic, and locally shallow-water deposition on the swells occurred during the Middle and Late Jurassic, with resulting deposits of marl, limestone, nodular limestone, and locally shallow shelf carbonate facies (Garcia-Hernandez et al., 1980; Molina, 1987).

The typical Jurassic stratigraphic sequence of the middle Subbetic (Fig. 8.2-2) began with shallow-water shelf limestone and dolostone of early Liassic age, which are overlaid successively by (1) limestone-marl rhythmites and marly "ammonitico rosso" of late Liassic age; (2) a radiolarite sequence, which may locally be underlain by submarine volcanic rocks; and (3) carbonate turbidites of Late Jurassic age. The precise age of the radiolarites is not known for each outcrop because of the lack of characteristic fossils. The radiolarites are generally Bajocian and Kimmeridgian. In some middle Subbetic Jurassic sequences, such as in the Navalcan area, only claystone without radiolarian-rich intercalations occur and is also of Bajocian and Kimmeridgian age.

Generally two episodes of siliceous radiolarian-rich sedimentation, separated by a phase of pelagic carbonate deposition, can be distinguished in the radiolarite sections. This has also been observed in other alpine ranges, such as the Carpathian Mountains (Jenkyns and Winterer, 1982). When the two siliceous intervals occur, gray or green is typical of the stratigraphically lower unit, whereas the upper unit is commonly red. The sequence thickness ranges from 25 to 230 m, averaging 90 m (Fig. 8.2).

We studied seven localities (Figs. 8.1B, 8.2): (1) Sierra Elvira (SE); (2) Algarinejo (AG); (3) Sierra Pelada (SP); (4) Zegrí (ZE); (5) Jarandilla-Hachazo (JH); (6) Navalcán (NAV); and (7) Arbuniel (AB). Some features of these seven Jurassic stratigraphic sequences differ from the general series mentioned. In ZG, SE, and in some sections of the AG area, sedimentary rocks of the marly "ammonitico rosso" facies immediately underlie the radiolarite sequence. The condensed marly ammonitico rosso facies is interpreted as typical of deposition on submarine swells and adjacent slopes (Aubouin, 1964; Braga et al., 1981). In SP, a 1-m-thick bed of oolitic grainstone occurs between the two episodes of the radiolarite section. This grainstone is probably of turbidity current origin and is locally silicified. In JH and locally in AG, the radiolarite sequence directly overlies submarine volcanic rocks (Garcia-Yebra et al., 1972; Puga and Ruiz-Cruz, 1980) and breccia with volcanic rock clasts (Fig. 8.3A). The thickness of the volcanic rocks is variable, but in Algarinejo (AG), volcanic rocks average 10 m thick. Volcanic rocks include andesite, hawaiite, and tephrite with ophitic or hyalopilitic intersertal textures and locally abundant vesicles (Puga and Ruiz-Cruz, 1980). Pillows that generally range in size from 10 cm to 1 m occur with pahoehoe or massive lavas. Locally hypabyssal sills occur. In sections from JH, the transition from pillow lava to pillow breccia at the bottom of some lava flows is common. In sections from AG and JH, irregular-shaped blocks of carbonate rocks are included in the lavas and are locally metamorphosed. The blocks are similar to overlying rocks, and they probably represent pieces of rock from the conduit walls (Puga and Ruiz-Cruz, 1980).

Petrology and Sedimentary Structures

The radiolarite sequence is composed of different types of rocks. Three main types have been distinguished: radiolarite, siliceous mudstone and marl, and pelagic limestone containing radiolarians. Nev-

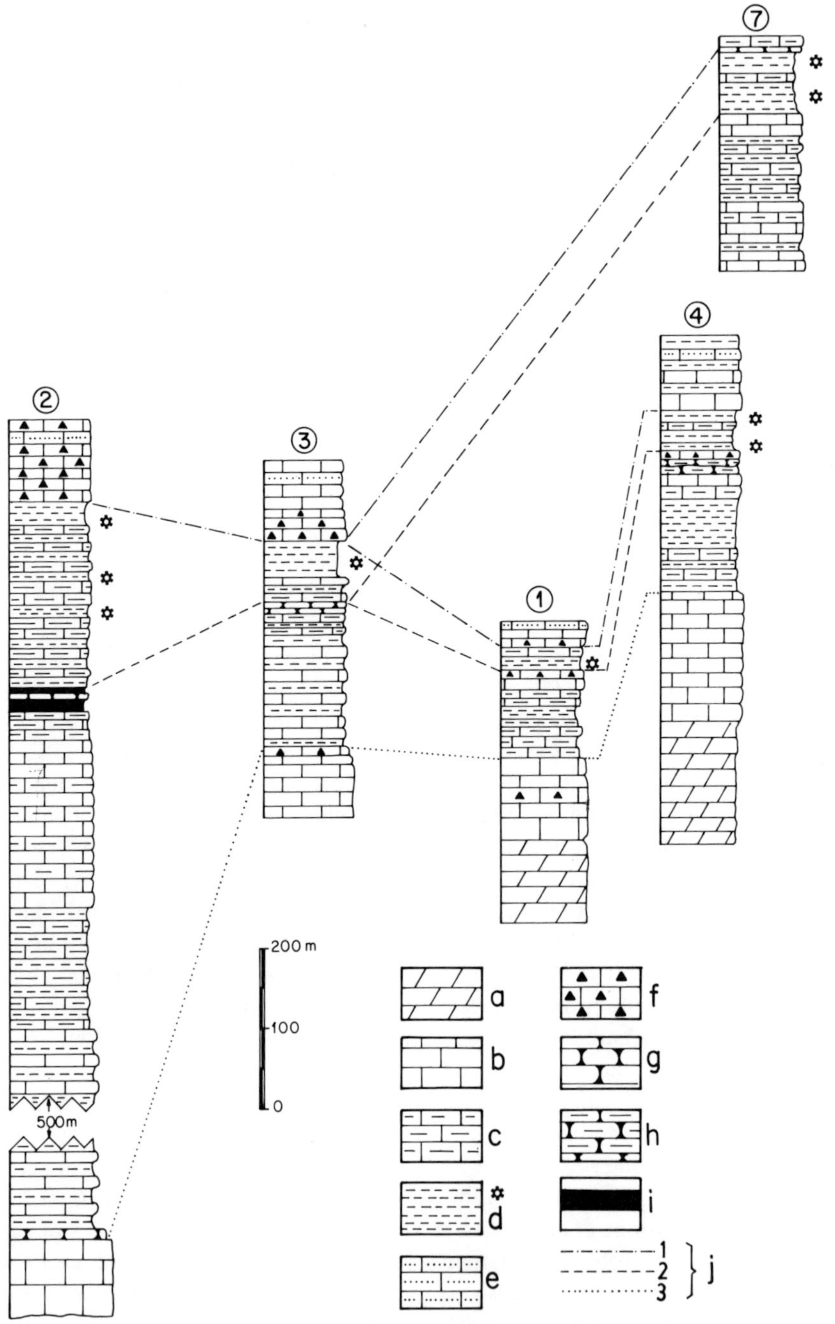

FIGURE 8.2. Jurassic stratigraphic sections of some of the studied localities. The numbers correspond to those in Figure 8.1. Key: (a) dolostone; (b) limestone; (c) marly-limestone; (d) marl and radiolarite (symbol); (e) calcarenite; (f) cherty limestone; (g) calcareous nodular limestone; (h) marly nodular limestone; (i) volcanic rock; (j) correlation lines: (1) top of the radiolarite sequence; (2) base of the radiolarite sequence; (3) top of the shallow-water shelf Liassic limestones.

FIGURE 8.3. (A) Volcanic breccia intercalated in siliceous pelagic limestone, JH section. (B) General view of radiolarite in Sierra Pelada (SP) section; thin clay or shale partings occur. (C) Rhythmically bedded sequence of radiolarite-siliceous mudstone (or marl), Algarinejo (AG) section. (D) Diagenetic layering of a radiolarite bed; Sierra Elvira (SE) section.

ertheless, a range of intermediate lithologic types occur between these three lithologies.

Radiolarite

The absence of quantitative boundaries in the definition of radiolarite and the difficulty in determining the composition in the field have necessitated the use of "radiolarite" as a general term. Nevertheless, the widespread use of radiolarite in Alpine geology makes it advisable to continue using the term, but also, the chemical and mineralogical compositions of each rock should be determined and the appropriate modifying terms used with radiolarite. Here we use radiolarite in a broad sense, to refer to hard, compact beds with more than 50% nondetrital quartz, most of which is biogenic, being derived from radiolarian tests. No bed having more than 85% nondetrital silica has been found. Other constituents include calcite, clay minerals, and hematite.

Consequently, rocks we studied from the Betic Cordillera are impure radiolarites. Thus, locally the terms argillaceous-, calcareous-, and argillaceous-calcareous radiolarites are used according to the most abundant impurity in each rock. This is a binomial classification system similar to that proposed by Jones and Murchey (1986). According to their classification, our samples are included in radiolarite types 1, 1A, 2, and 2A (Jones and Murchey, 1986, p 457).

The radiolarite beds have planar tops and bottoms and are mainly black, greenish-gray, or red (Fig. 8.3B). The bed thickness ranges from 2 to 20 cm, averaging 9 cm. Values from 4 to 9 cm are common, and locally, in sections from JH, some beds are up to 50 cm thick. Apart from some sections at SE, JH, and AG (Fig. 8.3C), no rhythmic radiolarite-mudstone sequences occur. Generally a very low correlation coefficient (0.22–0.26) exists for the thickness of the hard and softer beds of each couplet. The contacts between the radiolarites and the under-

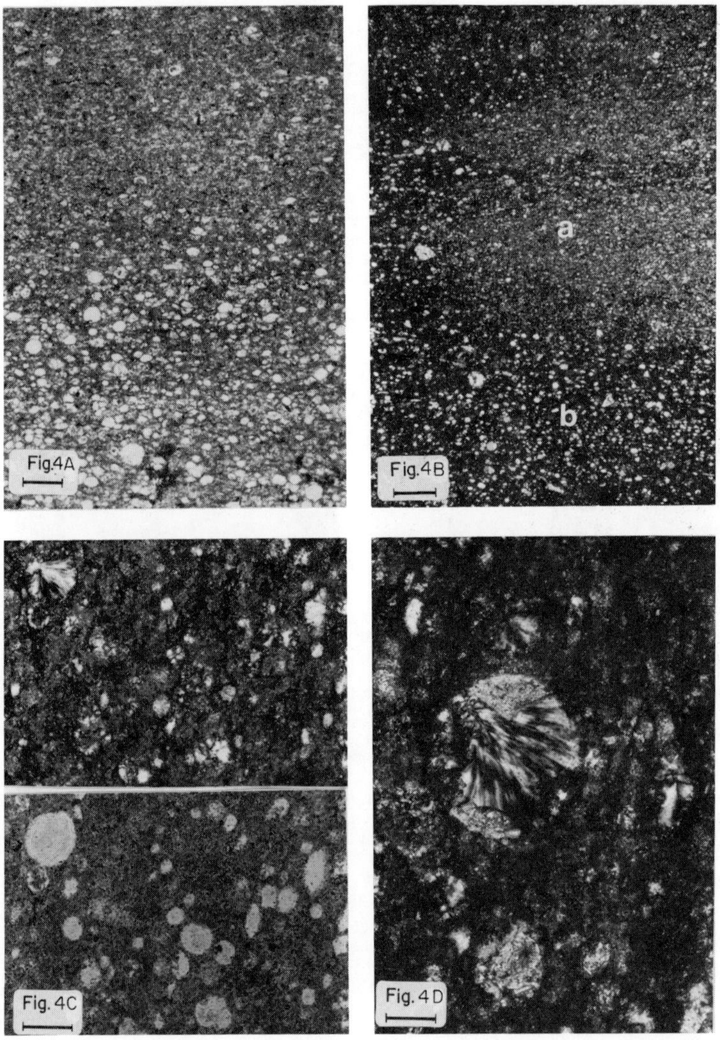

FIGURE 8.4. Photomicrographs of radiolarite sequences: (A) Radiolarite showing graded bedding and sedimentary layers; lower part is radiolarian wackestone, upper part is matrix-rich. Scale bar represents 0.25 mm. (B) Radiolarite showing diagenetic layering: (a) siliceous casts of radiolarians in a carbonate matrix; (b) calcite casts of radiolarians in a silicified matrix (crossed polarizers). Scale bar represents 0.5 mm. (C) Close-up of the (a) in Fig. 8.4B; radiolarians with length-fast chalcedony and megaquartz in a carbonate matrix. Upper, crossed polarizers; lower, parallel polarizers. Scale bar represents 0.125 mm. (D) Euhedral late-stage calcite crystals replacing length-fast chalcedony in a radiolarian cast; crossed polarizers. Scale bar represents 0.062 mm.

and overlying siliceous mudstones are usually gradational, although sharp bottoms of radiolarite beds are also present. Extensive bioturbation always occurs, though locally it may be masked by a diagenetic overprint (Fig. 8.3D). Radiolarite beds are chiefly massive and locally show a fine parallel lamination, especially evident in sections from SE, AG, and JH. Synsedimentary fractures and minor faults affected these deposits.

A detailed petrographic study was done on a sequence of rocks from SE, where the radiolarites are best preserved. Radiolarian tests, 40–270 μm in size, make up 50–80% by volume of the radiolarite beds. The matrix is fine-grained carbonate and clay minerals. Some radiolarite beds show a layering resulting from variable radiolarian contents. Intervals showing grading in the number and size of radiolarians are overlain by matrix-rich intervals with a lower radiolarian content (Fig. 8.4A). The vertical matrix enrichment in Figure 8.4A is about 2 cm thick and recurs within a single radiolarite bed. In the mud-rich intervals, the radiolarians show ovoidal and compaction-flattened forms, while in the radiolarian-rich, locally grain-supported intervals, radiolarians are better preserved. Fine parallel laminations result from a preferred orientation of grains.

Diagenetic layering (Fig. 8.4B) resulting from the redistribution of carbonate, silica, and Fe oxides is superimposed on the sedimentary beds. Diagenetic layers consist of alternating dark and light layers. The light layers (Fig. 8.4C) are composed of siliceous (microcrystalline quartz and length-fast chalcedony) radiolarian casts in a calcareous or argillaceous-calcareous matrix. Some radiolarian casts show later, frequently idiomorphic, calcite crystals (Fig. 8.4D). The dark reddish-green layers (Fig. 8.5) include only calcite radiolarian casts, whereas the matrix is isotropic silica and iron oxides or silica, iron oxides, and argillaceous material. We conclude that the light layers are primary, whereas the dark ones resulted from calcitization of radiolarians. The calcitization took place at the expense of the matrix carbonates, and the silica released from dissolution and replacement of the radiolarians was incorporated into the matrix. This process of diagenetic differentiation is better developed when the matrix is composed solely of carbonates. When the matrix is argillaceous-calcareous, the differentiation of light and dark layers is poorly developed, and siliceous and calcite radiolarian casts coexist. If the differentiation process was completed, all the radiolarian casts are calcite, and the rock has a uniform dark appearance.

In only a few places does silica occur as chert nodules and thin beds, where the silica radiolarian test infillings are contained in a matrix of cryptocrystalline quartz. At JH, radiolarite beds have a high carbonate content and are a transitional facies between pelagic limestone containing radiolarians and calcareous radiolarite. At the remaining localities—AG, AB, SP, and ZG—radiolarites consist of cryptocrystalline quartz mosaics and some ovoid radiolarian ghosts. Because of the low calcite content of these sequences, no diagenetic layering is present. The presence of Fe oxides and clay minerals and the intensive recrystallization and cementation by cryptocrystalline quartz make more detailed petrographic observations difficult.

Compaction, as determined by the deformation of radiolarians, varies depending on the composition of the radiolarite matrix. The higher the clay/carbonate ratio, the higher the degree of compaction.

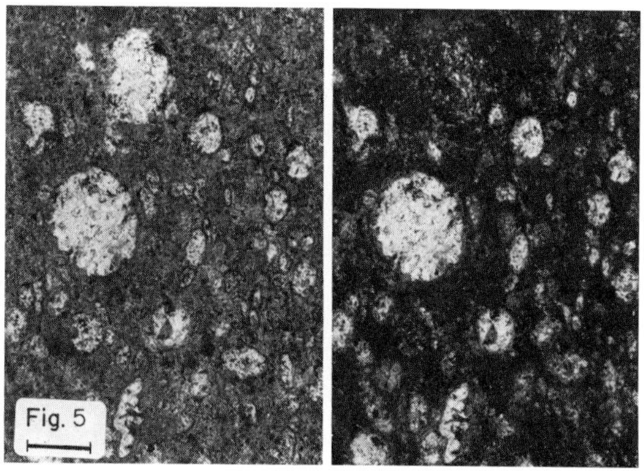

FIGURE 8.5. Close-up of the (b) in Figure 8.4B. Single crystals of calcite infilled the radiolarian molds, which occur in a silicified carbonate matrix. Left, parallel polarizers; right, crossed polarizers. Scale bar represents 0.125 mm.

Burrow structures occur in thin bedded radiolarite sections. Commonly, burrows were filled with quartz and argillaceous sediment, locally accompanied by Fe oxides.

Siliceous Mudstone and Marl

Siliceous mudstone and marl occur rhythmically interbedded with the radiolarite in parts of the SE, JH, and AG sections. Beds range in thickness from 0.5 to 120 cm, averaging about 6 cm. In only one section (JH) do some beds thicker than 30 cm occur. Greenish-gray and red beds are common. Beds have an earthy appearance and are locally fissile.

Siliceous mudstone and marl beds consist of clay minerals (35–80%), nondetrital quartz (10–50%), and minor amounts of calcite, hematite, and clastic quartz and feldspar. The beds richest in nondetrital quartz occur in sections from AG and SE and are in turn poorer in clay minerals. Calcite is less than 20% except in some beds from sections at AG, JH, ZG, and SP; these beds are classified as siliceous marls. Ovoid radiolarians always comprise less than 30% of the beds, and, when there is a micrite matrix, the radiolarian test infillings are composed of calcite. The siliciclastic grains are silt-size and comprise less than 10% of the beds. Clay mineral suites vary in different areas. In the sections from AB, only smectite is present; in the JH, AG, SP, and ZG sections, smectite and illite occur; and smectite, illite,

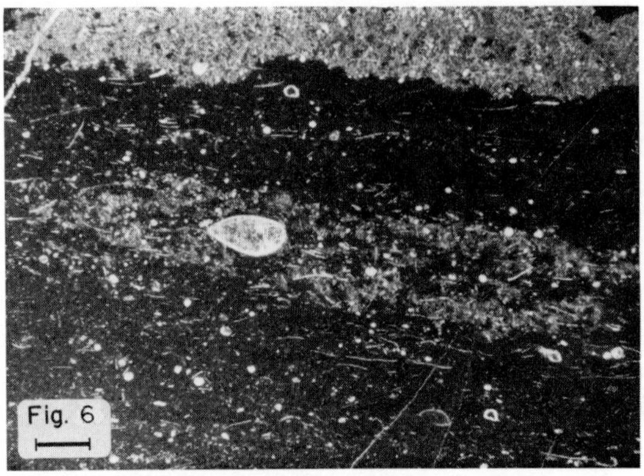

FIGURE 8.6. Wackestone of radiolarians and filaments (remains of thin-shelled pelecypods) with silicified laminae of cryptocrystalline quartz. Parallel polarizers. Scale bar represents 0.5 mm.

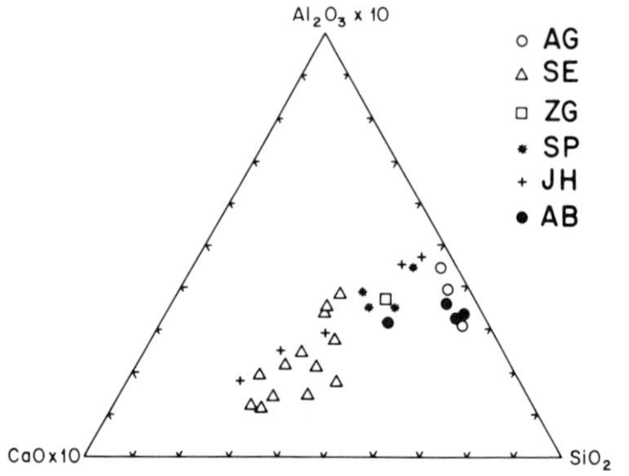

FIGURE 8.7. SiO_2-CaO(×10)-Al_2O_3(×10) ternary diagram. Radiolarite beds from the different areas studied are plotted.

and minor kaolinite occur in SE sections. Smectite is always greater than illite.

Radiolarian-Bearing Pelagic Limestone

Limestones are wackestones containing radiolarians, filaments (remains of thin-shelled pelecypods), and locally Saccocoma. Radiolarian abundances range from 5% to 50%, and most are calcified. When microfossils are abundant, the rock forms a packstone. Clay minerals are less than 10%.

This rock type is associated with the radiolarite and occurs in all the sequences. In JH sections, where they compose part of a rhythmic sequence, many intermediate lithologic types between radiolarian-bearing limestone and radiolarite occur. Variations in the carbonate content produce the rhythmicity and give the rock a compact appearance.

Silicification commonly occurred in the limestones (Fig. 8.6), although the silica (cryptocrystalline quartz), which is mainly intimately mixed with the micrite matrix, is commonly masked. Radiolarians composed of silica (microcrystalline mosaics and length-fast chalcedony) are included within silicified intervals which frequently form diffuse chert nodules. Calcite replaced radiolarians occur outside of these silicified intervals. Locally, certain carbonate fossils (e.g., belemnites) were silicified by quartzine as a consequence of the fossil structure. In sections from SE, highly calcareous facies occur as the result of extensive calcitization of radiolarians; no carbonate microfossils occur.

Geochemistry

The major elements of 55 samples were determined by X-ray fluorescence spectroscopy (Tables 8.1–8.5). Volatiles were measured by loss on ignition at 900°C. We also calculated the ratios Si/Si+Al+Fe+Ca = Ia, Si/Si+Al+Fe = Ib, Al/Al + Fe + Mn = Ic, Fe/Ti, and Fe/Al (Tables 8.1–8.5). Total iron is expressed as Fe_2O_3, because Fe^{+2}/Fe^{+3} is always very low, which suggests the prevalence of hematite over the Fe silicates.

SiO_2 content of the radiolarite beds ranges from 61% to 85%. Plotting the composition of radiolarites on a ternary diagram of SiO_2:Al_2O_3:CaO differentiates samples from different sequences (Fig. 8.7). Al_2O_3 represents the clay mineral fraction, and CaO the calcite; the amount of CaO related to the clay minerals is less than 2% using the method of Baltuck (1982). Radiolarites at AB and AG are the richest in silica, and those at JH and some at SE are the lowest in silica. The carbonate content in sections at AG and AB is low; at ZG and SP some calcareous radiolarites occur, and at SE and JH calcareous radiolarite is common. Uniform concentrations of alumina indicate a similar clay mineral content in the radiolarites from the different areas.

More insight can be gained about the nature and paleogeographic position of these deposits by using various geochemical ratios (Steinberg and Mpodozis Marin, 1978; Rangin et al., 1981; Hein et al., 1983). The ratio Ia (Si/Si+Al+Fe+Ca) gives information about the biogenic silica content in relation to aluminosilicates and ferruginous and calcareous minerals.

TABLE 8.1. Chemical composition in weight percent of rocks from Sierra Elvira (SE) by X-ray fluorescence.

	Radiolarite SE–2	Siliceous mudstone SE–3	Radiolarite SE–6	Siliceous mudstone SE–7	Radiolarite SE–9	Siliceous mudstone SE–10	Radiolarite SE–14	Radiolarite SE–15	Siliceous mudstone SE–2–3B	Radiolarite SE–2–3A
SiO_2	74.00	61.53	64.50	67.20	67.13	59.55	76.62	69.58	70.82	77.00
Al_2O_3	2.76	10.26	2.70	8.93	2.45	9.42	5.21	3.07	9.81	3.08
MnO	0.15	0.15	0.25	0.11	0.32	0.34	0.15	0.20	0.06	0.19
MgO	1.40	2.01	1.10	1.67	1.29	1.83	2.01	1.11	1.23	1.00
CaO	8.80	8.90	13.47	6.78	12.50	9.87	6.44	11.65	4.70	6.89
Na_2O	0.31	0.32	0.26	0.31	0.43	0.35	0.33	0.30	0.26	0.17
K_2O	0.24	1.40	0.33	1.25	0.31	1.35	0.70	0.37	1.48	0.30
TiO_2	0.11	0.52	0.11	0.49	0.11	0.47	0.27	0.14	0.33	0.10
P_2O_5	0.02	0.14	0.05	0.11	0.03	0.10	0.11	0.05	0.08	0.01
Fe_2O_3*	3.02	3.13	3.09	2.97	4.27	3.74	4.39	2.72	2.63	3.03
LOI	9.13	12.04	13.95	10.33	10.84	12.97	7.71	10.17	8.57	7.71
Total	99.94	100.04	99.81	100.15	99.68	99.99	99.94	99.36	99.97	99.48
Al/Al+Fe+Mn	0.396	0.702	0.378	0.686	0.286	0.634	0.464	0.441	0.734	0.418
Si/Si+Al+Fe	0.906	0.791	0.894	0.822	0.880	0.785	0.853	0.902	0.825	0.906
Si/Si+Al+Fe+Ca	0.778	0.673	0.695	0.729	0.704	0.655	0.765	0.733	0.761	0.806
Fe/Ti	32.030	7.022	32.773	7.071	45.288	9.284	18.969	22.667	9.298	35.350
Fe/Al	1.446	0.403	1.512	0.439	2.303	0.525	1.113	1.171	0.354	1.300

TABLE 8.1. *Continued.*

	Siliceous mudstone SE-2-4B	Radiolarite SE-2-4A	Siliceous mudstone SE-2-21B	Radiolarite SE-2-21A	Radiolarite SE-2-22A	Radiolarite SE-2-37A	Radiolarite SE-2-56A	Radiolarite SE-2-68A	Radiolarite SE-2-68B
SiO_2	72.00	70.00	69.18	70.80	64.60	71.79	72.66	69.97	70.80
Al_2O_3	9.30	5.21	9.63	7.41	4.60	7.84	3.98	4.91	7.13
MnO	0.06	0.14	0.17	0.09	0.21	0.06	0.20	0.22	0.20
MgO	0.27	1.11	1.03	0.77	0.89	1.15	1.22	0.65	0.93
CaO	4.41	9.14	5.46	6.92	13.01	5.87	7.93	10.87	7.06
Na_2O	0.19	0.25	0.12	0.20	0.11	0.15	0.25	0.13	0.18
K_2O	1.52	0.75	1.59	0.99	0.32	1.11	0.40	0.52	0.12
TiO_2	0.37	0.16	0.41	0.22	0.10	0.28	0.17	0.12	0.25
P_2O_5	0.15	0.04	0.09	0.05	0.04	0.04	0.06	0.04	0.09
Fe_2O_3*	2.50	2.92	2.74	2.76	3.37	2.93	4.12	1.91	1.94
LOI	7.91	10.18	8.85	9.60	12.32	8.38	8.92	10.48	9.49
Total	99.68	99.90	99.27	99.81	99.57	99.60	99.91	99.82	98.19
Al/Al+Fe+Mn	0.733	0.562	0.713	0.662	0.491	0.664	0.410	0.633	0.714
Si/Si+Al+Fe	0.835	0.872	0.822	0.850	0.863	0.844	0.872	0.893	0.866
Si/Si+Al+Fe+Ca	0.774	0.743	0.748	0.754	0.682	0.763	0.761	0.736	0.765
Fe/Ti	7.883	21.292	7.797	14.636	39.317	12.208	28.275	18.569	9.053
Fe/Al	0.355	0.740	0.376	0.492	0.968	0.494	1.368	0.511	0.359

*Total iron as Fe_2O_3

LOI: Loss on ignition at 900°C

TABLE 8.2. Chemical composition in weight percent of rocks from Algarinejo (AG).

	Siliceous marl AG-1-1	Siliceous marl AG-1-2	Siliceous mudstone AG-2-10	Siliceous marl AG-2-13	Siliceous mudstone AG-3-5	Siliceous mudstone AG-3-7	Radiolarite AG-3-8	Radiolarite AG-3-2	Radiolarite AG-3-11
SiO_2	50.73	56.88	59.13	59.65	71.15	65.49	85.76	80.78	84.10
Al_2O_3	5.75	5.82	8.12	6.51	9.47	9.41	4.14	6.77	5.87
MnO	0.11	0.17	0.15	0.15	0.10	0.20	0.26	0.01	0.17
MgO	0.96	1.02	1.27	0.91	1.18	1.14	0.40	0.56	0.43
CaO	18.43	13.82	10.83	11.19	0.92	5.09	0.77	0.53	0.68
Na_2O	0.19	0.18	0.26	0.20	0.26	0.35	0.12	0.18	0.30
K_2O	1.59	1.74	1.74	1.69	1.32	1.56	0.54	0.63	0.65
TiO_2	0.27	0.32	0.51	0.53	0.58	0.54	0.23	0.17	0.18
P_2O_5	0.07	0.13	0.09	0.08	0.12	0.10	0.14	0.04	0.08
Fe_2O_3 *	4.32	3.65	4.01	3.66	7.00	4.68	2.67	5.61	2.96
LOI	17.20	15.80	13.98	15.38	7.98	11.25	4.78	4.49	4.05
Total	99.62	99.53	100.09	99.95	100.08	99.81	99.81	99.77	99.47
Al/Al+Fe+Mn	0.495	0.534	0.595	0.563	0.502	0.592	0.514	0.477	0.585
Si/Si+Al+Fe	0.796	0.825	0.796	0.823	0.770	0.788	0.908	0.834	0.884
Si/Si+Al+Fe+Ca	0.552	0.632	0.651	0.666	0.759	0.720	0.897	0.827	0.874
Fe/Ti	18.667	13.307	9.173	8.057	14.080	10.111	13.543	38.500	19.185
Fe/Al	0.993	0.829	0.652	0.743	0.977	0.657	0.862	1.095	0.666

*Total iron as Fe_2O_3

LOI: Loss on ignition at 900°C

TABLE 8.3. Chemical composition in weight percent of rocks from Zegrí (ZG) and Sierra Pelada (SP) by X-ray fluorescence.

	Siliceous marl ZG-2	Radiolarite ZG-3	Siliceous mudstone ZG-4	Siliceous marl ZG-8	Siliceous marl SP-1	Radiolarite SP-2	Radiolarite SP-3	Radiolarite SP-4	Radiolarite SP-5	Siliceous mudstone SP-6	Siliceous mudstone SP-7
SiO_2	61.34	75.33	64.99	59.20	53.00	77.64	75.48	76.22	74.10	54.38	43.20
Al_2O_3	6.13	6.28	12.83	7.66	6.76	7.46	6.40	5.70	7.37	10.72	10.24
MnO	0.75	0.24	0.10	0.34	0.23	0.10	0.15	0.24	0.37	0.03	0.06
MgO	0.75	0.88	1.50	0.95	1.01	1.10	1.00	0.98	1.07	1.46	1.24
CaO	7.83	3.26	4.12	11.95	76.64	1.54	4.12	2.93	4.31	9.16	16.27
Na_2O	0.16	0.19	0.19	0.16	0.34	0.17	0.19	0.25	0.29	0.36	0.25
K_2O	1.06	1.30	1.79	1.74	1.47	1.49	1.11	0.92	1.40	2.35	1.81
TiO_2	0.32	1.33	0.68	0.36	0.31	0.42	0.31	0.26	0.40	0.60	0.47
P_2O_5	0.13	0.19	0.19	0.21	0.10	0.09	0.08	0.16	0.18	0.12	0.09
Fe_2O_3*	8.73	4.23	5.00	4.53	3.80	3.43	3.53	5.13	4.33	4.92	5.32
LOI	12.42	7.70	8.54	13.13	16.30	6.64	7.65	6.79	8.94	16.08	20.97
Total	99.62	100.93	99.93	100.23	99.96	100.08	100.02	99.58	102.76	100.18	99.92
Al/Al+Fe+Mn	0.327	0.514	0.655	0.542	0.558	0.615	0.567	0.444	0.541	0.621	0.590
Si/Si+Al+Fe	0.754	0.849	0.747	0.793	0.799	0.851	0.858	0.844	0.833	0.736	0.688
Si/Si+Al+Fe+Ca	0.657	0.803	0.697	0.637	0.577	0.830	0.800	0.804	0.776	0.619	0.493
Fe/Ti	31.828	3.711	8.578	14.681	14.301	9.528	13.285	23.019	12.629	9.567	13.206
Fe/Al	1.882	0.890	0.515	0.781	0.743	0.607	0.729	1.189	0.776	0.606	0.686

*Total iron as Fe_2O_3

LOI: Loss on ignition at 900°C

TABLE 8.4. Chemical composition in weight percent of rocks from Jarandilla (JH) by X-ray fluorescence.

	Siliceous mudstone JH-7	Radiolarite JH-9	Siliceous mudstone JH-10	Radiolarite JH-2-2A	Radiolarite JH-2-2B	Siliceous marly limestone JH-2-6A	Radiolarite JH-2-6B	Siliceous mudstone JH-2-10B	Siliceous limestone JH-2-10A	Siliceous limestone JH-2-15	Siliceous limestone JH-2-17A	Radiolarite JH-12
SiO_2	65.58	69.25	68.65	61.32	77.46	59.55	76.31	64.11	50.75	49.00	44.00	64.14
Al_2O_3	12.13	5.65	8.14	4.77	7.85	4.99	8.02	7.24	3.45	3.40	1.46	5.62
MnO	0.02	0.07	0.07	0.20	0.01	0.18	0.02	0.07	0.30	0.20	0.10	0.24
MgO	2.16	1.32	1.03	0.40	1.19	0.52	1.16	1.10	0.70	0.70	0.13	1.10
CaO	2.14	5.83	7.08	15.46	1.04	15.80	1.99	9.55	22.11	22.34	29.30	10.55
Na_2O	0.35	0.30	0.42	0.15	0.43	0.16	0.41	0.22	0.25	0.38	0.24	0.14
K_2O	1.63	0.77	1.34	0.64	1.45	0.61	1.58	1.15	0.56	0.68	0.24	0.48
TiO_2	0.71	0.27	0.42	0.15	0.27	0.11	0.23	0.19	0.06	0.05	0.03	0.15
P_2O_5	0.09	0.21	0.15	0.13	0.14	0.12	0.17	0.19	0.07	0.07	0.05	0.06
Fe_2O_3*	4.46	4.37	3.26	1.56	3.14	2.53	2.98	3.16	1.38	1.70	0.89	4.61
LOI	9.94	10.92	9.50	14.81	6.86	14.68	6.68	12.48	20.10	20.93	23.38	12.46
Total	99.21	100.01	100.06	99.59	99.84	99.25	99.55	99.46	99.73	99.45	99.82	99.55
Al/Al+Fe+Mn	0.672	0.490	0.649	0.670	0.653	0.581	0.669	0.629	0.604	0.573	0.504	0.466
Si/Si+Al+Fe	0.765	0.843	0.830	0.888	0.851	0.863	0.849	0.832	0.895	0.885	0.936	0.829
Si/Si+Al+Fe+Ca	0.738	0.747	0.734	0.661	0.836	0.639	0.821	0.700	0.561	0.547	0.479	0.686
Fe/Ti	7.329	18.883	9.056	12.133	13.568	26.833	15.116	19.404	26.833	39.667	34.611	35.856
Fe/Al	0.486	1.022	0.529	0.432	0.528	0.670	0.491	0.577	0.528	0.661	0.805	1.084

*Total iron as Fe_2O_3

LOI: Loss on ignition at 900°C

TABLE 8.5. Chemical composition in weight percent of rocks from Arbuniel (A).

	Radiolarite AB-5	Radiolarite AB-6	Radiolarite AB-3	Radiolarite AB-7
SiO_2	79.67	83.10	79.02	85.60
Al_2O_3	5.03	4.45	5.30	4.60
MnO	0.03	0.07	0.06	0.11
MgO	1.03	0.86	0.51	0.40
CaO	0.88	0.50	3.49	0.86
Na_2O	0.18	0.22	0.14	0.09
K_2O	0.51	0.46	0.84	0.70
TiO_2	0.28	0.30	0.30	0.26
P_2O_5	0.11	0.10	0.08	0.03
Fe_2O_3*	4.19	4.04	2.80	1.98
LOI	7.83	5.63	6.92	4.90
Total	99.74	99.73	99.46	99.53
Al/Al+Fe+Mn	0.474	0.450	0.583	0.624
Si/Si+Al+Fe	0.869	0.882	0.886	0.913
Si/Si+Al+Fe+Ca	0.857	0.875	0.836	0.900
Fe/Ti	17.458	15.711	10.889	8.885
Fe/Al	1.101	1.199	0.698	0.569

*Total iron as Fe_2O_3
LOI: Loss on ignition at 900°C

TABLE 8.6. Range of the Ic (Al/Al+Fe+Mn) ratio for different areas.

	$\bar{x}$	∂
JH	0.661	0.011
SE	0.577	0.150
SP	0.541	0.072
AG	0.519	0.046
AB	0.508	0.008

$\bar{x}$ = Mean (I_c)
∂ = Standard error mean

Typical biogenic silica-rich radiolarites have values of Ia equal to 0.8–0.9. In contrast, only 14 of the 55 samples analyzed in this study have such high Ia values; two samples described as siliceous mudstone having an earthy appearance are included in the 14 samples. The presence of calcareous minerals is usually responsible for the low Ia values (0.48–0.9) of the Subbetic radiolarites.

The ratio Ic (Al/Al+Fe+Mn) indicates the relative contribution of continental versus oceanic detritus or whether a large submarine exhalative contribution exists (Bostrom et al., 1973). Generally, the mean values for radiolarite beds from the same sequence (Table 8.6) are lower than 0.60, considered as the mean for biogenic hemipelagic chert (Adachi et al., 1986). The values typical of marine biogenic material (0.39; Baltuck, 1982) are obtained only from some radiolarite beds in sections from SE (Table 8.1). Lower values appear in some samples from the ZG and SE sequences (Tables 8.1, 8.3).

The Fe/Mn ratio of the radiolarites ranges from 4 to 56, which falls in the range (1–100) suggested by Audley-Charles (1956) as typical of radiolarites. Values near the mean of siliceous oozes (Fe/Mn = 10; Audley-Charles, 1956) are frequent for Subbetic radiolarites. Fe is not correlated with the aluminosilicate fraction. According to the correlation coefficients shown in Table 8.7, only JH and AG sequences can be used for a comparative study of the Fe-Al linear regression.

Plotting Fe/Ti against Ic (Al/Al + Fe + Mn) (Fig. 8.8), the radiolarites generally have a higher Fe/Ti ratio than the siliceous mudstones. The siliceous mudstones show higher values of the Ic ratio, but the radiolarites show a greater variability. The radiolarites that overlap with the siliceous mudstones on Figure 8.8 correspond to beds with a relatively high detrital component.

Comparisons of Couplets in Rhythmically Bedded Sequences

At SE, radiolarite beds (A) are overlain by siliceous mudstone beds (B), although some siliceous mudstones may be classified as radiolarites; however, the B-radiolarite always shows a greater clay mineral content than the adjacent A-radiolarite bed. At JH, the A-beds have a variable composition ranging from calcareous radiolarite to pelagic siliceous limestone with varying amounts of radiolarians. The B-beds are either a radiolarite or a siliceous mudstone. A-beds are enriched in carbonates, which gives a

TABLE 8.7. Data for Al(x) and Fe(y) linear regressions.

	y Intercept	Slope	Correlation coefficient	Number of samples
SE	2.62	−0.17	−0.35	13
SE[a]	2.57	−0.16	−0.45	16
JH	−0.65	1.06	0.82	7[b]
SP	2.52	−9.64	−0.03	4
AB	5.57	−0.75	−0.61	4
AG	0.82	1.15	0.93	4

[a] Rocks where Ia > 0.75.
[b] Radiolarites and siliceous limestones.

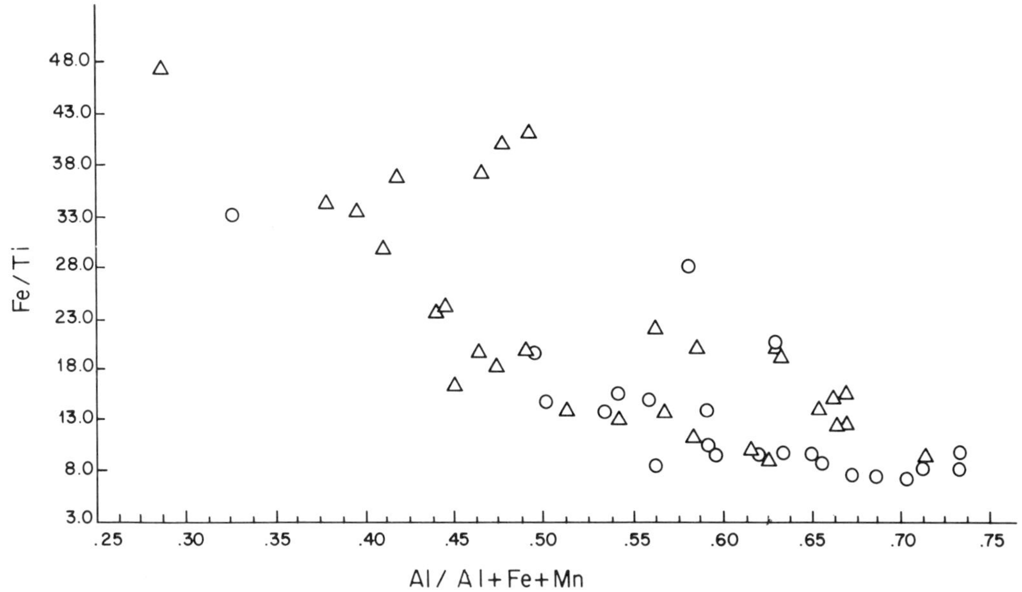

FIGURE 8.8. Fe/Ti versus Al/Al+Fe+Mn scatter plot for radiolarites (triangles) and siliceous mudstones and marls (circles).

hard and compact appearance to the beds and enhances the definition of the rhythmicity in both areas, but more so in area JH. In both areas, biogenic silica may be greater in the B-bed, as indicated by the Ia ratios which can be lower in A-beds than in adjacent B-beds (Table 8.8). This is a consequence of the enrichment in carbonates in the A-beds.

The terrigenous contribution as indicated by the Ic ratio is generally greater in the B-beds from sections at SE, and this terrigenous contribution is relatively constant (Ic = 0.634–0.734). Nevertheless, the values of this ratio in the radiolarites (A-beds) vary within a wide range. This variability may be a consequence of the sampling procedure, because the SE radiolarites have an internal sedimentary layering with radiolarian-rich and mud-rich layers and an irregular distribution of Fe oxides across the bed. In all the JH sequence, similar values of Ic occur in divisions A and B.

The K/Al ratio can be used as an indicator of the abundance of illite relative to the other clay minerals (smectites and kaolinite). In the siliceous mudstone, the ratio K/Al ranges from 0.214 to 0.311, whereas in the radiolarites it ranges from 0.026 to 0.226. K/Al is generally higher in the B-beds. In the SE sequence, this relation holds not only for specific couplets (Table 8.8) but for all couplets analyzed in the sections (Table 8.1).

Fe/Ti and Fe/Al of couplets from SE show that the radiolarite beds are relatively richer in Fe than the siliceous mudstone beds. This contrasts to what Barrett (1981) found in sections from the Apennines in Italy, where Fe is greater in the siliceous mudstones. On the other hand, the values of the Fe/Ti ratio are much greater in A-beds (14.63–35.35) than in B-beds (7.79–9.30) (Table 8.8).

Discussion

Geochemical and Mineralogical Considerations

Smectite is dominant over illite. Smectite is characteristic of some pelagic marine environments and may be authigenic or diagenetic (Griffin and Goldberg, 1963; Heath and Dymond, 1977; Hein et al., 1979). Smectite may also be of detrital origin or from the submarine alteration of basaltic rocks (Singer and Müller, 1983; Hein et al., 1983). In the sequences studied, intercalation of submarine volcanic rocks is frequent, and they mark the beginning of radiolarite sedimentation; weathering of these volcanic rocks is the most likely source of the smectite. Smectite is also more abundant in the siliceous mudstone, although the illite/smectite ratio is higher than in the radiolarite. Illite might be derived from

TABLE 8.8. Geochemical ratios of A and B beds of some couplets of rhythmically bedded sequences.

	Siliceous mudstone SE-2-3B	Radiolarite SE-2-3A	Siliceous mudstone SE-2-4B	Radiolarite SE-2-4A	Siliceous mudstone SE-2-21B	Radiolarite SE-2-21A	Radiolarite SE-2-68B
Si/Si+Al+Fe+Ca	0.716	0.806	0.774	0.743	0.748	0.754	0.765
Si/Si+Al+Fe	0.825	0.906	0.835	0.872	0.822	0.850	0.866
Al/Al+Fe+Mn	0.734	0.418	0.733	0.562	0.713	0.662	0.714
K/Al	0.237	0.153	0.256	0.226	0.258	0.209	0.026
TiO_2/Al_2O_3	0.034	0.032	0.040	0.031	0.042	0.030	0.035
Fe/Al	0.354	1.300	0.355	0.740	0.376	0.492	0.359
Fe/Ti	9.298	35.350	7.883	21.292	7.797	14.636	9.053
	Radiolarite SE-2-68A	**Radiolarite JH-2-2B**	**Radiolarite JH-2-2A**	**Radiolarite JH-2-6B**	**Siliceous marly limestone JH-2-6A**	**Siliceous mudstone JH-2-10B**	**Siliceous limestone JH-2-10A**
Si/Si+Al+Fe+Ca	0.736	0.836	0.661	0.821	0.639	0.700	0.561
Si/Si+Al+Fe	0.893	0.851	0.888	0.849	0.863	0.832	0.895
Al/Al+Fe+Mn	0.633	0.653	0.670	0.669	0.581	0.629	0.604
K/Al	0.166	0.289	0.210	0.309	0.191	0.249	0.254
TiO_2/Al_2O_3	0.024	0.034	0.031	0.028	0.022	0.026	0.017
Fe/Al	0.514	0.528	0.432	0.491	0.670	0.577	0.528
Fe/Ti	18.569	13.568	12.133	15.116	26.833	19.404	26.833

the diagenetic transformation of smectite (e.g., Singer and Müller, 1983; Palomo-Delgado et al., 1985; Iijima et al., 1985), but more likely, illite is of detrital origin derived from a continental source, as found in other studies (Weaver, 1978; Hein et al., 1983). A detrital origin is supported by an increase in the illite/smectite ratio in the beds richest in clay minerals, siliceous mudstone, and marl.

A close relation exists between the nondetrital silica (dominantly biogenic) content and the presence of smectite. The stratigraphic intervals with the greatest silica contribution (Ia = 0.89, 0.90) and the lowest terrigenous supply (Ic ratio; Table 8.6) occur in sections at AG and AB. In both areas, the smectite either is the only clay mineral present (AB) or it is strongly dominant (AG). On the other hand, in the Navalcan sequence, made up chiefly by clay minerals without radiolarians, only illite is present. Thus, the greater the oceanic influence the higher the biogenic silica and smectite contents of the sections, and as the continental influence increases, the illite content increases and the silica decreases. The occurrence of volcanic rocks in general increases with increasing amounts of smectite in the sections (AG area).

Kaolinite is derived from the weathering of edaphic profiles in warm and humid climates under hydrolyzing conditions (Biscaye, 1978; Deconinck et al., 1985). Because it occurs only in sections from SE, a different source area compared to the other areas may have contributed sediment to this section.

The radiolarite beds are systematically enriched in Fe in respect to the siliceous mudstone. This relative Fe enrichment in the radiolarite may result from low rates of sedimentation and therefore from long periods of exposure of the sediment to oxidizing bottom waters. Benthic organisms would thrive under these bottom-water conditions, resulting in intense sediment bioturbation, which is noted for the deposits we studied. Bioturbation in turn enhances sediment oxygenation.

The use of geochemical ratios to determine paleogeographic conditions (Steinberg and Mpodozis Marin, 1978; Rangin et al., 1981; Hein et al., 1983) poses problems when applied to lithologically heterogeneous sequences, such as those considered here. Rocks with only a similar nondetrital silica content should be compared. The Ia (Si/Si + Al + Fe + Ca) ratio is used to determine the nondetrital silica contribution. Only rocks with Ia greater than 0.75 are used in the following comparisons. The chosen Ia value is close to the minimum value (0.8) of silica-rich radiolarites. Diagenetic remobilizations could modify the value of this ratio, and we must assume that the beds acted as chemically closed systems in order for the ratio to be considered as representative of the sediment composition. For example, the carbonate filling radiolarian tests can only be considered in the denominator of the ratio (Ca) if the source of the carbonate was in the same bed, such as the micrite matrix.

Although Bostrom et al. (1973) only used the Ic index as a measure of the detrital (continental or oceanic) contribution, other authors (Rangin et al., 1981; Baltuck, 1982) believed it to indicate a continental provenance. The following succession from greater to lower terrigenous contribution is obtained from the mean of Ic indices of each sequence studied: JH–SE–SP–AG–AB (Table 8.6). Thus, AB and AG appear to be the sequences deposited farthest from a continent or topographically most isolated from a terrigenous supply. However, the use of mean indices for a comparative analysis poses problems for interpretation of results when a great variability in the ratio within a single sequence occurs. Thus, in the sequences from SE, the beds with lowest Ic values (0.286) occur, and according to Adachi et al. (1986), these values fall within the range (0.12–0.32) found when a hydrothermal contribution exists. Nevertheless, other factors must be taken into account, as these low Ic values might also be due to local accumulations of Fe minerals, as suggested by the high Fe/Ti ratios of the beds. However, in the JH and AG sequences, where intercalations of volcanic rocks occur, the radiolarites do not show low Ic indices, as would be expected if contamination with the basic igneous material were prevalent (Baltuck, 1982).

Using the Ib (Si/Si+Al+Fe) ratio as a measure of the proximity of the deposit to a detrital source (Rangin et al., 1981), the placement of the areas studied is different from that obtained with the Ic index. Because in the Ib ratio a continental provenance of the Fe is assumed (Rangin et al., 1981), whereas a greater part of the Fe in the samples studied is of oceanic origin, we consider that the Ic index provides the best ordering of sequences as to proximity of source areas (see also Hein et al., 1983).

The slope of the regression lines on Fe and Al plots has also been used (Steinberg and Mpodozis Marin, 1978; Baltuck, 1982; Steinberg et al., 1983; Hein et al., 1983) to determine the relative position of deposits to source terranes within a single depositional basin. According to the correlation coefficients in Table 8.1, only JH and AG can be used and deposits at AG can apparently be formed farther from a continental source than deposits at JH. Hein et al. (1983) also showed data with low correlation coefficients. We conclude that because the Fe and Al are not associated with the same minerals and because diagenetic remobilizations of Fe oxides occurred, the Fe/Al linear regressions are not useful for paleogeographic studies of Subbetic deposits.

Mechanisms of Sedimentation

Evidence favors an interpretation of the radiolarite sequences as resulting from pelagic settling in a marine basin. Evidence includes the gradational contact between the siliceous mudstone and radiolarite, the lack of a positive correlation between the thickness of A- and B-beds in couplets, complete bioturbation of beds, and Fe enrichment of radiolarite beds relative to siliceous mudstone. On the other hand, similar values of the Fe/Ti ratio in A- and B-beds of couplets have been interpreted as caused by the deposition of the two beds from a single mechanism of sedimentation, a turbidity current (Sugisaki et al., 1982). On the contrary, in Subbetic rhythmically bedded sequences, the values of the Fe/Ti ratio are much greater in A-beds than in B-beds of couplets. Generally, the lithologic change from siliceous mudstone to radiolarite is caused by a relative decrease in the terrigenous supply and by an increasing contribution to the sediment of carbonate and silica tests. Likely, climatic changes and the hydrodynamic regime of the basin partly caused these compositional differences which were later enhanced by diagenesis.

Rhythmically bedded radiolarite-argillite sequences cannot be explained by a single mechanism of deposition. At JH, where the terrigenous contribution expressed by the Ic ratio is very similar in A- and B-beds of each couplet, the rhythmic bedding results from the higher carbonate content of the lower A-beds. This is a typical limestone-marl rhythmite, which frequently occurs in middle Subbetic Jurassic sequences overlying and underlying radiolarite sections. The rhythmite was probably generated by climatic cycles which stimulated carbonate plankton productivity cycles in the surface waters (Einsele, 1982). These productivity cycles were coeval with increase in the abundance of radiolarians within the basin and in the Tethys generally. Volcanic effusions in this middle Subbetic topographically restricted basin (Fig. 8.9) (Garcia-Hernandez et al., 1980; Vera, 1981) could have increased the seawater pH, decreasing the carbonate production as a consequence. We conclude that these rhythmic sequences represent a constant terrigenous

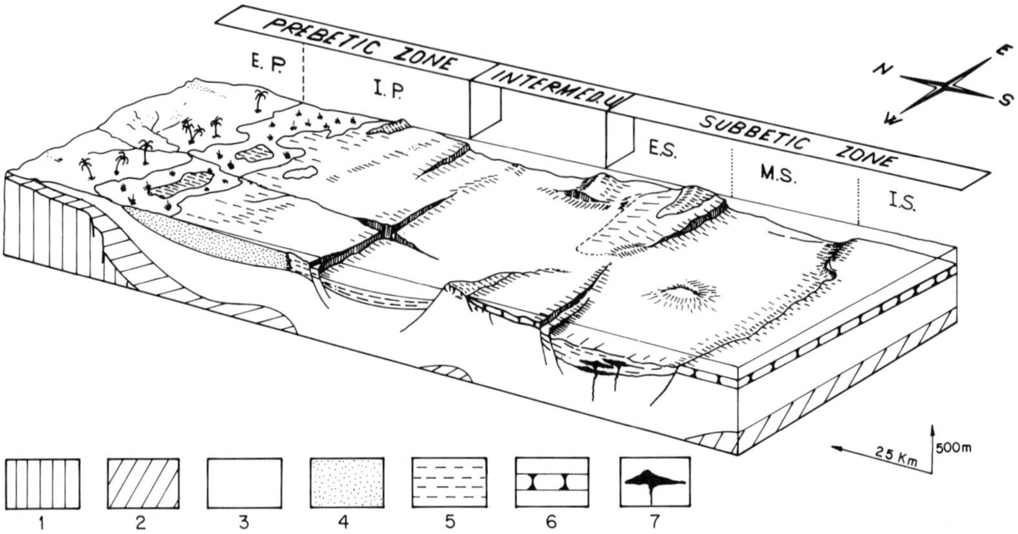

FIGURE 8.9. Paleogeographic reconstruction for the boundary between the Middle and Late Jurassic: (1) Paleozoic basement; (2) Triassic; (3) Lower and Middle Jurassic; (4) shelf carbonates of the Prebetic zone; (5) pelagic limestone and marl and, locally, radiolarite; (6) Ammonitico Rosso facies; (7) volcanic rock.

supply, a cyclic carbonate production, and a cyclic or constant radiolarian production. If radiolarian productivity were cyclic, the times of greater radiolarian productivity would have coincided with times of lower carbonate production, resulting in marls enriched in siliceous organisms.

In sections from SE, the terrigenous contribution is greater, the amount of clay minerals is greater, and the illite/smectite is higher in the B-beds relative to the A-beds. This variability in the terrigenous supply is responsible for the cyclicity of the sequence. Thus, terrigenous dilution cycles (Einsele, 1982) were generated in which the clay mineral-rich beds (B) are poorer in siliceous organisms and/or carbonate. Moreover, sedimentary layering, graded bedding, and reworking resulted from bottom currents and distal dilute turbidity currents. Graded sequences are composed of a lower, relatively matrix-poor unit showing grading of the number and size of radiolarian tests and an upper mud-rich unit. The orientation of the flattened radiolarian tests and other planar bioclasts, especially thin-shelled pelecypods, are responsible for the fine horizontal lamination locally observed in the radiolarite beds.

The sedimentation rates, averaged over entire sections of sequences from the Subbetic, range from 1 mm/1000 years to 9 mm/1000 years and average 3.5 mm/1000 years. These values are similar to those proposed by other authors for other Mesozoic radiolarites, such as radiolarites from the Alps and Apennines (3–9 mm/1000 years; Schlager and Schlager, 1973; Jenkyns and Winterer, 1982).

Paleogeographic Implications

The interpretation that some radiolarites of the northern limestone Alps and other regions were deposited below the CCD (Garrison and Fischer, 1969; Jenkyns and Winterer, 1982) cannot be applied to the Subbetic radiolarites because of the ubiquitous presence of carbonates. A few radiolarite beds at AG and AB are devoid of carbonates, but siliceous limestones and calcareous radiolarites are intercalated. Therefore, for deposition below the CCD to apply to the beds lacking carbonate, frequent recurrent fluctuations of the CCD must have taken place. Folk (in McBride and Folk, 1979) suggested deposition of the Lombardian cherts associated with evaporites in shallow intertidal flats. We favor an interpretation of fluctuations in radiolarian and carbonate plankton productivity to explain the lithologic variations. Estimates of the level of the pre–Late Jurassic CCD in the western Tethys range from 2.1 to 2.5 km (Bosellini and Winterer, 1975; Winterer and Bosellini, 1981). However, Schlager and Schlager (1973) suggested that the CCD during the deposition of the Alpine radiolarites cannot be precisely known, because variations in volcanic activity and plankton supply can produce substantial oscillations in the CCD. Formation of the alpine Subbetic radiolarites on continental crust also suggests deposition in relatively shallow water.

Except for possible differences in depth of water of deposition, the paleogeography of the external zones of the Betic Cordillera (Fig. 8.9) fits the small-basin model of Jenkyns and Winterer (1982). These small basins tend to have a shallow CCD, favoring accumulation of carbonate-poor sections, and are the loci of high plankton productivity and radiolarian sedimentation. We favor deposition of the Subbetic radiolarites above the CCD at a depth range of about 1000 to 1500 m—not as deep as originally thought, but in agreement with the depth estimates for other deposits of the Betic External Zones (Ruiz-Ortiz, 1980; Vera, 1981; Garcia-Hernandez et al., 1982).

Recent papers (Molina et al., 1985; Ruiz-Ortiz et al., 1985; Garcia-Hernandez et al., 1986; Vera et al., 1988) argue for the existence in the depositional basin of emergent swells during the Middle Jurassic (Fig. 8.9). These swells were the loci of deposition of the external and internal Subbetic Jurassic series. The swells were generated by listric faults, generally dipping toward the basin, which gave rise to a topography where the deeper-water areas were more protected from detritus originating from the northern continent; these structurally down-dropped blocks formed troughs to the south of the swells. Weathering of the carbonate islands formed by the swells, where karstification occurred (Garcia-Hernandez et al., 1986; Vera et al., 1988), could have been the source of detritus found in some sequences, such as the kaolinite found in some sections at SE. On the other hand, the topographic restrictions imposed by the swells to the north and south of the middle Subbetic basin (Fig. 8.9) likely determined the low sedimentation rate of the radiolarite sequence. Lack of detrital supply and episodes of radiolarian productivity produced the radiolarite beds on the down-dropped blocks of the listric faults. Similar environmental conditions have been suggested by De Wever and Dercourt (1985) for the deposition of radiolarites of Middle and Late Jurassic age of the Tethys region.

Complex basin geometry, topography, and thrust faulting hamper the reconstruction of the individual sequences studied and their relationship to the hypothetical northern continental source (Iberian meseta). For instance, from the geochemical indices, SE is one of the more proximal sequences, but it was probably deposited in the most distal southern part of the basin, as suggested by palinspastic reconstructions of the Betic External Zones (Garcia-Hernandez et al., 1980).

Conclusions

1. Radiolarite, siliceous mudstone, and pelagic limestone with radiolarians are the lithologic types making up the middle Subbetic radiolarite sequence; transitional lithologies among these three also occur.

2. Smectite and illite are the main clay minerals present, smectite always in greater amounts. Smectite and abundant radiolarians characterize the sequences with a greater oceanic influence. They were deposited in the deepest basin areas protected from input of continental detritus. Greater amounts of illite and lesser amounts of radiolarians make up the sequences with a large continental influence.

3. Various geochemical indices used as paleogeographic markers of distance from a terrigenous source show inconsistent results. Also, no correlation exists between the paleogeographic results derived from geochemical techniques and the positions of the studied sequences as determined from palinspastic reconstructions of the Betic External Zones. Complex basin geometry and topography are partly responsible for these inconsistencies.

4. Sedimentary structures and petrographic textures suggest that deposition was by pelagic settling. Productivity cycles and dilution by terrigenous debris gave rise to rhythmic sequences. Low sedimentation rates (average 3.5 mm/1000 years) allowed for enrichment in Fe in the radiolarite beds and for intensive bioturbation of the sediments; sediments were locally reworked and deposited by contour currents and distal dilute turbidity currents.

5. Deposition above a relatively shallow CCD (1000–1500 m) is suggested, a depth range shallower than previously thought for these deposits.

Acknowledgments. We thank J.R. Hein, U.S. Geological Survey and J.A. Vera, University of Granada for critical review of the paper; their comments improved the text and are much appreciated. We also thank J. Arroyo, M. Castillejo, J.L. Casaseca, M.A. Fernandez, and A. Viloria for technical assistance. Rocks were analyzed for their chemical composition by R. Garcia, M.I. Ruiz, and M. Vallejo. Financial support was provided by the CAICYT (project 84/0065) and CSIC (project ID-200) (Spain).

References

Adachi M, Yamamoto K, Sugisaki F (1986) Hydrothermal chert and associated siliceous rocks from the Northern Pacific: their geological significance as indication of Oceanic Ridge activity. Sedimentary Geology 47:125–148.

Aubouin J (1964) Reflexions sur les facies "Ammonitico Rosso." Bulletin de la Société Géologique de France 7(6):475–501.

Audley-Charles MG (1956) Some aspects of the chemistry of Cretaceous siliceous sedimentary rocks from eastern Timor. Geochimica et Cosmochimica Acta 29:1175–1192.

Azema J, Foucault A, Fourcade E, Garcia-Hernandez M, Gonzalez-Donoso JM, Linares A, Linares D, Lopez-Garrido AC, Rivas P, Vera JA (1979) Las microfacies del Jurásico y Cretácico de las Zonas Externas de las cordilleras Béticas. Secretariado de Publicaciones de la Universidad, Granada, 83 pp.

Baltuck M (1982) Provenance and distribution of Tethyan pelagic and hemipelagic siliceous sediments, Pindos Mountains, Greece. Sedimentary Geology 31:63–88.

Barrett TJ (1981) Chemistry and mineralogy of Jurassic bedded chert overlying ophiolites in the North Apennines, Italy. Chemical Geology 34:289–317.

Bernoulli D, Jenkyns HC (1974) Alpine, Mediterranean and central Atlantic Mesozoic facies in relation to the early evolution of the Tethys. In: Dott RL Jr, Shaver RH (eds) Modern and Ancient Geosynclinal Sedimentation: A Symposium. Society of Economic Paleontologists and Mineralogists Special Publication 19:129–160.

Biscaye PE (1978) Clays, deep sea. In: Fairbridge RW, Bourgeois J (eds) The Encyclopedia of Sedimentology. Dowden, Hutchinson and Ross, Stroudsburg, PA, pp 157–158.

Bosellini A, Winterer EL (1975) Pelagic limestone and radiolarite of the Tethyan Mesozoic: A genetic model. Geology 3:279–282.

Bostrom K, Kraemer T, Gartner S (1973) Provenance and accumulation rates of opaline silica, Al, Ti, Fe, Mn, Cu, Ni and Co in Pacific pelagic sediments. Chemical Geology 11:123–148.

Braga JC, Comas MC, Delgado F, Garcia-Hernandez M, Jimenez AP, Linares A, Rivas P, Vera JA (1981) The liassic Rosso Ammonitico facies in the Subbetic zone (Spain). Genetic considerations. In: Farinacci A, Elmi S (eds) Rosso Ammonitico Symposium Proceedings. Tecnoscienza, Rome, pp 61–76.

Bustillo MA, Molina JM, Ruiz-Ortiz PA (1986) Radiolarites and associated sedimentary facies in the Betic Cordillera. Abstracts, Third International Conference on Siliceous Deposits, IGCP project 187, Belgrade, p 99.

Deconinck JF, Beaudoin B, Chamley H, Joseph P, Raoult JF (1985) Controles tectonique, eustatique et climatique de la sédimentation argileuse du domaine subalpin français au Malm-Crétacé. Revue de Géologie Dynamique et de Géographie Physique 26:311–320.

De Wever P, Dercourt J (1985) Les radiolaires triasico-jurassiques marqueurs stratigraphiques et paléogéographiques dans les chaines alpines périméditerranéennes: une revue. Bulletin de la Société Géologique de France 8(I):653–662.

Einsele G (1982) Limestone-marl cycles (periodites): Diagnosis, significance, causes—A review. In: Einsele G, Seilacher A (eds) Cyclic and Event Stratification. Springer-Verlag, Berlin, pp 8–53.

Garcia-Dueñas V (1967) La zona Subbética al norte de Granada. Doctoral Thesis, University of Granada, Granada, Spain, 417 pp.

Garcia-Hernandez M, Vera JA (1986) La sedimentación liásica en el sector central del Subbético medio: Una clave para la interpretación de la evolución de un rift intracontinental. Abstract, XI Congreso Español de Sedimentología, Barcelona, p. 83. (Acta Geológica Hispánica, in press.)

Garcia-Hernandez M, Lopez-Garrido AC, Rivas P, Sanz de Galdeano C, Vera JA (1980) Mesozoic palaeogeographic evolution in the Externes zones of the Betic Cordillera (Spain). Geologie en Mijnbouw 59:155–168.

Garcia-Hernandez M, Lopez-Garrido AC, Ruiz-Ortiz PA, Vera JA (1982) Turbiditas carbonatadas del Cretácico inferior en el Arroyo Bercho, Prebético de Jaén: Interpretación genética e implicaciones paleogeográficas. Cuadernos de Geología Ibérica 8:433–447.

Garcia-Hernandez M, Lupiani E, Vera JA (1986) Discontinuidades estratigráficas en el Jurásico de Sierra Gorda (Subbético, provincia de Granada). Acta Geológica Hispánica (in press).

Garcia-Yebra R, Rivas P, Vera JA (1972) Precisiones sobre la edad de las coladas volcánicas Jurásicas en la región de Algarinejo-Lojilla (zona Subbética). Acta Geológica Hispánica 7:133–137.

Garrison RE, Fischer AG (1969) Deep-water limestones and radiolarites of the Alpine Jurassic. In: Friedman GM (ed) Depositional Environments in Carbonate Rocks. Society of Economic Paleontologists and Mineralogists, Special Publication 14:20–56.

Griffin JJ, Goldberg ED (1963) Clay mineral distributions in the Pacific Ocean. In: Hill MN (ed) The Sea, Vol 3. John Wiley, New York, pp 728–741.

Heath GR, Dymond J (1977) Genesis and transformation of metalliferous sediments from the East Pacific Rise, Bauer Deep, and Central Basin, northwest Nazca Plate. Geological Society of America Bulletin 88:723–733.

Hein JR, Yeh H-W, Alexander E (1979) Origin of iron-rich montmorillonite from the manganese nodule belt of the north equatorial Pacific. Clays and Clay Minerals 27:185–194.

Hein JR, Kuijpers EP, Denyer P, Sliney R (1983) Petrology and geochemistry of Cretaceous and Paleogene cherts from western Costa Rica. In: Iijima A, Hein JR,

Siever R (eds) Siliceous Deposits in the Pacific Region. Elsevier, Amsterdam, pp 143–174.

Iijima A, Matsumoto R, Tada R (1985) Mechanism of sedimentation of rhythmically bedded chert. Sedimentary Geology 41:221–233.

Jenkyns HC, Winterer EL (1982) Palaeoceanography of Mesozoic ribbon radiolarites. Earth and Planetary Science Letters 60:351–375.

Jones DL, Murchey B (1986) Geologic significance of Paleozoic and Mesozoic radiolarian chert. Annual Review Earth Planetary Sciences 14:455–492.

McBride EF, Folk RL (1979) Features and origin of Italian Jurassic radiolarites deposited on continental crust. Journal Sedimentary Petrology 49:837–868.

Molina JM (1987) Análisis de facies del Mesozoico en el Subbético externo (provincia de Córdoba y sur de Jaén). Doctoral Thesis, University of Granada, Granada, Spain, 518 pp.

Molina JM, Ruiz-Ortiz PA, Vera JA (1985) Sedimentación marina somera entre sedimentos pelágicos en el Dogger del Subbético externo (Sierras de Cabra y de Puente Genil, Prov. de Córdoba). Trabajos de Geologia, Universidad de Oviedo 15:127–146.

Palomo-Delgado I, Ortega Huertas M, Fenoll Hach-Ali P (1985) The significance of clay minerals in studies of the evolution of the Jurassic deposits of the Betic cordillera, SE Spain. Clay Minerals 20:39–52.

Puga E, Ruiz-Cruz MD (1980) Observaciones petrológicas y geoquímicas sobre el magmatismo básico mesozoico del sector central de la Cordillera subbética. Cuadernos de Geología, Universidad de Granada 11: 127–155.

Rangin C, Steinberg M, Bonnot-Courtois C (1981) Geochemistry of the Mesozoic bedded cherts of Central Baja California (Vizcaino–Cedros–San Benito): Implications for paleogeographic reconstruction of an old oceanic basin. Earth and Planetary Science Letters 54: 313–322.

Ruiz-Ortiz PA (1980) Análisis de facies del Mesozoico de las Unidades Intermedias (Entre Castril – prov. de Granada – y Jaén). Doctoral Thesis, Doctoral Thesis, University of Granada, Granada, Spain, 272 pp.

Ruiz-Ortiz PA, Molina JM, Vera JA (1985) Coral-ooid-oncoid facies in a shallowing-upward sequence of the middle Jurassic (external Subbetic, southern Spain). Sixth European Regional Meeting IAS, Lleida, Spain, pp 403–406.

Schlager W, Schlager M (1973) Clastic sediments associated with radiolarites (Tauglboden Schïchten, upper Jurassic, eastern Alps). Sedimentology 20:65–89.

Singer A, Müller G (1983) Diagenesis in argillaceous sediments. In: Larsen G, Chillingar GV (eds) Diagenesis in Sediments and Sedimentary Rocks. Elsevier, Amsterdam, pp 115–212.

Steinberg M, Mpodozis Marin C (1978) Classification géochimique des radiolarites et des sédiments siliceux océaniques, signification paléoceanographique. Oceanologica Acta 1:359–367.

Steinberg M, Bonnot-Courtois C, Tlig S (1983) Geochemical contribution to the understanding of bedded chert. In: Iijima A, Hein JR, Siever R (eds) Siliceous Deposits in the Pacific Region. Elsevier, Amsterdam, pp 193–210.

Sugisaki R, Yamamoto K, Adachi M (1982) Triassic bedded cherts in central Japan are not pelagic. Nature 298: 644–647.

Vera JA (1966) Estudio geológico de la zona Subbética en la transversal de Loja y sectores adyacentes. Doctoral Thesis, University of Granada, Granada, Spain. Memorias del Instituto Geológico y Minero de España LXXII, 192 pp.

Vera JA (1981) Correlación entre las Cordilleras Béticas y otras cordilleras alpinas durante el Mesozoico. Real Academia de Ciencias, Madrid PICG 2:129–162.

Vera JA, Ruiz-Ortiz PA, García-Hernandez M, Molina JM (1988) Paleokarst and related pelagic sediments in the Jurassic of the Subbetic zone, southern Spain. In: James NP, Choquette PW (eds) Paleokarst. Springer-Verlag, New York, pp 364–384.

Weaver CH (1978) Clay sedimentation facies. In: Fairbridge RW, Bourgeois J (eds) Encyclopedia of Sedimentology. Dowden, Hutchinson and Ross, Stoudsburg, Pennsylvania, U.S.A., pp 159–165.

Winterer EL, Bosellini A (1981) Subsidence and sedimentation on a Jurassic passive continental margin, southern Alps, Italy. Bulletin American Association Petroleum Geologists 65:394–421.

9
Petrology and Diagenetic Changes in Miocene Marine Diatomaceous Deposits from Zakynthos Island, Greece

M. Stamatakis, M. Dermitzakis, A. Economou-Amilli, and A. Magganas

Overview

Silica diagenesis in a diatomaceous section of marly limestone and chert from central Zakynthos Island, Ionian sea, was investigated. This section, underlain by Cretaceous-Eocene reefal limestone and overlain by a middle-upper Miocene terrigenous–evaporitic succession, is defined as the Romiri Formation. Diatom and planktonic foraminiferal assemblages occur, the latter determining the biostratigraphic interval N_4–N_7 of the Blow zonation (Aquitanian-Burdigalian).

To determine the distribution of authigenic minerals in the Romiri Formation we used SEM, XRD, and LM techniques. A gradual diagenetic transformation of the biogenic silica (opal-A) to opal-CT and clinoptilolite, and then to chalcedonic quartz was determined. The clinoptilolite plus opal-CT and the chalcedonic quartz form two distinct diagenetic zones, while carbonates, detrital quartz, smectite, and illite are widely distributed throughout the formation. The chalcedonic quartz, appearing as spherical chert nodules in a carbonate matrix, occurs at the base of the formation, while clinoptilolite and opal-CT occur at higher stratigraphic levels, but below the layers rich in opal-A. However, because of tectonic disruption, the distinct zones of silica polymorphs are commonly not found in their normal stratigraphic position. Highly etched valves of diatoms are found scattered through lithologies other than diatomaceous marl, including the zones characterized by clinoptilolite plus opal-CT and chalcedonic quartz.

We conclude that 1) opal-CT, clinoptilolite, chalcedonic quartz, and probably part of the clay minerals are of authigenic origin derived in part from the dissolution of diatom frustules, radiolarians, and sponge spicules and 2) the rate of silica diagenesis is mainly controlled by burial depth and chemistry of the pore fluids.

Introduction

The terminology used in this paper is that of Jones and Segnit (1971). Opal-A is X-ray amorphous biogenic silica, and opal-CT is disordered intermixed cristobalite and tridymite. Opal-CT and chalcedonic quartz associated with opal-A in chert, porcelanite, and cherty or porcelaneous limestone are interpreted to have formed from dissolution of siliceous organisms (Murata and Larson, 1975; Kastner, 1979; Iijima and Tada, 1981; Pisciotto, 1981). The diagenetic transformation of opal-A to opal-CT and then to chalcedonic quartz is considered to result from a dissolution-reprecipitation process which can take place in marine, lacustrine, and terrestrial environments (Heath and Moberly, 1971; Murata et al., 1977; Kastner et al., 1977; Tada and Iijima, 1983). In some deposits, formation of authigenic minerals such as zeolite and clay minerals accompany the transformation of opal-A to opal-CT (Papavasiliou, 1979; Iijima and Tada, 1981; Mann and Müller, 1985). Diagenesis of silica is influenced by many factors, such as burial depth, time, temperature, host sediment lithology, pore solution chemistry, specific surface area of opal-A particles, presence of detrital minerals, and sediment permeability (Siever, 1962; Keene, 1975; Kastner et al. 1977; Pisciotto, 1981; Hein et al., 1983; Williams et al., 1985; Mann and Müller, 1985).

Chert layers, lenses, and nodules in limestone and marl from Zakynthos Island occur in a lower-middle Miocene succession belonging to the Paxi geotectonic zone of Greece (Dermitzakis, 1978). Similar

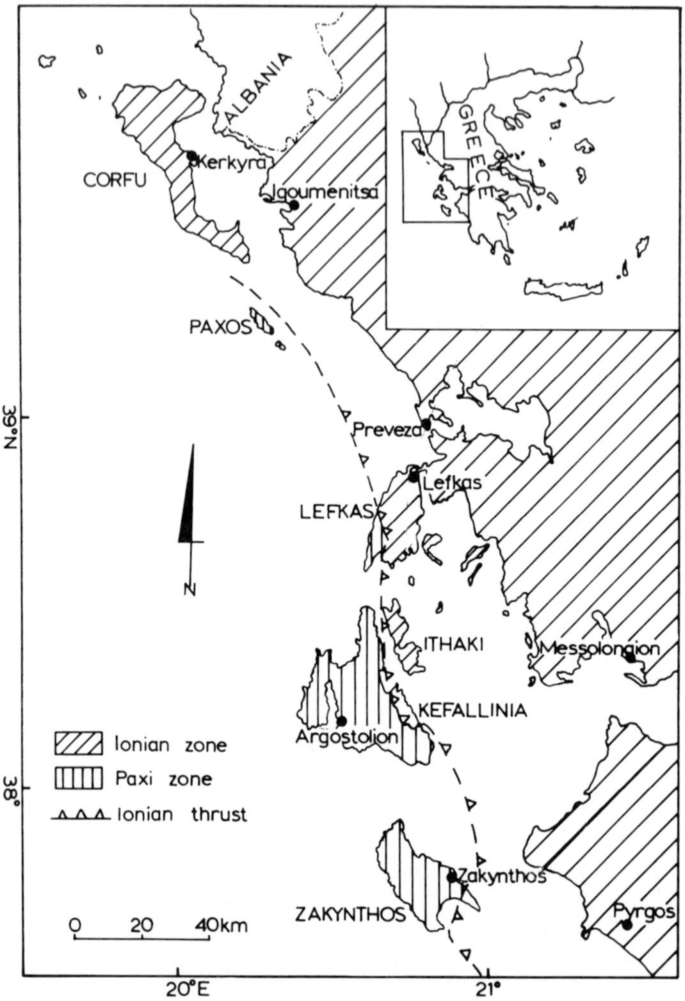

FIGURE 9.1. Location map of the Ionian Islands and the Paxi and Ionian geotectonic zones, western Greece.

cherty sections overlying Eocene limestone also occur in the same tectonic zone of the neighboring islands of Lefkas and Kefallinia, where the chert succession is overlain by marls containing opal-CT and clinoptilolite. Diatomaceous layers (opal-A) of middle Miocene age are found mainly on Zakynthos Island.

We report on the geology, petrology, and texture of deposits from Zakynthos Island containing silica polymorphs. We discuss the conditions that control the transformation rate of opal-A to opal-CT and then to quartz.

Mineralogical and textural studies were carried out with X-ray diffraction, light microscopy, and scanning electron microscopy (XRD, LM, and SEM, respectively). In most samples, mineralogical analysis was done after treatment of samples with 1 N HC1 to remove the carbonate.

Geological Setting

Zakynthos Island is one of the Ionian islands occurring off the west coast of Greece (Fig. 9.1). It comprises part of the external Hellenides, which form a NNW-SSE trending orogenic belt. According to Aubouin (1959), Zakynthos Island belongs to the Ionian and Paxi geotectonic zones, the latter comprising the eastern slope of the foreland block, the Apulian ridge. The Ionian and Paxi zones are separated by a major thrust that transects several Ionian islands (Fig. 9.1).

In the Paxi zone, pre-Neogene rocks consist of pelagic to reefal limestone and range in age from Jurassic through Paleogene. Neogene sedimentary rocks of the same zone occupy the central-southeastern part of Zakynthos Island, forming a monoclinal succession which deepens to the east. The Neogene

FIGURE 9.2. Simplified geological map of the southern part of Zakynthos Island showing sampling localities. (Modified from Dermitzakis, 1978.)

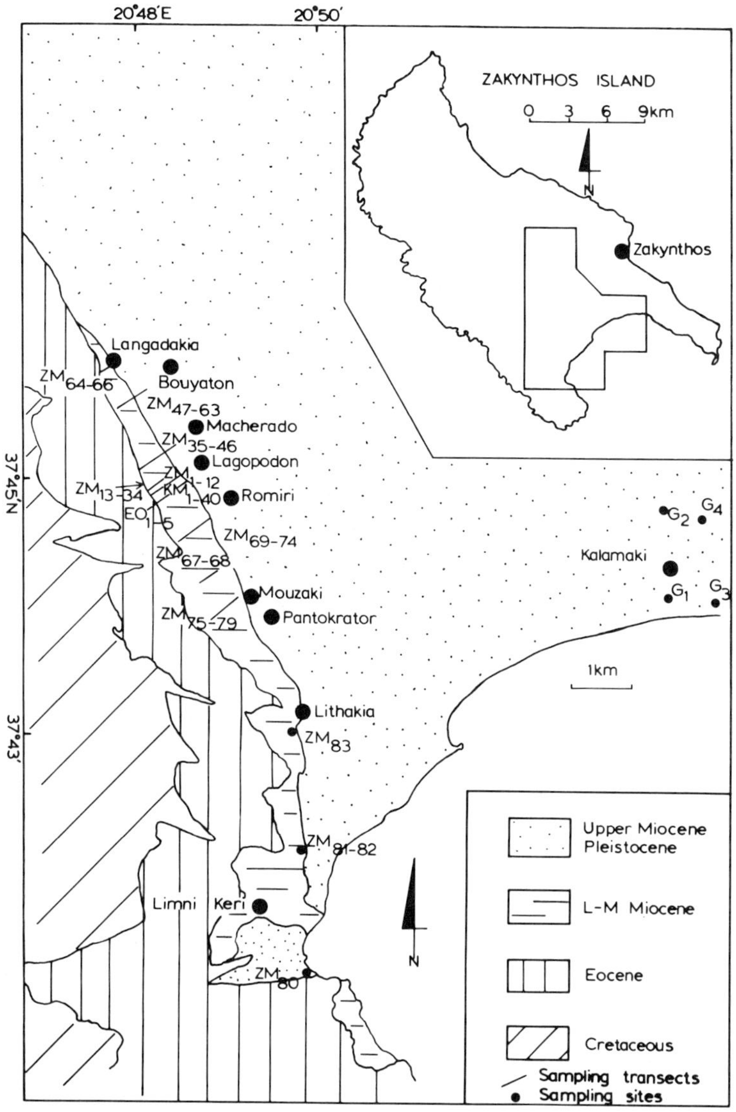

deposits can roughly be subdivided into two units: a lower unit is predominantly composed of calcareous and siliceous sedimentary rocks of early to middle Miocene age, and an upper unit consists of terrigeneous clastic deposits of late Miocene to Pliocene age with local occurrences of evaporites (Horstmann, 1977; De Mülder, 1975; Dermitzakis, 1978). The Neogene sedimentary rocks overlie middle to upper Eocene reefal limestone, which crops out to the west of the Neogene successions (Fig. 9.2). Immediately overlying the reefal limestone is a strongly faulted carbonate-siliceous section which in turn is overlain by marl, claystone, and evaporite of Laghian-Serravalian to Pliocene age (Heimann, 1977; Dermitzakis, 1978; Heimann et al., 1979). We name this succession the Romiri Formation.

Romiri Formation

The Romiri Formation consists of well-bedded limestone, marl, marly diatomite, and porous unstratified limestone with chert in the form of nodules, lenses, and thin laminated intercalations. Some porcelanite nodules are found in marls and in diatom-bearing layers. Large foraminifera (Fig. 9.3(1–9)), bryozoa, echinoids, and algae are common in the marl and the coarse intercalations; in the diatom-bearing layers, diatom frustules, coccoliths, radiolarians, sponge spicules, and silicoflagellates are abundant (Fig. 9.4(1–6)). The presence of diatom frustules has also been reported in the Messinian strata of the Kalamaki area (Heimann et al., 1979). All the fossil types present suggest a depth of water

FIGURE 9.3. Characteristic foraminifera species from marls of the Romiri Formation: (1) *Globorotalia kugleri* Bolli; (2) *Globigerinoides sicanus* De Stafani; (3) *Globorotalia peripheroroda* Blow and Banner; (4) *Praeorbulina glomerosa* Blow; (5) *Globorotalia peripheroroda* Blow and Banner; (6) *Globigerina bulloides* D'Orbigny; (7) *Globigerinoides sicanus* De Stefani; (8) *Globigerinoides sicanus* De Stefani; (9) *Globigerina dehiscens* (Chapman, Parr, and Collins) Parker. Scale bar: 1–9 = 0.1 mm.

of deposition between 300 and 500 m (Mircou-Peripopoulou, 1974).

Foraminifera species (Fig. 9.3(1–9)) define the following biostratigraphic intervals using the Blow zonation: *Globorotalia kugleri, Globigerinoides primordious* (zone N_4), and *Globigeriniodes sicanus* (zone N_6). These zones correspond to Aquitanian to late Burdigalian age. The upper part of the formation is assigned to the late Burdigalian (zone N_7), since the base of the Langhian (zone N_8) coincides with the appearance of *Praeorbulina glomerosa*, which is absent in the Romiri Formation but present at higher stratigraphic levels.

Strata rich in diatoms occur in a continuous NNW-SSE belt from Langadakia to Lithakia (Fig. 9.2). These strata are overlain by a porous carbonate succession which contains abundant chert in the form of nodules, lenses, and discontinuous layers. The contact of the cherty limestone with the underlying marly diatomite is tectonically disrupted at all outcrops where it occurs. Dermitzakis (1978) noted some discontinuity in the faunal succession at this contact.

The entire lithological succession is thickest near Lagopodon and thins to the north and south, near Bouyaton and Mouzaki, respectively, where chert is nearly absent. The cherty limestone is overlain by marl containing clinoptilolite and opal-CT. The distribution of the authigenic minerals follows the distribution of chert. In places porcelanite occurs in

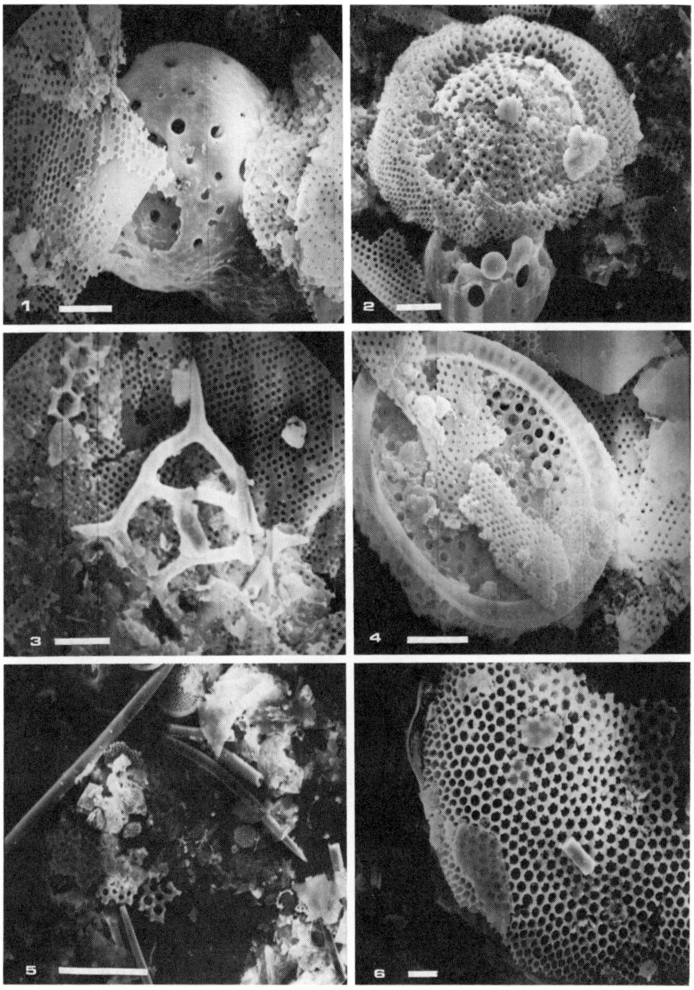

FIGURE 9.4. SEM photomicrographs of microfossils in diatomaceous rocks of zone A. (1) Very slightly dissolved radiolarian behind broken and partially dissolved diatom frustules (KM18 HCl). (2) Extensive dissolution of a diatom frustule. A broken radiolarian is observed at the lower left (KM18 HCl). (3) Silicoflagellate skeleton on diatom fragments. Slight dissolution in a small fragment of a diatom valve (upper left) (KM18 HCl). (4) Inner valve of a partly corroded diatom frustule (KM15 HCl). (5) Unaltered sponge spicules and fragments of diatoms and radiolarians (KM15 HCl). (6) Diatom valve showing noticeable dissolution of the fine structure (KM15 HCl). Scale bars: 1-4 and 6 = 10 µm; 5 = 100 µm.

marly diatomite as fracture filling up to 50 cm wide, or as nodular concretions with a waxy appearance and dark brown to gray color [Fig. 9.5(2–4)].

Finally, above the Romiri Formation there occurs grayish-green to whitish-yellow claystone and marl of middle Miocene to Pliocene age. Near Kalamaki, the Messinian strata (Heimann et al., 1979) contain gypsum bands and diatomaceous beds.

Thus, based on the distribution of silica polymorphs, the Romiri Formation can be subdivided into three zones, from the base to the top: C–zone of cherty limestone, presence of chalcedony. Maximum estimated thickness is 50 m. B–zone of porcelaneous marl, presence of clinoptilolite and opal-CT. Maximum estimated thickness is 100 m. A–zone of diatomaceous layers, presence of opal-A. Maximum estimated thickness is 50 m.

Zone C

This zone consists of a porous unstratified limestone where detrital quartz is the dominant silica polymorph. White-yellow to pink siliceous concretions occur in the limestone [Fig. 9.6(1–4)].

Various amounts of calcite are present within siliceous nodules, represented mainly by foraminiferal

FIGURE 9.5. Silica lithologies of zone A. (1) Thin, laminated diatomite near the village of Bouyaton. (2) Porcelanite nodule (arrow) in diatomaceous marlstone near the village Bouyaton. (3) Detailed view of the nodule from Fig. 9.5(2). Gradual transition of opaline porcelanite (dark area) (ZM70 nod.) to calcareous marlstone (white area). (4) Hand specimen of cherty material developed in a fracture perpendicular to the diatomaceous layers. The different shades are due to the predominance of chalcedony or opal-CT (ZM61 fract.).

tests. Chalcedonic quartz infills tests and is a massive pore-filling cement. In some places, such as near Bouyaton, relicts of diatoms are found in the chert nodules. Also, a few prismatic crystals (clinoptilolite?) are pseudomorphed by calcite and chalcedonic quartz. Near Romiri, carbonate material with sharp outlines is present within large chert concretions [Fig. 9.6(1)]. Most chert concretions are flattened and asymmetric, but a few are spherical [Fig. 9.6(2–4)]. Internally, the chert nodules commonly show an alternating repetition of colors.

Neither opal-CT nor clinoptilolite has been found in limestone-hosted chert concretions (Table 9.1). Chert nodules are more abundant in zone C, which tectonically overlies zone A. Zone C is stratigraphically thickest around Lagopodon area, which represents the central outcrop area of the Romiri Formation.

Zone B

Zone B is characterized by the presence of authigenic clinoptilolite and opal-CT in marly limestone and porcelaneous marl. In addition to calcite and detrital quartz, marls contain illite, smectite, chlorite, feldspars, clinoptilolite, and opal-CT (Table 9.1, Fig. 9.7). The opal-CT d(101)-spacing is sharp and is located at 4.07 Å, indicating a relatively well-ordered opal-CT (see also Iijima and Tada, 1981).

Some clinoptilolite has a stalky prismatic form, and some is found in vugs with radially oriented crystals. Opal-CT occurs as pore filling with asymmetric or spherical form. Foraminifera tests are commonly filled with rhombohedral calcite, prismatic clinoptilolite, and spherical opal-CT. In some places, prismatic clinoptilolite encloses calcite microspar, which probably acted as a nucleus for clinoptilolite crystallization.

In zone B, clinoptilolite is more abundant than opal-CT. Opal-CT occurs almost always accompanied by clinoptilolite. Chalcedonic quartz is absent. Strongly etched diatom frustules, radiolarian tests, and sponge spicules occur in zone B. These biogenic components probably transformed to opal-CT and were the source of silica for clinoptilolite. Radiolarian tests and sponge spicules are more abundant than diatom frustules because of their higher resistance to dissolution (Tada and Iijima, 1983).

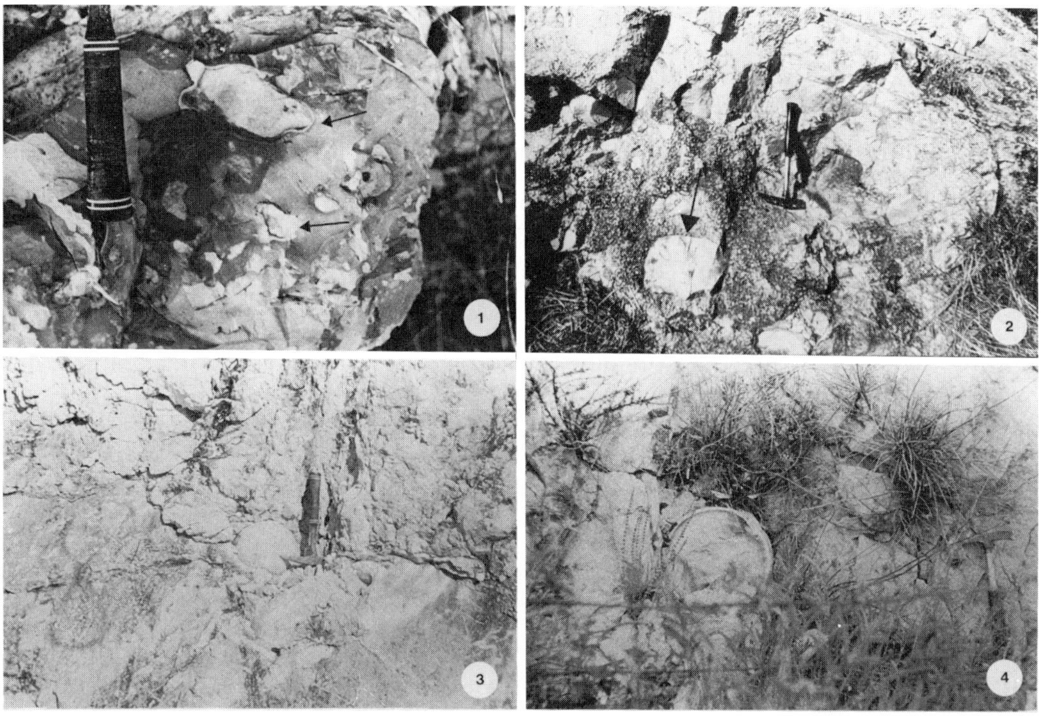

FIGURE 9.6. Diagenetic macrotextures in the porous limestone of zone C near the villages of Romiri and Lagopodon. (1) An asymmetric chert concretion, with nonreplaced carbonate inclusions, marked by the arrows. (2) Asymmetric siliceous nodule conchoidally fractured (arrow). (3) Spherical siliceous concretion in the carbonate matrix (left of the hammer). (4) Spherical siliceous concretion with an earthy rind.

Zone A

Zone A, characterized by the presence of diatomaceous rocks [Fig. 9.5(1,2)], has an irregular outcrop occurrence, because it is tectonically sandwiched between the strata of zones C and B. Zone C tectonically overlies zone A.

XRD patterns of samples treated with 1 N HC1 show a broad peak, characteristic of opal-A, and also reflections for the minerals quartz, illite, smectite, chlorite, and traces of feldspar (Table 9.1). Quartz and feldspar are of detrital origin and occur as broken crystals in the groundmass of samples.

Slightly to strongly etched large diatom frustules and well-preserved radiolarian tests, sponge spicules, and silicoflagellate skeletons [Fig. 9.4(1–6)] occur in zone A. Because of the significant degree of diatom dissolution, the only easily recognizable species is *Coscinodiscus* sp. In some diatom frustules from other sections, a noticeable dissolution of fine-scale structures occurs [Fig. 9.4(6)], indicating an early transformation stage of opal-A to opal-CT (Hein et al., 1978; Mann and Müller, 1985).

An interesting local occurrence in this zone (near the villages of Lagopodon and Bouyaton) is the presence of waxy brown porcelanite nodules in the diatomaceous layers. Some fractures in the same layers are also filled with pink to yellow-red porcelaneous material [Fig. 9.5(2–4)], both containing opal-CT, chalcedonic quartz, and detrital quartz. Botryoidal aggregates of quartz with a porcelaneous texture also fill cavities. The opal-CT d(101) peak is somewhat rounded and occurs at 4.11 Å, indicating a relatively disordered opal-CT that may never have been deeply buried (Mitsui and Taguchi, 1977; Carver, 1979; Pisciotto, 1981; Hein et al., 1981).

Controls on Diagenesis

In the Romiri Formation on Zakynthos Island, the diagenetic transformation of opal-A to opal-CT took place in at least two stages. In the first stage, opal-CT in the form of foraminifera and pore space infillings shows a relatively high degree of ordering, d(101) 4.07 Å. The sedimentary rocks of the lower-

TABLE 9.1. Mineralogical composition of samples taken from different stratigraphic levels of the Romiri Formation.

Sample No.	Zone	Mineralogical analysis
EO1	Underlying Romiri Formation	Calcite
EO4	Underlying Romiri Formation	Calcite
EO5	Underlying Romiri Formation	Calcite
G3	Overlying Romiri Formation	Gypsum, quartz, illite, chlorite
G4	Overlying Romiri Formation	Calcite, gypsum, illite, quartz, chlorite, magnesite
KM1	Underlying zone B	Calcite, quartz, chlorite
KM2 HCl	Zone B	Quartz, clinoptilolite, smectite, albite
KM5 HCl	Zone A	Opal-A, quartz, illite, smectite
KM8 HCl	Zone A	Opal-A, quartz, illite, smectite
KM15 HCl	Zone A	Opal-A, quartz, illite, smectite, albite
KM17 HCl	Zone A	Opal-A, quartz, smectite, feldspar
KM18 HCl	Zone A	Opal-A, quartz, smectite
KM19 HCl	Overlying zone A	Quartz, smectite
KM20	Overlying zone A	Calcite, quartz, smectite
KM22	Overlying zone A	Calcite, quartz, smectite
KM24 nod.	Zone C	Quartz
KM25 nod.	Zone C	Quartz, calcite
KM25 host HCl	Zone C	Quartz, smectite, illite
KM27 host HCl	Zone C	Quartz, smectite
KM27 nod.	Zone C	Quartz
KM28 HCl	Zone B	Quartz, clinoptilolite, smectite, illite, chlorite
KM29 HCl	Zone B	Quartz, clinoptilolite, smectite, feldspar
KM30 HCl	Zone B	Quartz, opal-CT, clinoptilolite, smectite, anorthoclase, illite
KM31 HCl	Zone B	Quartz, smectite, clinoptilolite, high sanidine, opal-CT
KM32 HCl	Zone B	Quartz, clinoptilolite, smectite, illite
KM33 HCl	Zone B	Quartz, smectite, high sanidine, illite, chlorite
KM36 HCl	Zone B	Quartz, smectite, illite
KM37 HCl	Zone B	Quartz, clinoptilolite, smectite
KM38 HCl	Zone B	Quartz, clinoptilolite, albite, smectite
KM39 HCl	Zone B	Quartz, clinoptilolite, illite
KM40 HCl	Zone B?	Quartz, smectite, illite, chlorite, plagioclase
ZM11 HCl	Zone B	Quartz, clinoptilolite, illite, smectite
ZM18 nod.	Zone B to A	Opal-CT, quartz
ZM35 HCl	Zone B	Quartz, smectite, chlorite, illite, albite
ZM40 nod.	Zone A	Opal-CT, quartz
ZM41 nod.	Zone A	Opal-CT, quartz
ZM42	Zone A	Calcite, opal-A, quartz
ZM46 HCl	Overlying Romiri Formation	Quartz, illite, chlorite, albite
ZM50 HCl	Zone B	Quartz, illite, smectite, clinoptilolite, anorthoclase
ZM52 HCl	Zone B	Quartz, illite, chlorite
ZM54	Overlying Romiri Formation	Calcite, quartz
ZM55 HCl	Overlying Romiri Formation	Quartz, illite, smectite, anorthoclase
ZM56 HCl	Zone C	Quartz, smectite, illite, chlorite, albite
ZM59 fract.	Zone A	Opal-CT, quartz
ZM60 HCl	Zone A	Opal-CT, quartz, clay minerals
ZM61 fract.	Zone A	Opal-CT, quartz
ZM62 nod.	Zone A	Opal-CT, quartz
ZM70 nod.	Zone A	Quartz, opal-CT
ZM70	Zone A	Calcite, quartz, opal-A, opal-CT, illite
ZM76	Zone A	Calcite, quartz
ZM76 HCl	Zone B	Quartz, illite, albite, smectite, clinoptilolite
ZM78 HCl	Zone B	Quartz, smectite, illite, chlorite
ZM79 HCl	Zone B	Quartz, smectite, illite
ZM80 HCl	Overlying Romiri Formation	Quartz, albite, illite, smectite, chlorite
ZM81 HCl	Zone B	Quartz, illite, smectite

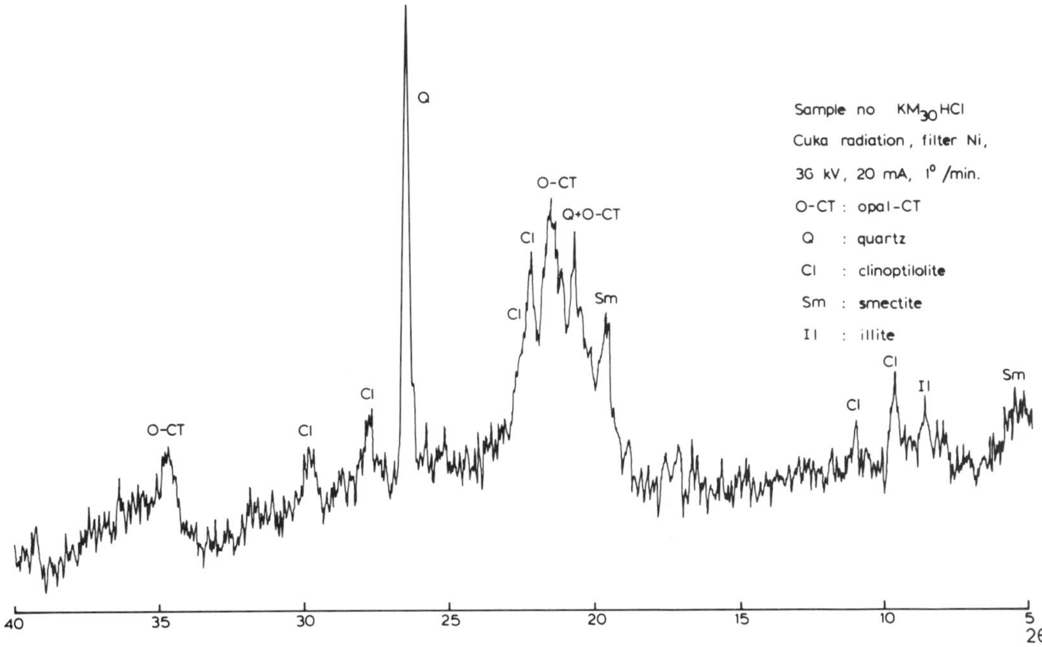

FIGURE 9.7. X-ray diffractogram of HCl-treated samples from zone B. Sharp peak of moderately ordered opal-CT.

middle Miocene Romiri Formation were deposited in relatively deep water, where precipitation of these early-formed silica infillings occurred. In the second stage, opal-CT fracture fillings have a d(101)-spacing of 4.11 Å; this relatively disordered opal-CT suggests formation during shallow burial, probably during uplift (Iijima and Tada, 1981) or during semiarid surface conditions (Dapples, 1979). These two stages of opal-CT formation are attributed to a major tectonic event that took place between early-middle Miocene and middle-late Miocene to Pliocene in Zakynthos and resulted in a fundamental change in the island's paleogeography (Dermitzakis, 1978). During the Messinian stage, semiarid climatic conditions were developed in Zakynthos and resulted in local deposition of evaporites (Dermitzakis, 1978; Heimann et al., 1979).

The occurrence of clinoptilolite similar to that found in zone B has been reported many times (Hein et al., 1978; Kidd et al., 1978; Melieres et al., 1978; Papavasiliou, 1979; Tada and Iijima, 1983; Mann and Müller, 1985). Its formation is due to the reaction of smectite with interstitial waters rich in Na^+, K^+, $HSiO_4^{3-}$, high Si/Al ratio, and pH lower than 8. Na^+ and K^+ might derive from seawater as well as from partial dissolution of feldspar and illite. $HSiO_4^{3-}$ was derived from opal-A dissolution (Papavasiliou, 1979). Smectite can be transformed directly to clinoptilolite (Papavasiliou, 1979) and vice versa (Mann and Müller, 1985). At least a part of the smectite in the Romiri Formation may be of authigenic origin, being derived from another authigenic mineral, clinoptilolite. Volcanic debris is absent in post-Miocene deposits in both the Paxi and Ionian zones and precludes the derivation of smectite from volcanic glass. Furthermore, the presence of only small amounts of opal-CT in zone B could be explained by the simultaneous presence of Mg^{2+}, OH^- (from seawater), and smectite, which all prevent the formation of opal-CT (Papavasiliou, 1979). The absence of clinoptilolite in zone C should be attributed to the presence of calcareous sediment, which does not contain the necessary reactive material for formation of clinoptilolite.

Chalcedonic quartz, formed during late diagenesis in terrigenous and siliceous sections of the Romiri Formation, occurs along with opal-CT in nodules and fracture infillings; in carbonate sections, chalcedonic quartz occurs as nodules and is formed during early diagenesis. Chalk inclusions, like those in chert of zone C in the Romiri Formation, have been reported in similar deposits of the Pacific (Dapples, 1979). Foraminiferal silty limestone with chert and zeolites were also found in the Atlantic Basin (Calvert, 1971; Dapples, 1979). All the previous observations on opal-CT, zeolites, and chalcedony found in the Pacific and Atlantic Oceans are supported by our studies of the Romiri Formation.

The diagenetic silica transformations in the Romiri Formation can be summarized as follows. During early diagenesis, transformation of opal-A to opal-CT in carbonate layers and to clinoptilolite in clayey sediment took place; clinoptilolite formation may also have involved a contribution from smectite degradation. In a later stage of diagenesis, the transformation of opal-CT to chalcedonic quartz took place in the deeper parts of the carbonate sections. The authigenic silica minerals formed during the early and late stages, infilling pores in the host rock and chambers of tests, and replacing preexisting mineral phases. Finally, when the Romiri Formation was exposed to near surface conditions, a final stage of opal-CT and quartz formation occurred. In this stage, opal-CT and quartz formed nodules and veins that are found in fractures and microfaults.

In the Ionian Islands, no evidence of volcanic activity or metamorphism has been recognized. Also, the heat flow in the area is low, ≤ 1.0 HFU (Ericson et al., 1977; Čermak et al., 1976), suggesting that high temperature was not an important factor in silica diagenesis. The subsidence that the Romiri Formation underwent was about 600 m, as estimated from the thickness of overlying sedimentary rocks. The depth of burial and time (age of the Romiri Formation) are the main controls for the silica transformations.

Burial of diatomaceous layers to a sufficient depth was responsible for the dissolution of opal-A. The chemistry of pore fluids enriched in $HSiO_4^{3-}$, K^+, and Na^+ and the type of host sediment were responsible for the formation of clinoptilolite and opal-CT. The development of chert nodules in lower stratigraphic levels floating in a carbonate matrix resulted from replacement of the host sediment. Tectonic disturbances and percolating groundwater are responsible for the late-stage opal-CT and chalcedony occurrences.

Acknowledgments. We are grateful to J.R. Hein and J. Obradović for critically reviewing the manuscript. Thanks are also due Z. Karotsieris for suggestions on the tectonic interpretations.

References

Aubouin J (1959) Contribution a l' étude géologique de la Grèce septentrionale: Les confins de l' Épire et la Thessalie. Annales Géologiques des Pays Helléniques 10:1–485.

Blow WH (1969) Late middle Eocene to recent planktonic biostratigraphy. Proceedings First International Conference on Planktonic Microfauna, Geneva 1967, pp 199–241.

Calvert CE (1971) Composition and origin of North Atlantic deep sea cherts. Contributions to Mineralogy and Petrology 33:273–288.

Carver RE (1979) Petrology of Paleocene-Eocene and Miocene opaline sediments, southern Atlantic coastal plain. Journal of Sedimentary Petrology 50:569–582.

Čermak V, Hartig E, Kutas R, Loddo M, Lubimova E, Mougelli F, Morgan P, Smirnov Y, Tezcan A (1976) Heat flow map of southern Europe and Mediterranean region. International Congress on Thermal Waters, Geothermal Energy and Volcanism of Mediterranean Area, Athens 1976, 1:149–168.

Dapples EC (1979) Silica as an agent in diagenesis. In: Larsen G, Chillingar GV (eds) Diagenesis in Sediments and Sedimentary Rocks. Elsevier, Amsterdam, pp 99–142.

De Mülder JE (1975) Microfauna and sedimentary-tectonic history of the Oligo-Miocene of the Ionian Islands and western Epirus (Greece). Utrecht Micropalaeontology Bulletin, Utrecht 13:138.

Dermitzakis MD (1978) Stratigraphy and sedimentary history of the Miocene of Zakynthos (Ionian Islands, Greece). Annales Géologiques des Pays Helléniques 29:47–186.

Ericson AG, Simmons G, Ryan WBF (1977) Review of heat flow data from Mediterranean and Aegean seas. In: Biju-Duval B, Montadert L (eds) Structural History of the Mediterranean Basins. Editions Technip, Paris, pp 263–280.

Heath GR, Moberly R (1971) Cherts from the western Pacific, leg 7, DSDP. In: Winterer EL, Riedel WR, et al. (eds) Initial Reports of the Deep Sea Drilling Project 7. U.S. Government Printing Office, Washington, pp 991–1007.

Heimann KO (1977) Die Fazies des Messins und untersten Pliozäns auf den Ionischen und auf Sizilien. Ph.D. Dissertation, Munich, 158 pp.

Heimann KO, Just J, Muller C (1979) Sedimentological and palaeontological aspects of Messinian strata on the Ionian Islands (Greece). Annales Géologiques des Pays Helléniques, Hors Serie 1:501–514.

Hein JR, Scholl DW, Barron JA, Jones MG, Miller J (1978) Diagenesis of late Cenozoic diatomaceous deposits and formation of the bottom simulating reflector in the southern Bering sea. Sedimentology 25:155–181.

Hein JR, Vallier TL, Allan MA (1981) Chert petrology and geochemistry, mid-Pacific Mountains and Hess Rise, leg 62, DSDP. In: Vallier TL, Thiede J, et al. (eds) Initial Reports of the Deep Sea Drilling Project 62, U.S. Government Printing Office, Washington, pp 711–748.

Hein JR, Sancetta C, Morgenson LA (1983) Petrology and

geochemistry of silicified upper Miocene chalk, Costa Rica Rift, DSDP leg 69. In: Cann JR, Langseth MG, et al. (eds) Initial Reports of the Deep Sea Drilling Project 69. U.S. Government Printing Office, Washington, pp 395–422.

Horstmann G (1977) Géologie de la partie meridionale de l'île de Zante (Grèce). Thèse, University of Paris, 127 pp.

Iijima A, Tada R (1981) Silica diagenesis of Neogene diatomaceous and volcaniclastic sediments in northern Japan. Sedimentology 28:185–200.

Jones JB, Segnit ER (1971) The nature of opal. I. Nomenclature and constituent phases. Journal of Geological Society of Australia 18:57–68.

Kastner M (1979) Silica polymorphs. In: Marine Minerals. Mineralogical Society of America (Short Course Notes) 6:99–107.

Kastner M, Keene JB, Gieskes JM (1977) Diagenesis of siliceous oozes. I. Chemical controls on the rate of opal-A to opal-CT transformation—an experimental study. Geochimica et Cosmochimica Acta 41:1041–1059.

Keene JB (1975) Cherts and porcelanites from the north Pacific, DSDP leg 32. In: Larson RL, Moberly R, et al. (eds) Initial Reports of the Deep Sea Drilling Project 32. U.S. Government Printing Office, Washington, pp 429–507.

Kidd RB, Bernoulli D, Garrison RE, Fabricius F, Melieres F (1978) Lithologic findings of DSDP leg 42a, Mediterranean Sea. In: Hsu KJ, Montadert L, et al. (eds) Initial Reports of the Deep Sea Drilling Project 42. U.S. Government Printing Office, Washington, pp 421–443.

Mann U, Müller G (1985) Early diagenesis of biogenic siliceous constituents in silty clays and claystones, Japan Trench. Neues Jahrbuch für Mineralogie Abhandlungen 153:33–57.

Melieres F, Chamley H, Coumes F, Rouge P (1978) X-ray mineralogy studies, leg 42a DSDP, Mediterranean Sea. In: Hsu KJ, Montadert L, et al. (eds) Initial Reports of the Deep Sea Drilling Proejct 42. U.S. Government Printing Office, Washington, pp 361–387.

Mircou-Peripopoulou RM (1974) Stratigraphy and geology of the northern part of the Zakynthos Island. Annales Géologiques des Pays Helléniques 26:35–108.

Mitsui K, Taguchi K (1977) Silica mineral diagenesis in Neogene Tertiary shales in the Tempoku district, Hokkaido, Japan. Journal of Sedimentary Petrology 47:158–167.

Murata KJ, Larson RR (1975) Diagenesis of Miocene siliceous shales, Templor Range, California. Journal Research U.S. Geological Survey 3:553–566.

Murata KJ, Friedman I, Gleason JD (1977) Oxygen isotope relations between diagenetic silica minerals in Monterey Shale, Templor Range, California. American Journal of Science 277:259–272.

Papavasiliou C (1979) The geochemistry and mineralogy of some NW Indian Ocean DSDP cores (sites 222 and 223, leg 23 DSDP). Ph.D. Thesis, University of Southampton, U.K., 520 pp.

Pisciotto KA (1981) Diagenetic trends in the siliceous facies of the Monterey Shale in the Santa Maria region, California. Sedimentology 28:547–571.

Siever R (1962) Silica solubility, 0°C–200°C and the diagenesis of siliceous sediments. Journal of Geology 70:127–150.

Tada R, Iijima A (1983) Petrology and diagenetic changes of Neogene siliceous rocks in northern Japan. Journal of Sedimentary Petrology 53:911–930.

Von Rad U, Rosch H (1974) Petrography and diagenesis of deep-sea cherts from the central Atlantic. In: Hsu KJ, Jenkyns HC (eds) Pelagic Sediments on Land and Under the Sea. Special Publication of International Association of Sedimentologists 1:327–347.

Williams LA, Parks GA, Crerar DA (1985) Silica diagenesis. I. Solubility controls. Journal of Sedimentary Petrology 55:301–311.

10 Thermally Induced Silica Transformation in Pliocene Diatomaceous Layers from Aegina Island, Greece

Michael G. Stamatakis and Andreas C. Magganas

Overview

The Pliocene diatomaceous and porcellaneous marine sediments of St. Thomas hill, Aegina Island, Greece contain opal-A and authigenic opal-CT. XRD, SEM, and LM analyses show that opal-A is partly transformed to opal-CT. The age of the sedimentary rocks, the absence of distinct zones of opal-A and opal-CT, and the shape and position of opal-CT (101) peak in XRD patterns show that silica diagenesis took place at shallow burial depths. The transformation of opal-A to opal-CT was primarily controlled by the high heat flow in the region since the Pliocene. This silica transformation is uncommon in deposits of such a young geologic age and that were not deeply buried.

Introduction

Opal-CT (disordered cristobalite-tridymite; Jones and Segnit, 1971) has been shown to be derived from tests of siliceous organisms (opal-A) (Heath, 1973; Friedman and Sanders, 1978; Iijima, 1980; Iijima and Tada, 1981). The diagenetic transformation of opal-A to opal-CT takes place through a dissolution-reprecipitation mechanism (Heath and Moberly, 1971; Murata and Larson, 1975; Murata et al., 1977; Tada and Iijima, 1983). This silica transformation occurs under common temperature conditions in various environments, such as in marine sediments, in lake deposits with saline-alkaline pore waters, and in sediments subjected to high temperatures (Sheppard and Gude, 1969; Murata and Larson, 1975; Keene, 1975; Dapples, 1979; Pisciotto, 1981; Tada and Iijima, 1983; Hein et al., 1983; Hein and Yeh, 1983; Kastner and Siever, 1983). The most common occurrence of opal-CT is in thick to moderately thick sections of marine siliceous rock. In deeper parts of the sediment column, opal-CT may transform to quartz (Ernst and Calvert, 1969; Murata and Nakata, 1974; Hein et al., 1978; Iijima and Tada, 1981), whereas under very high thermal gradients, opal-A transforms directly to quartz (Kastner and Siever, 1983).

In the St. Thomas hill area of Aegina Island, Greece, porcellanite is interbedded at various stratigraphic levels with diatomaceous layers in a Pliocene marine succession (Livaditis, 1974; Benda et al., 1979). In this paper, we discuss the petrology, texture, and conditions that control the rate of opal-A to opal-CT transformation.

Location and Geological Setting

Aegina Island, 12 miles from Athens, is situated in the western part of the Pliocene and Quaternary southern Aegean volcanic arc (Fig. 10.1). The island is composed of Cretaceous limestone, sandstone, and chert; Miocene and Pliocene marl, marly limestone, diatomaceous marlstone, sandstone, tuff and tuffite, and lava; and Pleistocene sandy limestone (Fig. 10.2) (Van Leyden, 1940; Davis, 1957; Pe, 1973; Livaditis, 1974; Tsoli-Kataga, 1977; Benda et al., 1979).

Diatomaceous rocks have been reported in the northern part of the island on St. Thomas hill. Some

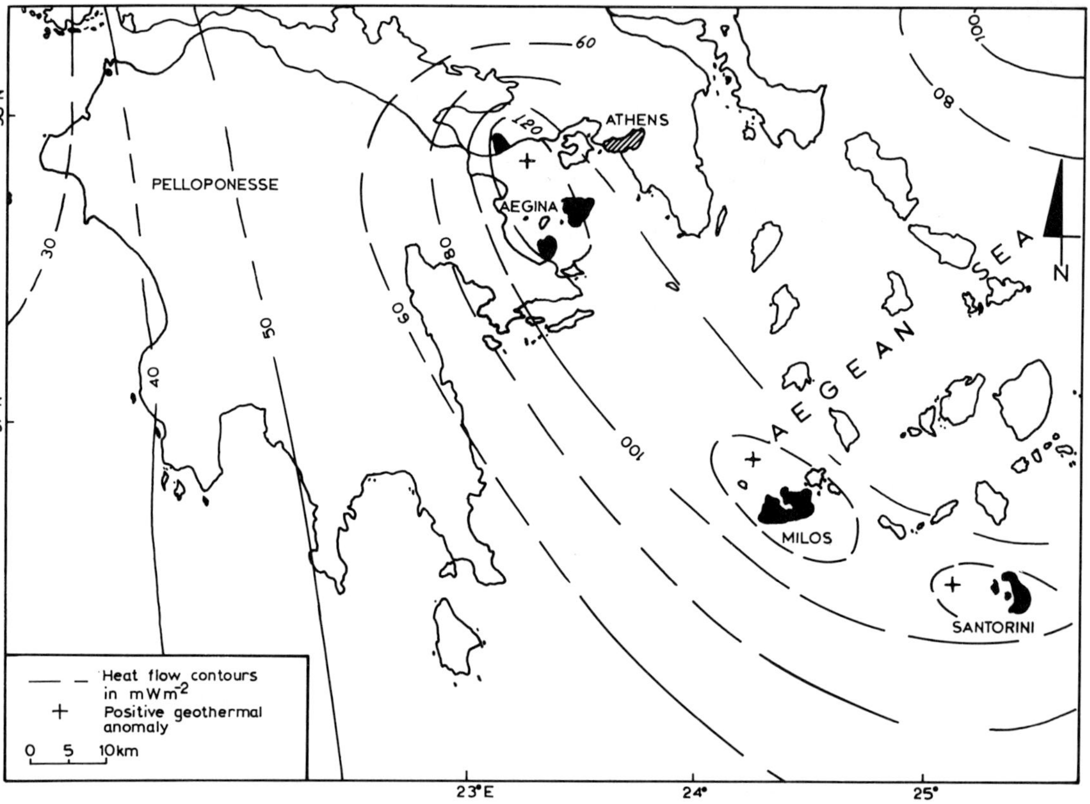

FIGURE 10.1. Location map showing the position of Aegina, the heat flow, and the positive geothermal anomalies as well as the distribution of the Pliocene and Quaternary volcanism (black areas) in the western part of the Aegean volcanic arc. (Modified from Fyticas and Kolios, 1979).

intercalations of porcellanite occur in these rocks (Benda et al., 1979). The age of this marine, diatom-rich formation is late early Pliocene (4.4 $\pm$ 0.2 my) (Benda et al., 1979; Müller et al., 1979). Samples were studied from this site from various stratigraphic levels (Fig. 10.3).

Petrology of the St. Thomas Section

Petrological and textural determinations were carried out using LM, SEM, and XRD analytical methods. XRD of silica phases and the d-spacings of opal-CT were determined using the techniques of Murata and Larson (1975) and Hein et al. (1978).

The St. Thomas section (Fig. 10.3) can be subdivided with respect to the occurrence of silica polymorphs (from the base): (1) The lower zone contains detrital quartz and no opal phases; (2) the middle zone contains intermixed opal-A and opal-CT and detrital quartz; (3) the upper zone contains volcanic glass, volcanic and/or detrital quartz, and no opal phases.

Lower Zone

This zone consists of argillaceous marlstone, calcareous marlstone, and thinly laminated gypsiferous claystone. The opal polymorphs are absent, but detrital quartz is present. The presence of micritic gypsum and dolomite in the claystone (Table 10.1) indicates deposition during evaporative conditions.

Middle Zone

We subdivide this zone into two subzones of unequal thickness. The opal-A subzone consists of diatomaceous-dolomitic marlstone, and the opal-CT subzone consists of opaline porcellanite. The opal-CT subzone is less extensive than the opal-A

FIGURE 10.2. Simplified geological map of Aegina Island showing the location of the St. Thomas section. (Modified from Benda et al., 1979).

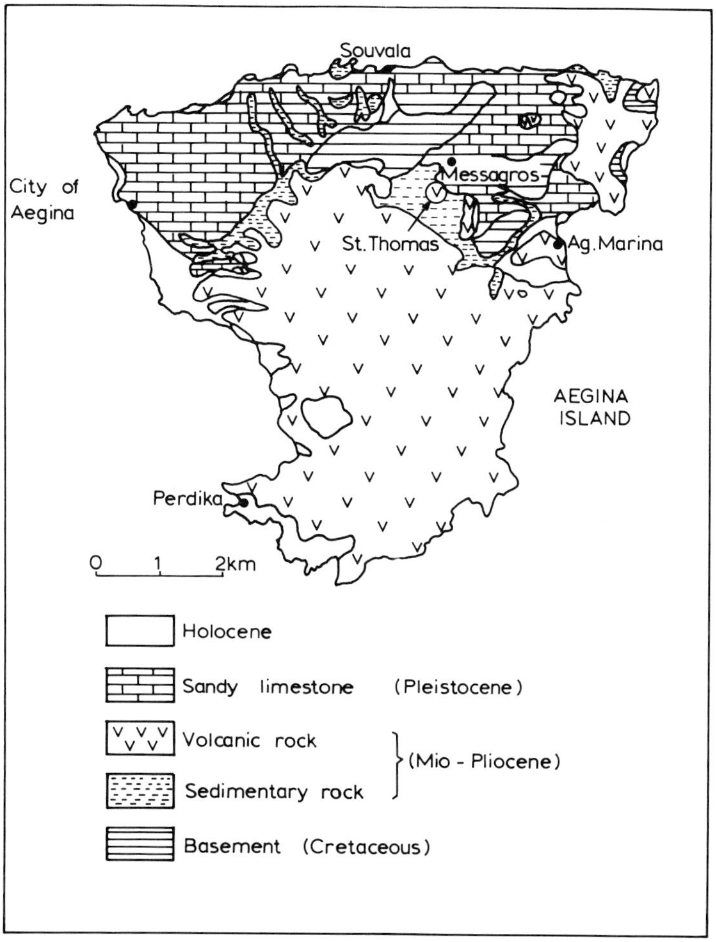

subzone. The total thickness of both subzones is about 50 m.

Opal-A Subzone

The white to pale yellow diatomaceous marlstone and marly diatomite that comprise this zone consist mainly of opal-A, dolomite, quartz, albite, smectite, calcite, chlorite, and illite (Table 10.1). The degree of preservation of diatom frustules is poor. Most of the frustules are broken and/or dissolved, and the delicate structure of frustules has been destroyed (Fig. 10.4a,b). Detrital quartz crystals are broken, and authigenic quartz crystals were not identified. Dolomite occurs as idiomorphic, rhombohedral crystals, reflecting its authigenic origin.

Opal-CT Subzone

Opaline porcellanite occurs in the form of asymmetric, massive nodules, lenses, and boudinaged layers of waxy brown color. In places, relict diatomaceous material occurs as patches in the nodules. A gradational transition zone from opal-A to opal-CT is common. The brownish opaline porcellanite consists mainly of opal-CT and dolomite; detrital quartz and albite are minor components. In a few samples, calcite coexists with chlorite, white mica, and amphibole (Table 10.1).

The d(101)-spacing of opal-CT is 4.115 ± 0.005 Å, and the peak is broad and somewhat rounded (Fig. 10.5). The opaline porcellanite displays a spongy texture that, after 6 N HCl treatment, yields scattered rhombohedral vugs where dolomitic crystals occurred (Fig. 10.4c,d).

The groundmass of samples consists of an isotropic to slightly anisotropic silica mineral (opal-CT by XRD). Dolomite rhombohedra (nonstained with Alizarin Red S) occur in discrete laminae in this silica groundmass. Opal-CT is colorless but in some samples is pale brown owing to the presence of

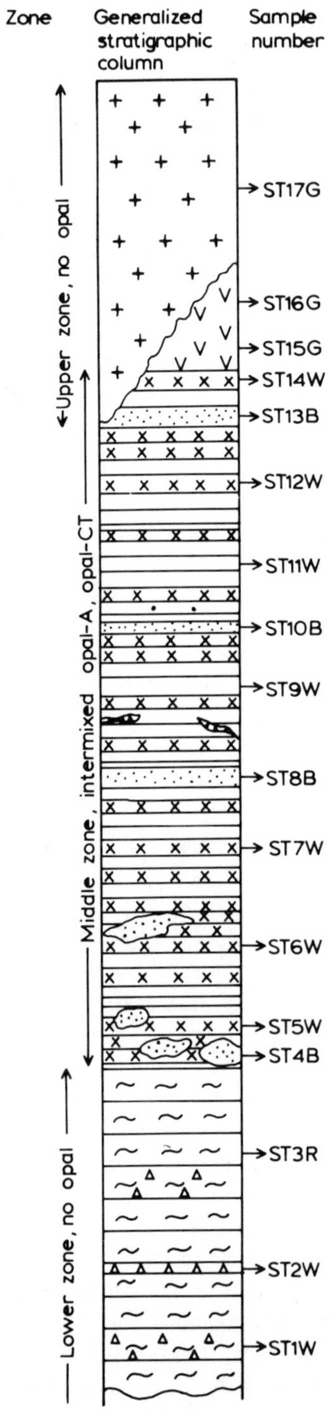

FIGURE 10.3. Generalized stratigraphic column of the St. Thomas section showing zones defined by silica polymorphs and sample locations. The column represents about 50 m of the section, and it is not to scale. The age of the entire section is late early Pliocene. See Table 10.1 for a description of the lithologies.

organic material. Scattered clastic grains of albite and quartz are enclosed by oriented crystallites of brown opal-CT. Relict forms of diatom frustules, foraminiferal tests, and mollusk shells are noted in the opaline porcellanite. The primary bioclasts have been replaced by opal-CT (Fig. 10.6) showing a slight anisotropy at the edges. Authigenic dolomite crystals formed simultaneously with or earlier than opal-CT, as dolomite occurs as idiomorphic crystals in the opal-CT matrix.

Upper Zone

The rocks from the upper zone are tuff and tuffite overlain by andesitic breccia. The only silica polymorph in this zone is volcanic and/or detrital quartz. Volcanic glass is also present. The zone has a maximum thickness of 80 m, but it is only 10 m thick in the St. Thomas section.

Diagenetic Transformation of Opal-A to Opal-CT

Previous work has shown that the main factors controlling the diagenetic transformation of opal-A to opal-CT and then to quartz are burial depth, temperature, time, morphology and size of the particles, composition of the host sediment, and pore solution chemistry (presence of Mg^{2+}, electrolytes, and alkalinity) (Murata and Larson, 1975; Keene 1975; Kastner et al., 1977; Kastner, 1979; Pisciotto, 1981; Rubin and Friedman, 1981; Iijima and Tada, 1981; Tada and Iijima, 1982; Hein et al., 1983; Williams et al., 1985). The diagenetic transformation results in the creation of three distinct diagenetic zones, where opal-A, opal-CT, and quartz predominate respectively from the top to the bottom of a section (Kastner, 1979). The transformation from opal-A to opal-CT begins within the purest layers of diatomite, forming alternations of diatomaceous marlstone or claystone and chert (Murata and Nakata, 1974). The complete transformation of diatomaceous sediment into porcellanite takes place at a later diagenetic stage (Bramlette, 1946; Murata and Nakata, 1974).

In the St. Thomas section on Aegina Island, an unusual diagenetic transformation of opal-A to opal-CT occurred. The silica transformation occurred in siliceous deposits that are young (early Pliocene) and stratigraphically thin (up to 50 m) and that were

TABLE 10.1. Lithology and mineralogical composition of samples from the St. Thomas section.[a]

Sample	Lithology	Bulk X-ray mineralogy (MJ = major, MD = moderate, TR = trace)
ST17G	Andesitic breccia	Plagioclase (MJ), glass (MJ), biotite (MD), hornblende (MD)
ST16G	Tuff	Quartz (MJ), albite (MJ), glass (MD), smectite (MD)
ST15G	Tuffite	Albite (MJ), quartz (MJ), glass (MD), white mica (TR), amphibole (TR), illite (TR), smectite (TR)
ST14W	Clayey diatomite	Opal-A (MJ), quartz (MD), dolomite (MJ), albite (TR), smectite (TR)
ST13W	Opaline porcellanite	Opal-CT (MJ), dolomite (MD), quartz (MD), calcite (TR), albite (TR)
ST12W	Diatomaceous marlstone	Quartz (MJ), opal-A (MD), dolomite (MD), albite (TR), illite (TR), smectite (TR)
ST11W	Calcareous marlstone	Calcite (MJ), quartz (MD), dolomite (MD), albite (TR), chlorite (TR), illite (TR), smectite (TR)
ST10B	Opaline porcellanite	Opal-CT (MJ), dolomite (MJ), quartz (MD)
ST9W	Dolomitic marlstone	Dolomite (MJ), quartz (MJ), calcite (MD), albite (TR), smectite (TR)
ST8B	Opaline porcellanite	Opal-CT (MJ), dolomite (MD), quartz (MD)
ST7W	Diatomaceous marlstone	Quartz (MJ), opal-A (MD), calcite (MD), dolomite (MD), albite (TR), chlorite (TR), illite (TR), smectite (TR)
ST6W	Dolomitic diatomite	Dolomite (MJ), opal-A (MJ), quartz (MD), albite (TR)
ST5W	Dolomitic diatomite	Dolomite (MJ), opal-A (MJ), quartz (MD), opal-CT (MD), albite (TR)
ST4B	Opaline porcellanite	Opal-CT (MJ), quartz (MD), dolomite (MD)
ST3R	Dolomitic marlstone	Dolomite (MJ), quartz (MJ), calcite (MD), chlorite (TR), white mica (TR), albite (TR), illite (TR), smectite (TR)
ST2W	Gypsum	Gypsum (MJ)
ST1W	Gypsiferous dolomitic marlstone	Dolomite (MJ), quartz (MD), gypsum (MD), white mica (TR), albite (TR), chlorite (TR), illite (TR), smectite (TR)

[a]Sampling sites are shown in Figure 10.3.

never deeply buried. A shallow burial depth is also indicated by the shape and the position of the opal-CT d(101)-spacing which is a broad, rounded peak centered at 4.115 ± 0.005 Å, indicating an embryonic form of opal-CT (Pisciotto, 1981; Iijima and Tada, 1981; Hein et al., 1981).

Clay and carbonate minerals are common host-rock mineralogies, but host-rock composition did not control silica diagenesis, because no correspondence is noted between the two components within the sediment section. The lack of zeolites, authigenic K feldspar, and evaporitic minerals in the siliceous deposits of the St. Thomas section suggest that sediments probably accumulated in a normal seawater environment. The presence of abundant authigenic dolomite in both lower and middle zones attests to the high activity of Mg^{2+} in the pore water.

Apparently, the principal factor that controlled the transformation of opal-A to opal-CT was the high heat flow in the area. The presence of Mg^{2+} as well as the size and morphology of opal-A particles may also have been important.

Heat Flow

Temperature is the most important factor controlling diagenetic transformations of opal in young sedi-

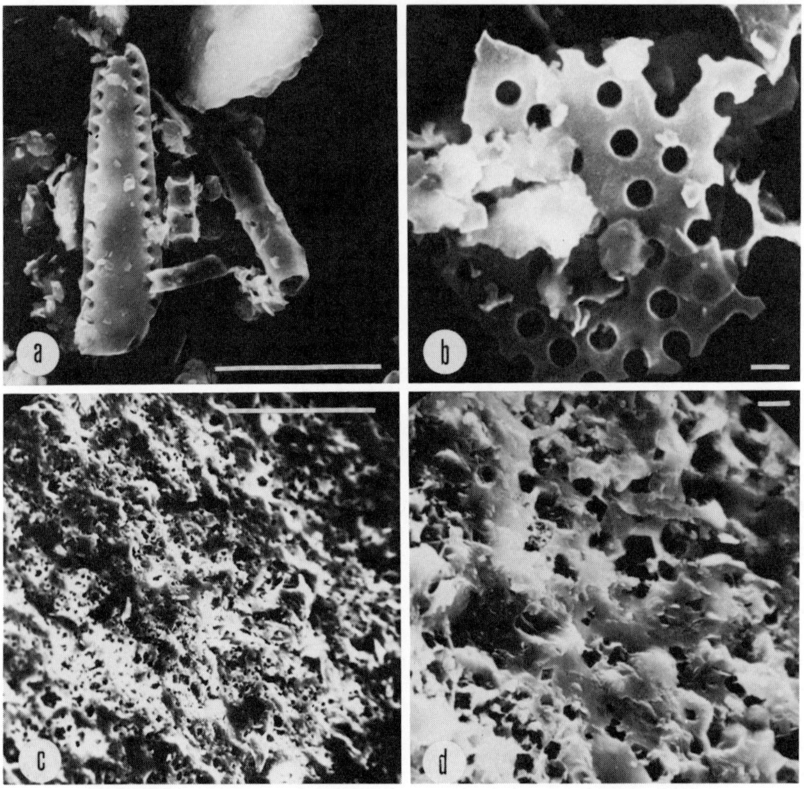

FIGURE 10.4. SEM photomicrographs (a,b). Dissolved and broken diatoms in the opal-A subzone (ST14W). The diatoms have lost their delicate structure after dissolution. (c) Opaline porcellanite after 6 N HCl treatment (ST10B). The matrix and cement, composed of opal-CT and detrital quartz, show a spongy texture. (d) Closeup of Figure 10.4c. Disseminated rhombohedral open spaces created by dissolution of dolomite crystals. Scale bars: a, 10 μm; b and d, 1 μm; c, 100 μm.

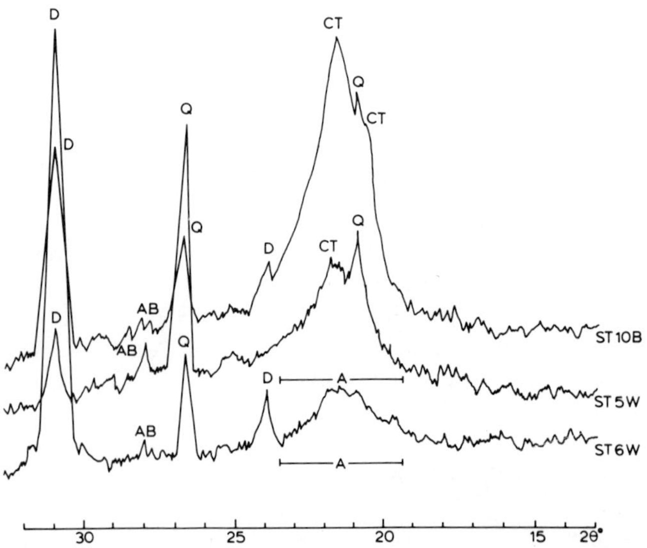

FIGURE 10.5. X-ray diffractograms of the progressive changes from biogenic opal-A to opal-CT in the intermixed opal-A and opal-CT zone. Part of the change in the opal-A to opal-CT pattern is probably due to an increasing crystallinity of the opal-CT, and part is due to the mixing of discrete opal-A and opal-CT particles. Q = quartz; CT = opal-CT; A = opal-A; AB = albite; D = dolomite.

ment. Keene (1975) suggested that chert formation is unusual in sediment younger than 10 my (late Miocene) except in areas with high in situ temperatures. Hein et al. (1978) pointed out that in continental margin and island arc environments, the opal-A transforms into opal-CT faster and at higher temperatures than in open ocean environments (Fig. 10.7). In these marginal environments, temperature is the dominant control of the silica transformation. It is well known that in the southern Aegean area, high heat flow and an active volcanic arc, which includes Aegina Island, have existed since the Pliocene (Fig. 10.1) (Jongsma, 1974; Fytikas et al., 1976; Erickson et al., 1977; Fytikas and Kolios, 1979; Dermitzakis and Papanicolaou, 1981; Innocenti et al., 1979, 1981, 1982; Papadopoulos, 1982). The highest heat flow value (2.73 HFU) in the Aegean Sea occurs in the active volcanic arc between the islands of Aegina and Milos (Jongsma, 1974). The high heat flow of the Aegean region is also due to the intense extensional tectonism occurring since the Pliocene (Mercier, 1976; Angelier, 1979).

High heat flow causes crystallization of opal-CT in a highly disordered state (Hein et al., 1983). So, the relatively highly disordered opal-CT (d-spacing at 4.115 ± 0.005 Å) in the young and not deeply buried deposits of the St. Thomas section supports high heat flow as the controlling mechanism for the transformation of silica.

The temperature of transformation of opal-A to opal-CT usually ranges from about 20 to 55°C (Murata and Larson, 1975; Hein et al., 1978). At very high heat flow (between ~4 and >30 HFU) and at elevated temperatures (150–200°C), opal-A may transform to opal-CT in a closed system, where rapid convective dissipation of silica is prevented; otherwise, opal-A transformed directly to quartz (Kastner and Siever, 1983). The heat flow value in the Aegean area (~3 HFU), the absence of silica deposits sandwiched between volcanic sills, which would create a closed system, and the absence of authigenic quartz clearly suggest a transformation temperature much lower than 150°C. Using the age of St. Thomas deposits (4.4 ± 0.2 my) and Kastner's (1979) equations, we estimate the conversion temperature of opal-A to opal-CT to have been about 51°C (50.6 ± 0.2°C). A similar conversion temperature was reported for silicified deposits at DSDP site 504, where the transformation was mainly controlled by high heat flow (Hein et al., 1983; Hein and Yeh, 1983). The age and the estimated conversion temperature for opal-CT in the St. Thomas deposits, plotted on Hein et al.'s (1978) diagram, fall at the extreme upper left of the field of the transformation of opal-A to opal-CT (Fig. 10.7).

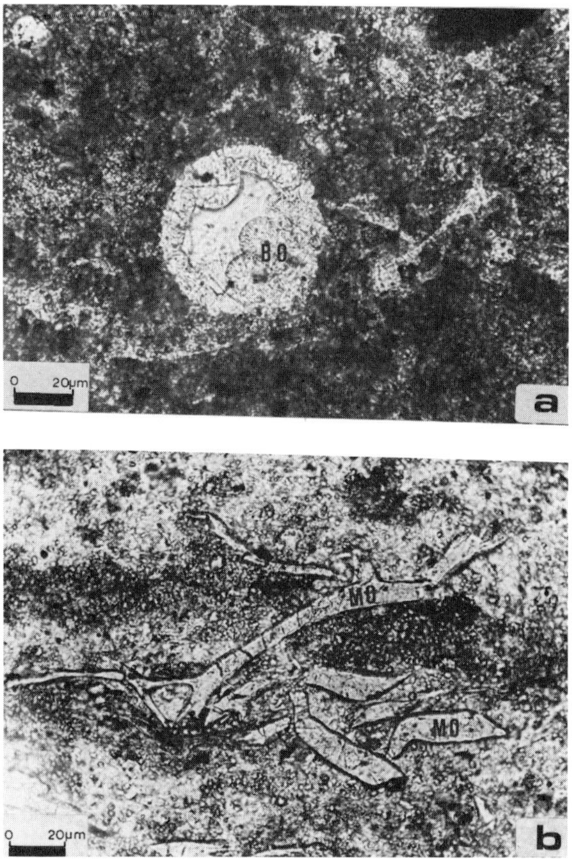

FIGURE 10.6. Thin-section photomicrographs of opaline porcellanite from the opal-CT subzone (ST4B). (a) Foraminferal test is filled with botryoidal (BO) opal-CT. (b) Mollusk shell (?) fragments are pseudomorphed by massive opal-CT. Plane polarized light.

We conclude that the diagenetic transformation of opal-A to opal-CT was mainly due to the high heat flow in the area of Aegina Island since the Pliocene. Pore water chemistry and sediment texture may have affected the transformation to a lesser extent.

Acknowledgments

Thanks are expressed by the authors to J.R. Hein (U.S. Geological Survey), K.A. Pisciotto (SOHIO), and R. Tada (University of Tokyo) for reviewing a version of the manuscript and providing valuable

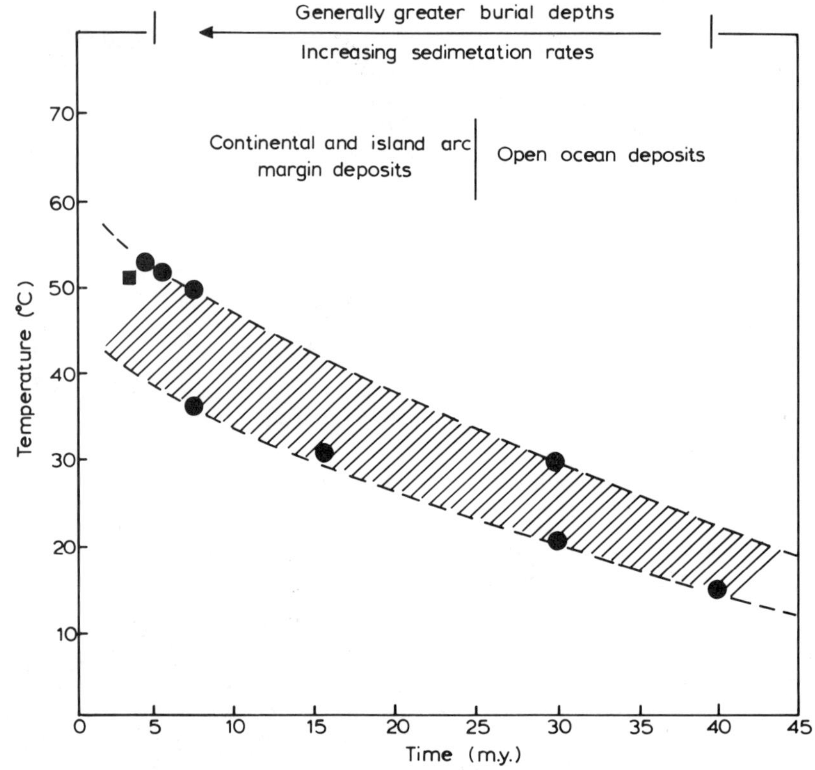

FIGURE 10.7. Field where the initiation of the transformation of opal-A to opal-CT occurs is defined by points (solid dots) taken from the literature (diagram from Hein et al., 1978, 1983). Datum point for the St. Thomas deposits is plotted as a solid square.

comments and suggestions. We are indebted to L. Benda (Hannover) for his critical remarks. We also thank A. Economou-Amilli (University of Athens) for helping in SEM analysis.

References

Angelier J (1979) Recent Quaternary tectonics in the Hellenic arc. Examples of geological observations on lands. Tectonophysics 52:267–275.

Benda L, Jonkers HA, Meulenkamp JE, Steffens P (1979) Biostratigraphic correlations in the eastern Mediterranean Neogene. Newsletter on Stratigraphy 8(1):61–69.

Bramlette MN (1946) The Monterey Formation of California and the origin of its siliceous rocks. U.S. Geological Survey Professional Paper 112:1–57.

Dapples EC (1979) Silica as an agent in diagenesis. In: Larsen G, Chillingar GV (eds) Diagenesis in Sediments and Sedimentary Rocks. Elsevier, Amsterdam, pp 99–142.

Davis E (1957) Die jungvulcanischen Gesteine von Aegina, Methana und Poros und deren Stellung in Rahmen der Kykladenprovinz. Publicationen Stiftung Vulcanistitut Immanuel Friedlaender 6, pp 74, Zurich.

Dermitzakis M, Papanicolaou D (1981) Paleogeography and geodynamics of the Aegean region during Neogene. Annales Géologiques des Pays Helléniques 32:245–289.

Erickson AG, Simmons G, Ryan WBF (1977) Review of heatflow data from the Mediterranean and Aegean Seas. In: Biju-Duval B, Montadert L (eds) Structural History of the Mediterranean Basins. Editions Technip, Paris, pp 263–280.

Ernst WG, Calvert SE (1979) An experimental study of the recrystallization of porcellanite and its bearing on the origin of some bedded cherts. American Journal of Science 267:114–133.

Friedman GM, Sanders JE (1978) Principles of Sedimentology. John Wiley, New York, 792 pp.

Fytikas M, Kolios N (1979) Preliminary Heat Flow Map of Greece. IGME, Athens.

Fytikas M, Giuliani O, Innocenti F, Marinelli G, Mazzuoli R (1976) Geochronological data on Recent magmatism of the Aegean Sea. Tectonophysics 31:29–34.

Heath GR (1973) Cherts from the eastern Pacific, leg 16, Deep Sea Drilling Project. In: Van Andel TG, Heath

GR, et al. (eds). Initial Reports of the Deep Sea Drilling Project 7. U.S. Government Printing Office, Washington, pp 609–613.

Heath GR, Moberly R (1971) Cherts from the western Pacific, leg 7, Deep Sea Drilling Project. In: Winterer EL, Riedel WR, et al. (eds). Initial Reports of the Deep Sea Drilling Project 7. U.S. Government Printing Office, Washington, pp 991–1007.

Hein JR, Yeh H-W (1983) Oxygen-isotope composition of secondary silica phases. Costa Rica Rift, Deep Sea Drilling Project leg 69. In: Cann JR, Langseth MG, et al. (eds). Initial Reports of the Deep Sea Drilling Project 69. U.S. Government Printing Office, Washington, pp 423–429.

Hein JR, Scholl DW, Barron JA, Jones MG, Miller J (1978) Diagenesis of late Cenozoic diatomaceous deposits and formation of the bottom simulating reflector in the southern Bering sea. Sedimentology 25:156–181.

Hein JR, Vallier TL, Allan MA (1981) Chert petrology and geochemistry, mid-Pacific Mountains and Hess Rice, leg 62, Deep Sea Drilling Project. In: Thiede J, Vallier TL, et al. (eds). Initial Reports of the Deep Sea Drilling Project 62. U.S. Government Printing Office, Washington, pp 711–748.

Hein JR, Sancetta C, Morgenson LA (1983) Petrology and geochemistry of silicified upper Miocene chalk, Costa Rica Rift, Deep Sea Drilling Project leg 69. In: Cann JR, Langseth MG, et al. (eds). Initial Reports of the Deep Sea Drilling Project 69. U.S. Government Printing Office, Washington, pp 395–422.

Iijima A (1980) Geology of natural zeolites and zeolitic rocks. Proceedings of the 5th International Conference on Zeolites, 1980, Naples 1:103–118.

Iijima A, Tada R (1981) Silica diagenesis of Neogene diatomaceous and volcaniclastic sediments in northern Japan. Sedimentology 28:185–200.

Innocenti F, Manetti P, Peccerillo A, Poli G (1979) Inner arc volcanism in NW Aegean Arc. Geochemical and geochronological data. Neues Jahrbuch für Mineralogie Monatshefte H4:145–158.

Innocenti F, Manetti P, Peccerillo A, Poli G (1981) South Aegean volcanic arc. Geochemical variations and geotectonic implications. Bulletin Volcanologique 44:376–391.

Innocenti F, Kolios N, Manetti P, Rita F, Villari L (1982) Acid and basic late Neogene volcanism in central Aegean sea. Its nature and geotectonic significance. Bulletin Volcanologique 45:86–97.

Jones JG, Segnit ER (1971) The nature of opal. I. Nomenclature and constituent phases. Journal Geological Society of Australia 18:57–68.

Jongsma D (1974) Heat flow in the Aegean Sea. Geophysics Journal Royal Astronomical Society 37:337–346.

Kastner M (1979) Silica polymorphs. In: Marine Minerals. Mineralogical Society of America. Short Course Notes, 6:99–107.

Kastner M, Siever R (1983) Siliceous sediments of the Guayamas basin: The effect of high thermal gradients on diagenesis. Journal of Geology 91:629–641.

Kastner M, Keene JB, Gieskes JM (1977) Diagenesis of siliceous ooze. I. Chemical controls on the rate of opal-A to opal-CT transformation—an experimental study. Geochimica et Cosmochimica Acta 41:1041–1059.

Keene JB (1975) Cherts and porcelanites from the North Pacific. Deep Sea Drilling Project leg 32. In: Larson RL, Moberly R, et al. (eds). Initial Reports of the Deep Sea Drilling Project 32. U.S. Government Printing Office, Washington, pp 429–507.

Keene JB (1976) The distribution, mineralogy and petrography of biogenic and authigenic silica from the Pacific Basin. Ph.D. Dissertation, University of California, San Diego, 264 pp.

Livaditis GJ (1974) Geological and geomorphological observations on Aegina Island. Ph.D. Thesis, University of Athens, 64 pp.

Mercier JL (1976) La néotectonique, ses methods es ses buts. Un example: l'arc Egéen (Mediteranée Oriental). Revue de Géographie Physique et Géologie Dynamique 18:323–346.

Müller P, Kreuzer H, Lenz H, Harre W (1979) Radiometric dating of two extrusives from a lower Pliocene marine section on Aegina Island, Greece. Newsletter on Stratigraphy 8:70–78.

Murata KJ, Nakata JK (1974) Cristobalite stage in diagenesis of diatomaceous shale. Science 184:567–586.

Murata KJ, Larson RR (1975) Diagenesis of Miocene siliceous shale, Temblor Range, California. Journal Research U.S. Geological Survey 3:553–556.

Murata KJ, Friedman I, Gleason JD (1977) Oxygen isotope relations between diagenetic silica minerals in Monterey shale, Temblor Range, California. American Journal of Science 277:259–272.

Papadopoulos G (1982) Active deep tectonics of the Aegean and surrounding area. Ph.D. Thesis, University of Thessaloniki, Greece, 176 pp.

Pe G (1973) Petrology and geochemistry of Aegina, Greece. Bulletin Volcanologique 37:491–514.

Pisciotto KA (1981) Diagenetic trends in the siliceous facies of the Monterey Shale in the Santa Maria region, California. Sedimentology 28:547–571.

Rubin DM, Friedman GM (1981) Origin of chert grains and a halite-silcrete bed in the Cambrian and Ordovician Whitehall Formation of eastern New York State. Journal of Sedimentary Petrology 51:69–72.

Sheppard RA, Gude AJ 3d (1969) Diagenesis of tuffs in the Barstow Formation, Mud Hills, San Bernardino County, California. U.S. Geological Survey Professional Paper 634, 35 pp.

Tada R, Iijima A (1982) Identification of mixtures of opaline silica phases and its implication for silica diagenesis. In: Iijima A, Hein JR, Siever R (eds). Siliceous deposits of the Pacific region. Elsevier, Amsterdam, pp 229–245.

Tada R, Iijima A (1983) Petrology and diagenetic changes of Neogene siliceous rocks in northern Japan. Journal Sedimentary Petrology 53:911–930.

Tsoli-Kataga E (1977) Geochemistry of clay formations from the island of Aegina. Bulletin of Geological Society of Greece 13:71–96.

Van Leyden R (1940) Der Vulcanismus des Golfes von Aegina und seine Beziehungen zur Tectonik. Publicationen Stiftung Vulcanistitut Immanuel Friedlander 1, p 151, Zurich.

Williams LA, Parks GA, Crerar DA (1985) Silica diagenesis. I. Solubility controls. Journal Sedimentary Petrology 55:301–311.

11
Silicified Wood from the Adoyama Chert, Kuzuh, Central Honshu, and Its Bearing on Compaction and Depositional Environment of Radiolarian Bedded Chert

A. Iijima, Y. Kakuwa, and H. Matsuda

Overview

Silicified *Brachyoxylon* wood measuring 11 × 17 × 22 cm was found in the uppermost part of the Triassic and Lower Jurassic Adoyama Chert from Kuzuh, northern Kanto, central Honshu. Sedimentary laminae bent around the silicified wood allowed us to estimate minimum compaction of the quartzose bedded chert to be 60%. Assuming an initial porosity for the sediment of 80%, the maximum compaction would have been 80%. The 3.9 cm average thickness of chert beds and the 67 m total stratigraphic thickness of the Adoyama Chert translate into precompaction thicknesses of 19.5 cm and 335 m, respectively.

A sedimentation rate can be determined by a geochemical method where the relationship between the average sedimentation rate of Triassic bedded chert (Y) obtained from stratigraphic method and the average MnO/Al_2O_3 ratio of the chert (X) is represented by a regression line, $\log_2 Y = -2.27 - 1.59 \log X$ ($r = -0.70$). The average sedimentation rate of the Adoyama Chert estimated from this method uncorrected for compaction is 2.1 m/m.y., corresponding to an uncompacted sedimentation rate of 10.5 m/m.y. The average sedimentation rate of 10.5 m/m.y. is comparable within the limit of error to 16 m/m.y. determined by the geochemical method derived from analysis of Pacific pelagic sediments (Matsumoto and Iijima, 1983.).

The *Brachyoxylon* wood drifted from the Kurosegawa landmass which was located on the south side of the elongate Jurassic marginal sea. The wood was deposited on the upper slope or outer shelf along with siliceous organic remains, and slumped down the slope as a siliceous debris flow into a basin which had a probable water depth of around 1000 m. The distance from the land was only a few hundred kilometers. In Middle Jurassic time, the central part of the marginal sea began to be uplifted possibly due to the northwestward movement of the Kula plate. In Late Jurassic time, the Adoyama Chert together with the Middle Jurassic Maki Sandstone and the Permian Nabeyama Limestone were located on the north side of the uplifted high and were displaced northeastward about 30 to 50 km due to gravitational tectonics. These rocks eventually settled on the flat seafloor and formed the Kuzuh duplex.

Introduction

The origin of radiolarian bedded chert formations in orogenic belts has been debated for a long time. Debates have centered on the depositional environments of bedded chert, mechanism of sedimentation, diagenesis, and tectonic implications (Jenkyns and Winterer, 1982; Iijima et al., 1983, 1985). Radiolarian chert is widespread in the Upper Paleozoic and Mesozoic geosynclinal deposits of the Japanese Islands associated with or without submarine mafic volcanics (Iijima and Utada, 1983; Igo and Koike, 1983). From sedimentological, petrological, and geochemical evidence, Iijima et al. (1978, 1985) and Matsumoto and Iijima (1983) concluded that most Mesozoic radiolarian bedded cherts in the Inner Zone of the Chichibu geosyncline accumulated in the central part of an elongate marginal sea under the influence of nearby landmasses. The authors also concluded that the rhythmic bedding, composed of radiolarian chert beds and Radiolaria-poor or Radiolaria-free shale partings, resulted from periodic (10^3 to 10^4 years) supply of distal mud turbidites into

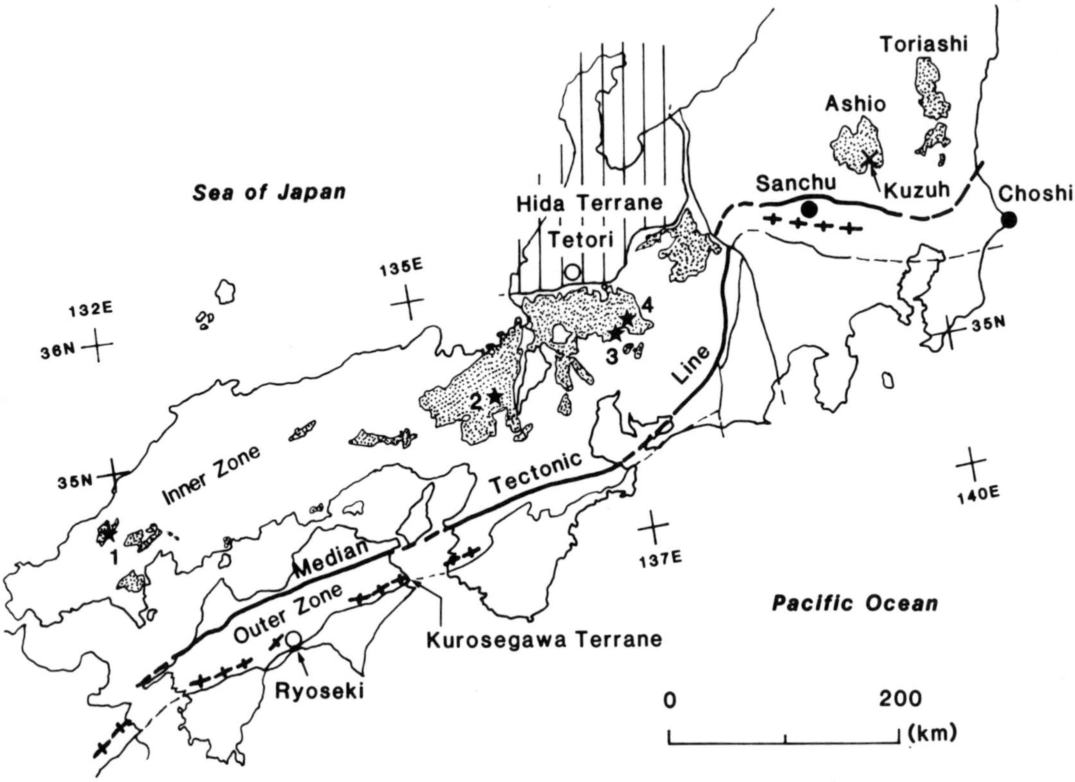

FIGURE 11.1. Index and geotectonic map of parts of Japan. ★ 1, Nichihara; ★ 2, Hozu; ★ 3, Inuyama; ★ 4, Hisui. •, Localities of *Brachyoxylon* excepting Kuzuh which is marked with *X*.

an area of slow and constant deposition of radiolarian skeletons.

Petrified *Brachyoxylon* wood was discovered in Lower Jurassic radiolarian bedded chert of the Adoyama Chert Formation. The wood locality is in the Karasawa mine, run by the Sumitomo Cement Company, and located in the Kuzuh area in the southeast of the Ashio Mountains, northern Kanto, central Honshu (Fig. 11.1). This is the first description of its mode of occurrence. Sedimentary laminae bend around the petrified log because of differential compaction, which allows us to estimate a compaction ratio for the radiolarian bedded chert. The depositional environment of the Adoyama Chert is discussed based on these new findings.

Geological Setting

Regional geological maps including the Kuzuh area were published by the Geological Survey of Japan (Fujimoto, 1961) and Tochigi-ken (Watanabe et al., 1957). However, the following geologic outline is mainly revised from Yanagimoto (1973) and Iijima et al. (1978), because the geologic ages of the noncarbonate rocks were not known in the older works.

The Kuzuh area is located in the southeastern part of the Ashio Mountains, where Upper Paleozoic and Mesozoic strata of the Inner Zone of the Chichibu geosyncline are widespread. The area comprises several imbricated thrust sheets, here called the Kuzuh duplex, which form, as a whole, an open SW-plunging syncline (Fig. 11.2). The stratigraphic sequence is well established within each thrust sheet, which vary in thickness from 0.3 to 2 km. Based on lithostratigraphic associations, the thrust sheets are grouped into carbonate-dominated sheets and chert-dominated sheets (Table 11.1). The carbonate-dominated sheets consist largely of Middle Permian (*Parafusulina* and *Neoschwagerian* zones) fusulinid-crinoidal limestone and dolostone of the Nabeyama Formation, which conformably overlies submarine basaltic tuff and lava of the Izuru Formation. The Nabeyama Formation represents an upward-shallowing sequence going from a lower basinal facies, to a middle dolostone, and finally to an upper shallow-water facies.

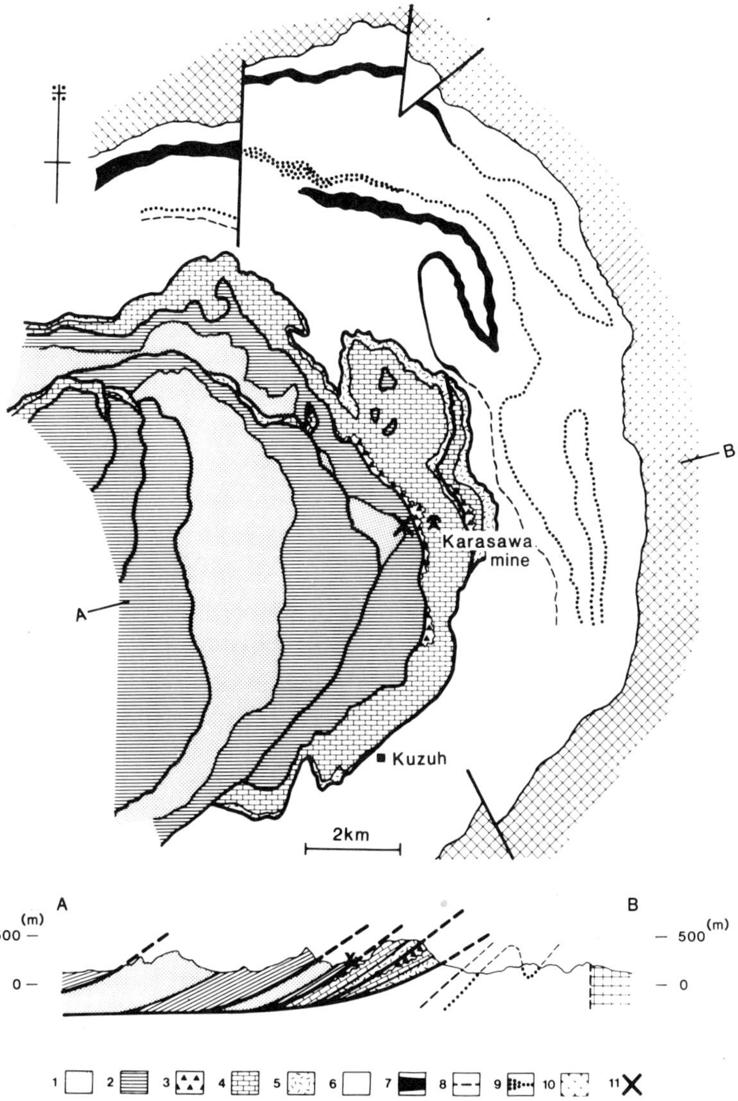

FIGURE 11.2. Geological map and profile of the Kuzuh area, showing the Kuzuh duplex and the "autochthon" (slightly modified from Iijima et al., 1978). 1, Maki Sandstone; 2, Adoyama Chert; 3, Karasawa Formation (dominantly radiolarian mudrock, sandstone, and chert); 4, Nabeyama Limestone, 5, Izuru Greenstone; 6, Aisawa Formation (dominantly sandstone, mudstone, and chert); 7, chert slump bed; 8, siliceous mudstone; 9, chert breccia; 10, Nakatsuma Chert; 11, locality of *Brachyoxylon*. A–B, position of cross-section.

The lower basinal facies is represented by thick-bedded, dark-grey, wackestone turbidites composed of fragments of fusulinids and crinoid stems in an argillaceous micrite matrix. The middle dolostone is of a mixed-meteoric-marine water type related to a slightly angular unconformity between the Nabeyama and Karasawa Formations (Matsuda and Iijima, 1985). The upper, shallow-water facies is made up of light gray, massive or cross-bedded, biocalcarenite, and small coral bioherms. Variegated radiolarian mudrock with intercalations of arkosic sandstone, chert-lithic sandstone, and argillaceous chert of the Karasawa Formation of Middle Jurassic age (Sashida et al., 1982) covers the karst surface of the Nabeyama Formation. Cave- and channel-fill calcareous deposits that yield Early to Late Triassic (Scythian to Carnian) conodonts (Koike et al., 1974, occur in places between the Nabeyama and Karasawa Formations. Clasts from the channel-fill limestone breccia indicate that the Late Permian part of the Nabeyama Formation was denuded by the pre-Triassic unconformity because the clasts contain *Lepidolina* and *Reichelina* limestone (Igo and Igo, 1977) as well as *Parafusulina* limestone, dolostone, and black nodular chert derived from the Nabeyama Formation.

The chert-dominated thrust sheets comprise the Adoyama Chert Formation of Triassic to Early Jurassic age and the Maki Sandstone Formation; the latter appears to conformably overlie the former,

TABLE 11.1. Stratigraphic correlation in the Kuzuh duplex and autochthon.

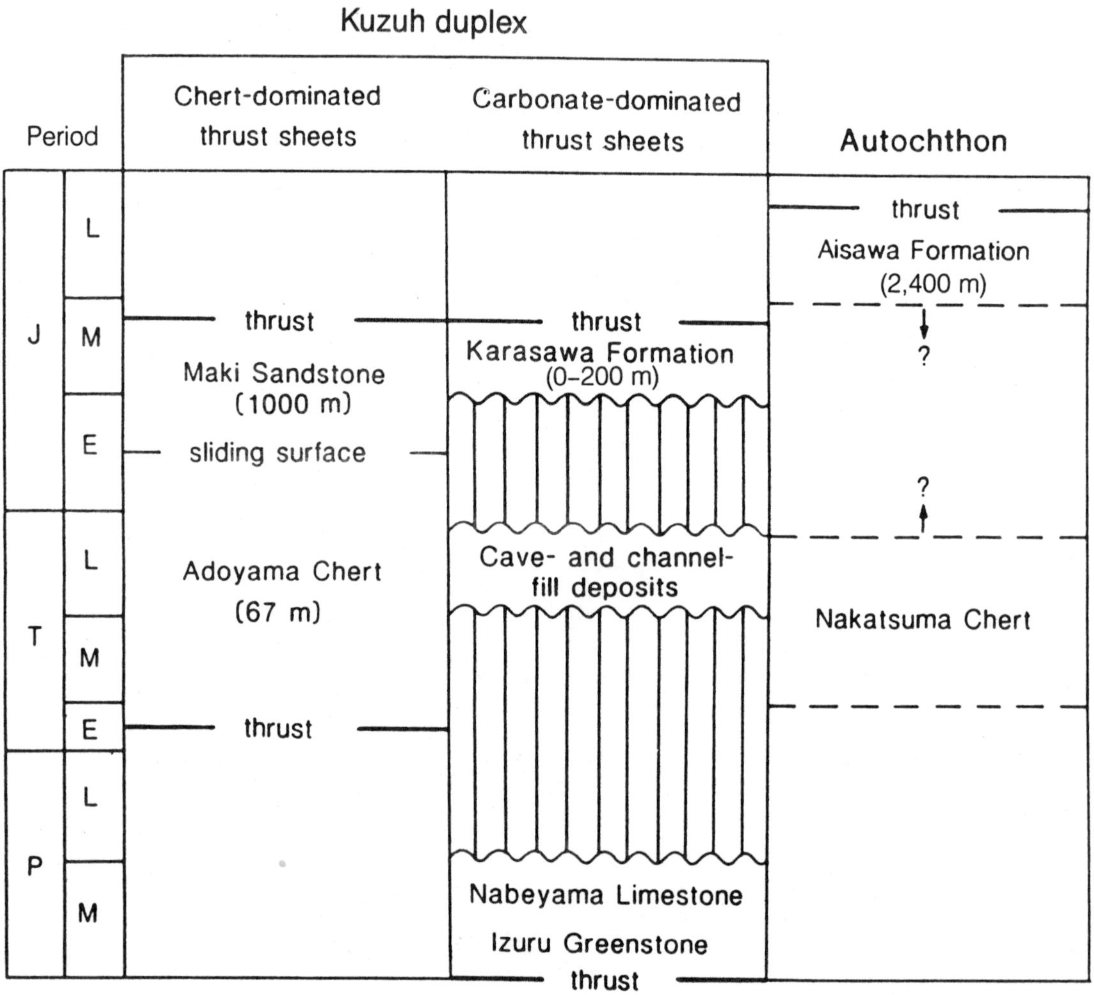

though a faultlike surface without a shear zone subparallel to the bedding plane was observed at an outcrop at the northwestern corner of the Karasawa mine (Iijima et al., 1978). The Maki Sandstone, at least 1000 m thick, is represented by light gray, thickly bedded, massive, medium- to very coarse-grained sandstone with some intercalations of cross-laminated, fine-grained sandstone and dark gray mudstone. These lithologies, and the absence of fossils except for carbonaceous fragments, suggest a shallow-water origin for the Maki Sandstone. The main components of the sandstone are quartz, graphic quartz-orthoclase, sericitized sodic plagioclase, rounded grains of granitic rock, meta-quartzite, pelitic hornfels, greenstone, and chert together with minor to rare biotite, muscovite, chlorite, tourmaline, and garnet. Moreover, angular intraformational fragments of chert, spicule-radiolarian siliceous shale, and silty shale are abundant in some very coarse-grained sandstone. Sandstone of the Karasawa Formation in the carbonate-dominated thrust sheets possesses the same components as the Maki Sandstone, suggesting that the two formations had the same provenance and are probably of the same geologic age.

The Kuzuh duplex was thrust over the Aisawa Formation, which consists mainly of massive, bioturbated, rippled, or laminated sandstone and some mudstone with frequent intercalations of slide-produced layers of Upper Triassic spicule-rich bedded

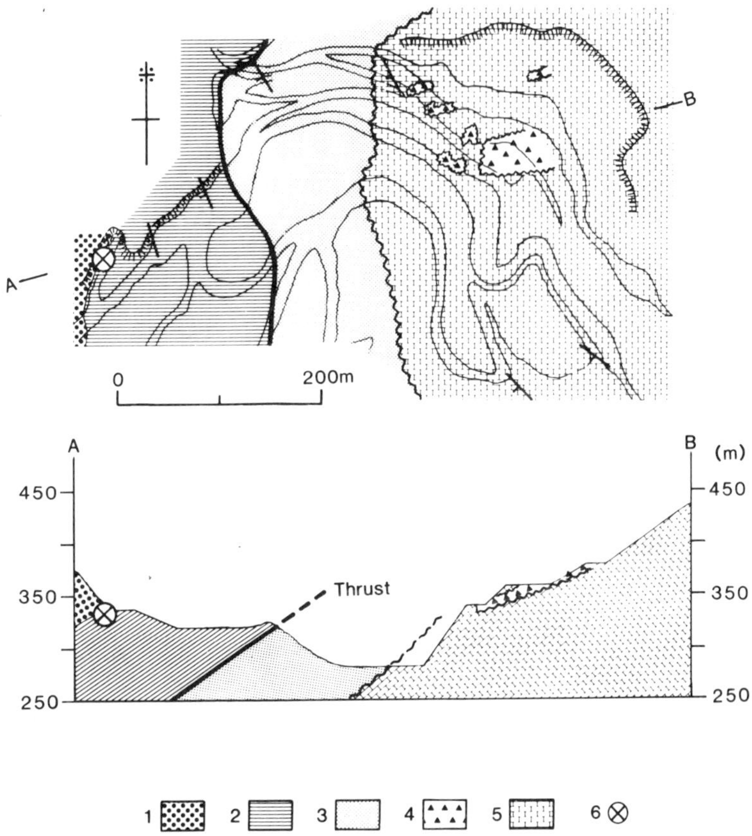

FIGURE 11.3. Geological map of the northern part of the Karasawa mine. 1, Maki Sandstone; 2, Adoyama Chert; 3, Karasawa Formation; 4, cave- and channel-fill deposits; 5, Nabeyama Limestone; 6, locality of *Brachyoxylon*. A–B, position of cross-section.

chert and slump-produced chert breccia. The Aisawa Formation, about 2400 m thick, is characterized by shallow-water micaceous sandstone and is likely correlative with the Late Jurassic *Perisphinctid*-bearing micaceous sandstone and mudstone of the Mashiko area in the Toriashi Mountains (Suzuki and Sato, 1972), 45 km east of Kuzuh; but it has not yielded any index fossils. Gray bedded chert of the Triassic Nakatsuma Chert underlies the Aisawa Formation, though the stratigraphic relation has not been definitely determined. Many stratabound manganese ore deposits are associated with the bedded chert.

Adoyama Chert Section of the Karasawa Mine

Two thrust sheets of the Kuzuh duplex appear in the northern part of the Karasawa limestone mine of the Sumitomo Cement Company: the upper is a chert-dominated sheet, and the lower a carbonate-dominated sheet (Fig. 11.3). A section of the Adoyama Chert containing the silicified wood was measured at the road-cut at the northwest corner of the mine. The measured chert section has a thickness of 67 m. Its bottom is cut by a thrust with a 5-m-wide shear zone between the two thrust sheets, whereas it is overlain by dark gray mudstone of the Maki Sandstone with a contact consisting of a minor fault without a shear zone and subparallel to the bedding plane. Minor faults of the same character as that of the upper fault are frequently seen within the chert section (Fig. 11.4). Several sections consisting of a few tens of chert beds show minor intraformational folding. The crests of anticlines of the intraformational folds are cut by the overlying shale parting, suggesting that the minor folds were produced by small-scale, contemporaneous, submarine slumps. Most axial planes of the minor folds are parallel to each other and overturned toward the northeast, indicating a paleoslope facing northeast (Fig. 11.5). Moreover, the minor faults are also probably caused by small-scale, contemporaneous, submarine slides.

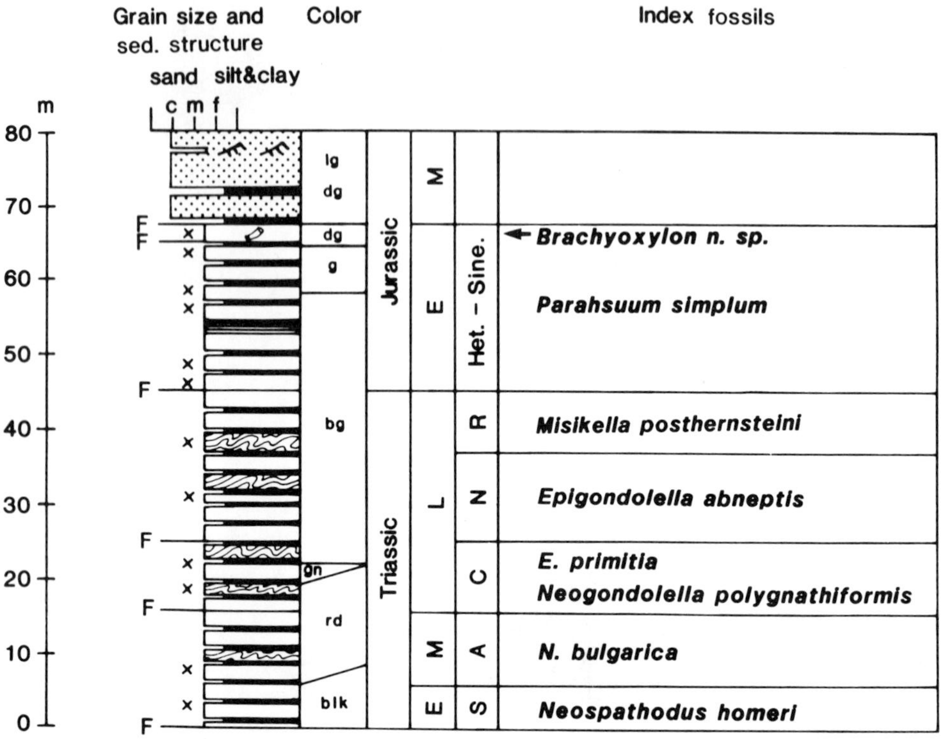

FIGURE 11.4. Columnar section of the Adoyama Chert in the Karasawa mine. X, Horizons of index fossils; F, fault; c, coarse-grained; m, medium-grained; f, fine-grained; dg, dark gray; lg, light gray; g, gray; bg, bluish gray; gn, green; rd, red; blk, black.

In spite of the frequent occurrence of submarine slump folds and slides, the measured chert section as a whole shows a uniform stratigraphic sequence form Early Triassic (Scythian) to Early Jurassic age, except for the late Middle Triassic (Ladinian), which was removed by a fault lacking a shear zone and subparallel to the bedding (Fig. 11.4). The *Parahsuum simplum* radiolarian assemblage in the upper part of the section is considered to correspond to the Hettangian-Sinemurian and possibly to the early Pliensbachian (Yao, 1986).

The Adoyama Chert consists of evenly bedded, rhythmic alternation of chert beds, 0.6 to 13.8 cm thick, averaging 3.9 cm, and shale partings, less than 1 cm thick (Iijima et al., 1978). Colors of associated chert beds and shale partings are commonly similar. The Karasawa chert section changes in color at different stratigraphic horizons. Black, due to carbonaceous matter and pyrite, occurs in the basal part: purplish to brownish red, due to hematite, occurs in the lower part: bluish-gray occurs in the middle part; gray occurs in the upper part; and dark gray due to carbonaceous matter occurs in the uppermost part of the section where the silicified wood was entombed. Chert beds are composed of radiolarian skeletons with appreciable amounts of spines and spicules. All siliceous organic remains were transformed into and cemented by microquartz. Most chert beds have the triple-layered type of layering (Iijima et al., 1978), which consists of a central, massive, more siliceous layer sandwiched by the lower and upper, clay-rich, less siliceous layers. The main constituents of the shale partings are illite, chlorite, and quartz. Radiolarian skeletons are rare, if found at all in the shale partings, and are as well preserved as those in associated chert beds. Microstylolites subparallel to the bedding are common in the clayey, less siliceous layers of the chert beds and result from pressure solution.

Occurrence of Silicified Wood in the Adoyama Chert

Silicified wood was found in the uppermost, Early Jurassic part of the measured chert section, 20 cm below the base of the Maki Sandstone. The wood is entombed in dark gray, radiolarian-bedded chert

(Fig. 11.6). The silicified wood rests on a shale parting, and its cross section is exposed. Its black, hard, massive appearance resembles the surrounding carbonaceous chert, and it was not recognized as wood until examination of thin sections. The silicified wood together with the surrounding bedded chert was excavated as a large block with the appearance of a concretion (Fig. 11.6). The excavated silicified wood is 11 × 17 × 22 cm in size. Thin-section observation reveals that the wood texture is well preserved in a clear aggregate of microquartz (Fig. 11.7) and that it belongs to the genus *Brachyoxylon* of Protopinaceae; it possesses brachyoxylic bordered pits (Nishida, Chiba University, written communication, 1986). It is noteworthy that there is no evidence of tectonic deformation of the silicified wood. Similarly, radiolarian tests of the Adoyama Chert are not significantly deformed.

A sketch of a cross section of the excavated block is synthesized from the mosaic of 70 thin sections (Fig. 11.8). Well-preserved annual rings definitely demonstrate that the *Brachyoxylon* is not a whole log but is a broken fragment of the trunk or a large branch. The diameter of the trunk or branch is estimated to be at least about 60 cm from the curvature of the rings. The radial cracks in the wood are probably caused by dehydration shrinkage during burial and are filled with clear, length-fast chalcedonic quartz. The wood texture in the peripheral part of the log, especially along the microstylolitic margin, is entirely destroyed owing to recrystallization of silica to coarser-grained quartz than that found in the places where the texture is preserved.

There are four chert beds (Ch-1, -2, -3, -4) and three shale partings (Sh-1, -2, -3) in the excavated block (Fig. 11.8). The *Brachyoxylon* wood is contained in the Ch-3 bed. The wedge-shaped hollow in the upper left of the wood is filled with crushed silicified wood chips. The bundle of rings disrupted by the hollow have a larger curvature than the other outer rings and appear to have been stretched along with creation of the hollow at that time when the waterlogged, easily breakable wood rested on the sea bottom. Also, the crushed chips probably formed along with the creation of the hollow. Two small silicified wood fragments float in the middle part of the Ch-3 bed. This distribution of wood fragments strongly suggests that the large and small wood fragments did not settle vertically in the water column but were transported downslope by a submarine slump. If so, the Ch-3 bed bearing the wood fragments should be a submarine slump deposit.

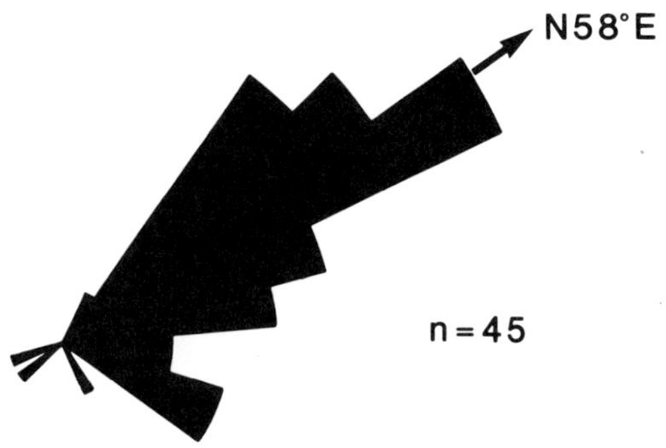

FIGURE 11.5. Azimuth diagram showing the directions of slopes estimated from orientation of axial planes of intraformational minor slump folds in the Adoyama Chert at the Karasawa section. Arrow indicates the mean direction.

The Ch-3 bed overlies the thin, discontinuous shale parting (Sh-2), which would have been disrupted by the slump. The Ch-3 bed underlies the black, carbonaceous siliceous claystone (A in Fig. 11.8) which has a quite different appearance from the other shale partings. The siliceous claystone (A) is massive and rich in minute carbonaceous particles which are likely derived from pulverized and decomposed *Brachyoxylon*; the bed A changes in thickness over short distances, covers the irregular surface of the wood, and is cut by small-scale faults oriented nearly perpendicular to the bedding plane; these faults, confined to the thicker part of the bed, were created during dehydration. The lower part of the bed A is locally fractured, and its fragments are scattered in the uppermost part of the Ch-3 bed (Fig. 11.8). We interpret the carbonaceous siliceous claystone (A) to have accumulated as the uppermost part of the Ch-3 slump bed. At that time, the top of the wood was covered with the carbonaceous clay (A) and would have formed a small high on the sea bottom.

The Ch-2 bed above the silicified wood shows vague parallel laminations which were probably caused by bottom currents of the upper flow regime. A cluster of small fragments of the carbonaceous siliceous claystone (A) are scattered within the Ch-2 bed just above the wedge-shaped hollow of the silicified wood (Fig. 11.8). This pattern can be explained if a bottom current flowing from right to left—i.e., southwesterly current—eroded the unconsolidated carbonaceous clay at the top of the small topographic high created by the wood and winnowed away the cluster of clay fragments from it. The orientation of the long axis of wood is NW-SE, perpendicular to

FIGURE 11.6. Photographs taken at the Karasawa section. (A) Radiolarian bedded chert of the Adoyama Chert. (B) Silicified *Brachyoxylon* wood entombed in the uppermost part of the Adoyama Chert. (C) Silicified wood partly excavated from the surrounding bedded chert.

FIGURE 11.6C.

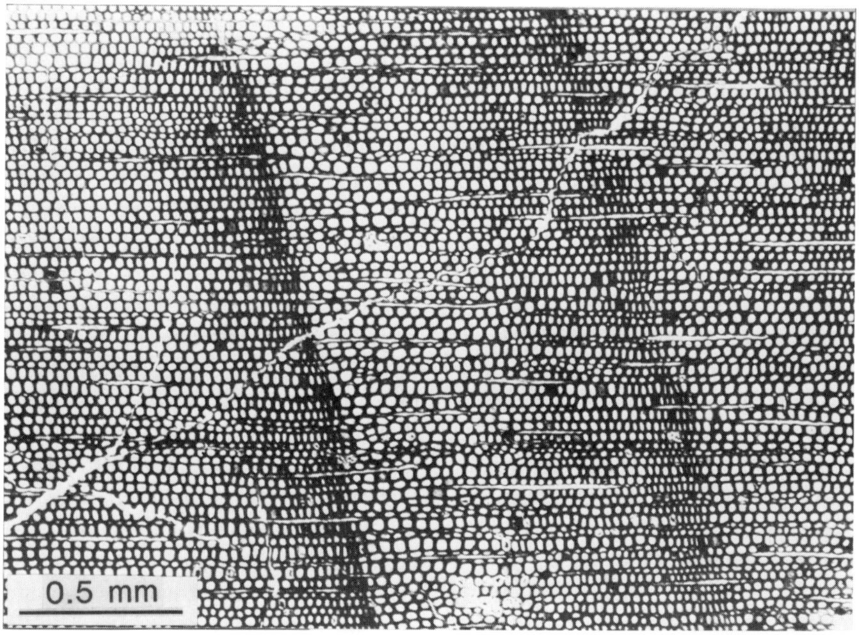

FIGURE 11.7. Photomicrograph of thin section showing a radial section of the silicified *Brachyoxylon* wood. Annual rings are recognized.

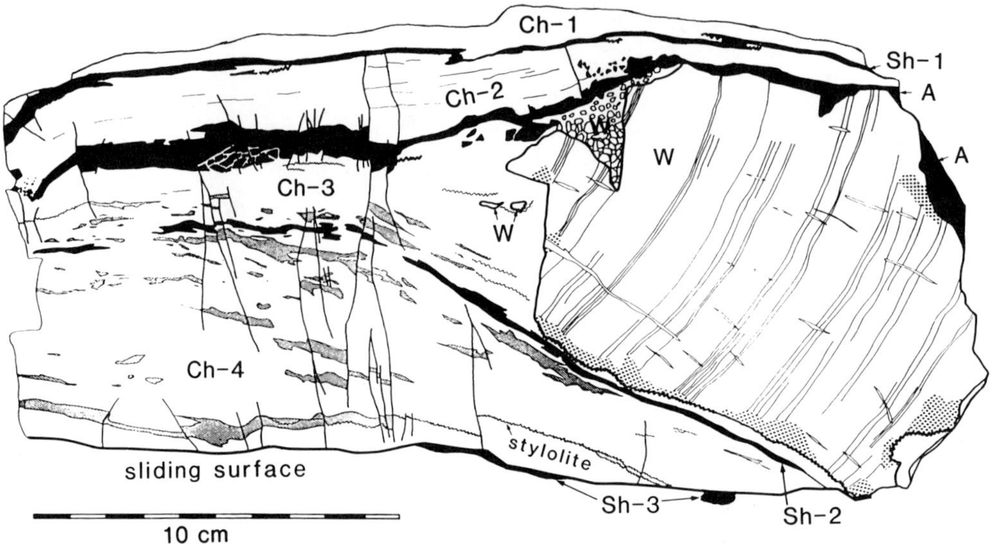

FIGURE 11.8. Sketch of a cross section synthesized from 70 thin sections of the *Brachyoxylon* wood-containing block of the Adoyama Chert. Note well-preserved annual rings. A = carbonaceous, siliceous claystone; Ch-1, -2, -3, -4 = chert beds; Gray stippled pattern = argillaceous laminae in Ch-3 and -4; Sh-1, -2, -3 = shale partings; W = silicified wood; Coarse stippled = completely destroyed wood texture by recrystallization of silica.

the paleoslope. The paleoslope indicated by the paleocurrent direction is consistent with the paleoslope estimated from the slump folds in the measured chert section (Figs. 11.4, 11.5). The Ch-2 bed filled the uneven surface of the slump bed (Ch-3 and A) so that the Sh-1 parting was deposited on a flat sea bottom.

The continuous Sh-3 parting is about 1 cm thick. There is a clear discordant relation between the Sh-3 and the Ch-4 +Sh-2 +Ch-3 beds (Fig. 11.8). The faultlike surface at the top of the Sh-3 parting is not associated with a shear zone or slickensides, suggesting that it is a sliding surface caused by a submarine slide before lithification of the sediment. In addition to the above-stated microfaults cutting the carbonaceous siliceous claystone (A), there are many minor, high-angle faults that are limited to the interval between the sliding surface and the base of the Sh-1 parting. Most, if not all, faults were likely caused by release of the stress around the silicified wood at the early stage of lithification.

Compaction Ratio of the Adoyama Chert

The petrified wood of the Adoyama Chert has a high silica content (99.3% SiO_2) as compared with its surrounding chert beds (93.9% SiO_2 based on an average of 8 samples; Table 11.2). The silica of the petrified wood undoubtedly originated from dissolved siliceous organic remains in the chert beds, migrated in pore waters of the porous siliceous deposits, and precipitated in and replaced the buried wood. Moreover, the well-preserved texture of the silicified tissue of the *Brachyoxylon* wood suggests that the silicification occurred at a relatively early stage in the burial history of the Adoyama Chert.

The shale partings Sh-1 and Sh-2 should have been originally horizontal and parallel to each other. In other words, the original thickness of the bed between the two shale partings was uniform. The present maximum thickness is 108 mm where the silicified wood occurs (a in Fig. 11.9), whereas the thickness decreases to 43 mm away from the wood (a′), where it remains uniform throughout the outcrop (Fig. 11.9). The difference in thickness is best explained by the bent laminae around the log indicating differential compaction. The silicified wood behaves like a concretion that formed at an early stage of burial diagenesis. The destruction of the wood texture at the margin of silicified wood was likely related to later stage silicification accompanied by microstylolitization (Fig. 11.8).

Thus, we can calculate a compaction ratio for this part of the Adoyama Chert from the difference in thickness of the interval between Sh-1 and Sh-2: a′/a = 43/108 = 0.40.* This value undoubtedly repre-

TABLE 11.2. Chemical composition of petrified wood, chert, and shale partings of the Adoyama Chert and mudrock of the Maki and Karasawa Formations.

	A	B	C	D	E	F	G	H
SiO_2 (%)	99.31	94.80	94.15	93.91	93.69	67.05	62.63	60.48
TiO_2	0.00	0.06	0.06	0.09	0.09	0.71	0.63	0.52
Al_2O_3	0.35	1.60	1.55	1.90	2.00	14.93	15.91	16.15
$Fe_2O_3^*$	0.05	0.26	0.40	0.90	0.90	4.93	6.83	4.95
MnO	0.00	0.00	0.00	0.01	0.01	0.02	0.15	0.07
MgO	0.00	0.17	0.20	0.30	0.31	1.95	3.14	1.72
CaO	0.00	0.17	0.16	0.07	0.01	0.16	1.04	1.78
Na_2O	0.00	0.06	0.25	0.05	0.02	0.01	1.30	0.00
K_2O	0.00	0.39	0.32	0.44	0.46	3.94	2.91	3.47
P_2O_5	0.00	0.11	0.00	0.01	0.00	0.02	0.17	0.64
Total	99.71	97.62	97.09	97.68	97.49	93.72	94.71	89.78
Ba (ppm)	82	217	260	239	441	789	862	1539
Cr	24	64	49	28	26	73	70	49
Cu	0	18	19	49	31	121	94	77
Mn	0	113	116	116	170	227	1139	598
Ni	0	16	0	10	2	54	66	101
Rb	8	19	26	33	36	163	142	144
Sr	36	50	85	91	70	78	107	117
V	9	0	17	21	23	111	101	96
Zn	0	0	17	25	38	159	152	160
Zr	0	8	0	6	3	114	134	158

X-ray fluorescent analysis by Y. Kakuwa. The correction is after Matsumoto and Urabe (1980).
*Total iron.
Key: A, petrified wood; B, chert bed Ch-3 containing the petrified wood; C, chert bed Ch-1; D, Early Jurassic chert, average of eight samples from the Karasawa section; E, Triassic chert, average of 16 samples from the Karasawa section; F, shale partings, average of five samples from the Karasawa section; G, variegated mudrock of the Karasawa Formation, average of two samples from the Karasawa mine; H, Mudstone at the base of the Maki Formation from the Karasawa section.

sents the minimum compaction of the bedded chert, because the thickest interval (a) does not necessarily show the initial thickness of the porous siliceous organic sediment but is somewhat thinned by both (1) the stylolite on the bottom of the silicified wood and (2) the compaction of the Ch-2 bed. Although the burial depth at which the wood has been silicified cannot be estimated, the apparently undeformed, well-preserved texture of the *Brachyoxylon* indicates an early stage of silicification at a relatively shallow depth. In Neogene tuffaceous strata of Japan, silicification of wood was completed in zone II (clinoptilolite–mordenite–low cristobalite) of the zeolitic burial diagenesis of silicic tuff (Iijima, 1988) and in the opal–CT diagenetic zone of siliceous deposits (Tada and Iijima, 1985). Consequently, if this pattern was the same in the Jurassic, the silicified *Brachyoxylon* wood would not have undergone significant physical compaction. If the thinning due to (1) and (2) were negligible, the Early Jurassic spicule-radiolarian sediment would have possessed a porosity of 60%. However, this value seems to be too small compared with over 80% porosity for modern diatomaceous earths. It is reasonable to assume that the initial porosity of the spicule-radiolarian sediment was about 80%. If so, the compaction ratio of the quartzose chert (porosity = 0%) to the initial sediment is 0.2, provided that additional silica cementation did not occur. This value, therefore, represents the maximum compaction of the bedded chert.

We can determine another compaction ratio from the chert bed, Ch-4. An argillaceous, Radiolaria-rich, parallel lamina is persistent in the lower part of the Ch-4 bed, though the lamina is cut by some high-angle microfaults (Fig. 11.8). Near the base of the silicified wood, a stylolite marks the bottom of the parallel-lamina, until it consumes the lamina. The compaction ratio is obtained by dividing the shortest interval (b′) between the shale parting Sh-2 and the stylolite by the interval (b) between the Sh-2 and the parallel lamina: $b'/b = 21/54 = 0.39$ (Fig. 11.9). This value is essentially the same as the 0.40 ratio for Ch-3, the small difference possibly resulting from pressure solution represented by the stylolite.

*In another cross section, the ratio is 0.43 (Iijima, 1988).

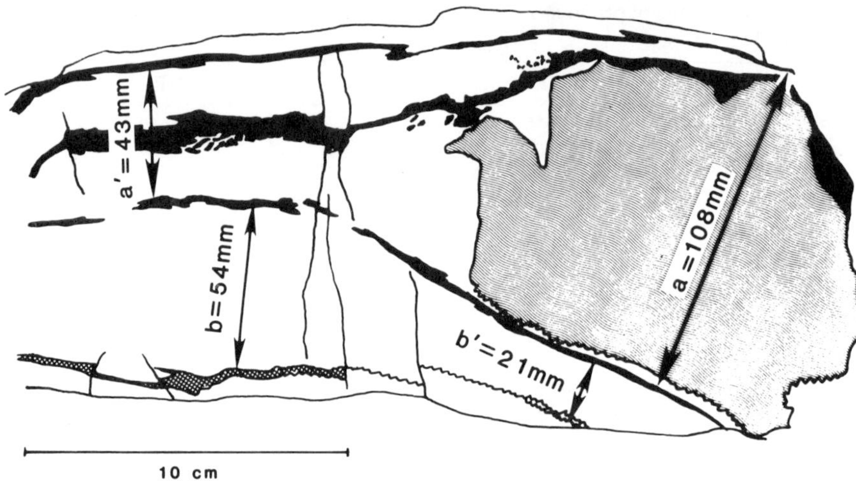

FIGURE 11.9. Schematic sketch drawn from Figure 11.8, showing the differential compaction of the bedded chert. Compaction ratio is calculated as a'/a = 0.40 and b'/b / 0.39.

Consequently, the thinning of the thickest interval (a) in Ch-3 is considered to be largely due to stylolitization at the base of the silicified wood. It is probable that the more massive character of the petrified wood, resulting from silicification at an early stage, accelerated the process of pressure solution.

The sedimentary overburden of the Adoyama Chert is, so far as is known, the Maki Sandstone with a thickness of 1,000 m. The silicification of the *Brachyoxylon* wood probably occurred during the deposition of the Maki Sandstone. The 1,000-m overburden alone is insufficient to cause the surrounding porous siliceous deposit to decrease by 60% of its original thickness and to become dense chert due to physical compaction. However, during the deposition of the Upper Jurassic Aisawa Formation, the Kuzuh duplex was emplaced over the Adoyama Chert. As a result, the *Brachyoxylon*-containing bedded chert was buried by the tectonic overburden with a thickness of 5,000 m or less, which represents the maximum burial depth (Fig. 11.10). The primary porosity of the bedded chert was then lost by physical and chemical processes.

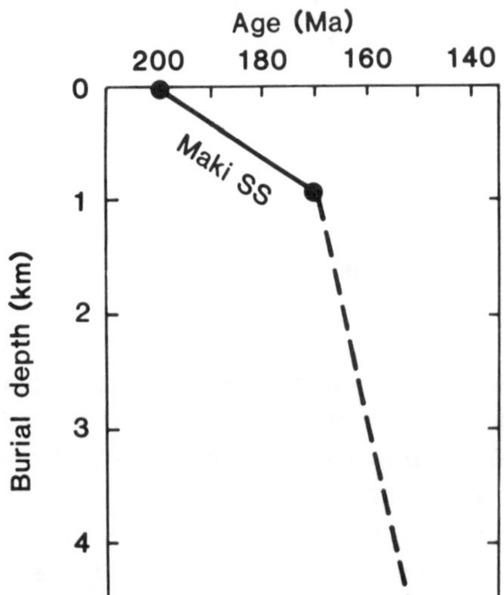

FIGURE 11.10. Schematic diagram showing the burial history of the *Brachyoxylon* wood in the Adoyama Chert.

Initial Thickness of the Adoyama Chert

The rhythmic bedding of the Adoyama Chert is a depositional feature. The chert beds and shale partings accumulated initially as spicule-radiolarian sediment and mud, respectively (Iijima et al., 1978). Most chert beds are of the triple-layered type caused by steady and slow sedimentation of siliceous organic remains (Iijima et al., 1978, 1985). In contrast, the *Brachyoxylon*-containing chert bed is a slump bed consisting mainly of siliceous organic remains. Moreover, its chemical composition cannot be distinguished from that of the Ch-1 bed, which is the common type of chert bed, or from the average composition of the Early Jurassic part of the

TABLE 11.3. Thickness of the Adoyama Chert corresponding to various compaction ratios.

Compaction ratios	Chert bed (average, cm)	Shale parting (average, cm)	Thickness (m)	Porosity (%)
0.2	3.9	1.0	67	0
0.4	9.8	2.5	168	?
1.0 (soft sediment)	19.5	5.0	335	80

Adoyama Chert (Table 11.2). Thus, we believe that the compaction ratio, 0.40, obtained from the differential compaction of the Ch-3 bed can be applied to the other chert beds in the section.

The initial thickness of the Adoyama Chert can be estimated by using the compaction ratio obtained for the Ch-3 bed (Table 11.3). The chert beds and shale partings are assumed to have the same porosity-versus-burial-depth relation. The chert beds and shale partings of the Adoyama Chert decrease in thickness to approximately one-fifth of the initial thickness; i.e., one chert-shale couplet was initially composed of a 19.5-cm-thick siliceous bed and a <5-cm-thick mud bed, on average. This translates into a total thickness for the 67-m-thick Adoyama Chert of 335 m in the unconsolidated state.

Sedimentation Rate of the Adoyama Chert

The Adoyama Chert in the Karasawa section has a total thickness of 67 m and covers the time span from Scythian (Early Triassic) to Sinemurian (Early Jurassic) excluding the Ladinian. This time span corresponds to 38 m.y. (Harland et al., 1982). The compacted average sedimentation rate of the Adoyama Chert is 1.76 m/m.y., which translates into 8.8 m/my uncompacted rate using the compaction ratio of 0.20. The Karasawa section is, however, cut by many low-angle faults, and most zones defined by fossil assemblages are bounded by them (Fig. 11.4). Therefore, sedimentation rates obtained by the stratigraphic method may not be accurate.

Matsumoto and Iijima (1983) devised a geochemical method to obtain sedimentation rates of siliceous deposits on the basis of a good correlation between the MnO/Al_2O_3 ratio and sedimentation rates of modern Pacific deep-sea sediment. This method contains the following assumptions: (1) Hydrogenous MnO in pelagic and hemipelagic sediment is primarily adsorbed from seawater, so that its content corresponds to the duration of time of interaction between the sediment and seawater; (2) the adsorbed MnO is immobile during diagenesis; and (3) the adsorption rate is almost constant in the normal marine environment throughout geologic time. Moreover, to use this technique, a reasonable compaction ratio must be estimated which is commonly difficult.

Chemical analysis of the silicified wood and surrounding chert beds indicates the immobility of MnO adsorbed on these siliceous deposits. Hydrogenous elements, such as Cu, Mn, and Ni are not detected in the silicified wood (Table 11.2). Above all, Mn is not detected in the silicified wood, whereas it is commonly more than 100 ppm in the surrounding chert beds. These chemical trends indicate that the hydrogenous elements were adsorbed by the sediment before the silicification of wood and that the adsorbed elements may have been immobile during burial diagenesis.

Biostratigraphy of bedded cherts of the Chichibu geosyncline has developed over the past decade (Igo and Koike, 1983). Based on the stratigraphic method, Kakuwa (MS) calculated average sedimentation rates of each series of Triassic bedded chert in the Nichihara, Hozu, Inuyama, and Hisui areas in the Inner Zone of southwest Japan belonging to the same tectonic terrane as the Kuzuh area. Furthermore, he constructs a diagram showing the relationship between the average sedimentation rate (Y) and average MnO/Al_2O_3 ratio (X) of the bedded chert and obtains a regression line of $\log_2 Y = -2.27 - 1.59 \log X$ ($r = -0.70$) (Fig. 11.11). On the diagram, we can obtain average sedimentation rates of each series of the Adoyama Chert at the Karasawa section by plotting the mean MnO/Al_2O_3 ratios on the regression line. The compacted sedimentation rates are 1.6 m/m.y. (standard error +0.7, −0.5) for the Carnian and 2.4 m/m.y. (SE +1.1, −0.7) for the Early Jurassic, and average 2.1 m/m.y. (SE +1.0, −0.7). These compacted sedimentation rates cor-

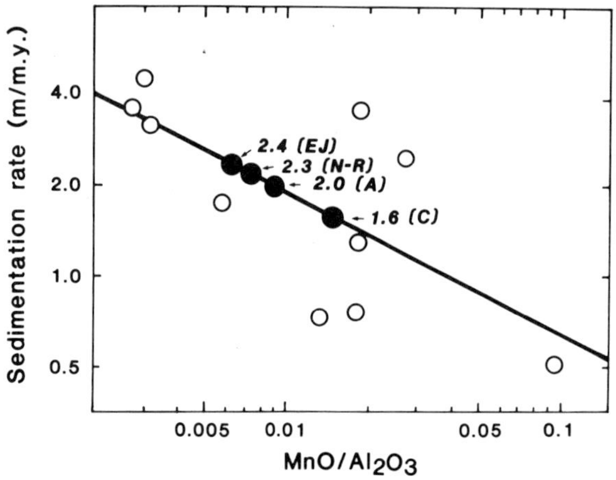

FIGURE 11.11. Diagram showing relationship between the average sedimentation rate and average MnO/Al_2O_3 ratio of some selected Triassic bedded chert sections from the Inner Zone of southwest Japan (Kakuwa, MS). The sedimentation rate of the Adoyama Chert at the Karasawa section is obtained from the diagram; (C), Carnian; (A), Anisian; (N-R), Norian and Rhaetian; (EJ), Early Jurassic.

respond to the range 8.0 to 12.0 m/m.y., average 10.5 m/m.y. (SE +5.0, −3.5) for the uncompacted depositional rates. In contrast, the geochemical method (Matsumoto and Iijima, 1983) produces an average sedimentation rate of the Adoyama Chert of 16 m/m.y. (SE +4.9, −3.8). However, 16 m/m.y. is not inconsistent with the average 10.5 m/m.y. and falls within the limit of error. The consistent results of the geochemical method with the stratigraphic method suggests that the concentration of Mn in the Mesozoic seawater is nearly equivalent to that in the present Pacific Ocean water.

The *Brachyoxylon*-bearing chert bed should have been laid down rapidly, because it is a slump deposit. Its MnO/Al_2O_3 rato is, however, indistinguishable from that of the other chert beds (see Table 11.2). This consistent composition is explained by the depositional setting, where the siliceous organic remains, which accumulated slowly on the upper slope or on the outer shelf, slumped down the slope together with the wood and were quickly deposited on the basin floor.

The MnO/Al_2O_3 ratios of five shale partings of the Adoyama Chert are within the limit of 0.61×10^{-3} to 4.07×10^{-3}, averaging 1.96×10^{-3} (Table 11.2). This mean value is much lower than 10.0×10^{-3} for the average of the chert beds and even lower than the 4.78×10^{-3} of a terrigenous mudstone of the Maki Sandstone. Consequently, most Mn of the shale partings appears to be of terrigenous origin, to which the sedimentation rate versus MnO/Al_2O_3 diagram (Fig. 11.11) is not applicable. Iijima et al. (1978, 1985) considered that the shale partings rapidly and periodically settled from bottom currents or, more likely, from distal turbidity currents, whereas the chert beds were slowly and continuously deposited.

The Middle Jurassic variegated, Radiolaria-bearing mudrock of the Karasawa Formation, covering the karst surface of the Nabeyama Limestone, is manganiferous with the MnO/Al_2O_3 ratio of 9.24×10^{-3} based on the average of two samples. The sedimentation rate of 2.0 m/m.y. (SE +0.9, −0.6) inferred from the ratio appears to be inconsistent with both the sedimentation rate estimated from the stratigraphic method and the frequent intercalations of arkosic sandstone. The variegated mudrock was derived from the weathered material (terra rossa) on the Nabeyama Limestone (Yanagimoto, 1973; Iijima et al., 1978); hence the Mn of the variegated mudrock was also of terrigenous origin.

Depositional Environment and Displacement of the Adoyama Chert

A marginal sea occurred in the southeast periphery of the Sino-Korean continental plate and was the site of deposition of the upper Paleozoic to Jurassic strata of the Chichibu geosyncline of the Japanese Islands (Kimura, 1983). The pre-Silurian metamorphic and plutonic rocks in the Hida and Kurosegawa terranes are considered to represent segments of the continental plate on either side of the marginal sea (Fig. 11.1). Iijima et al. (1985) discussed the depositional mechanism of radiolarian bedded cherts, including the Adoyama Chert, of the Chichibu geosyncline and concluded that the bedded cherts accumulated in the marginal sea under the influence of nearby landmasses. The discovery of *Brachyoxylon* wood in the slump-produced chert bed of the Adoyama Chert supports their theory of nearby landmasses.

The frequent intercalations of synsedimentary chert slump beds in the Adoyama Chert (Fig. 11.4) suggest a site of deposition at the foot of the northeast-facing slope. Sedimentary structures indicate that the *Brachyoxylon* wood was derived from a landmass on the southwest side of the basin. The Kurosegawa terrane bounding the southern margin

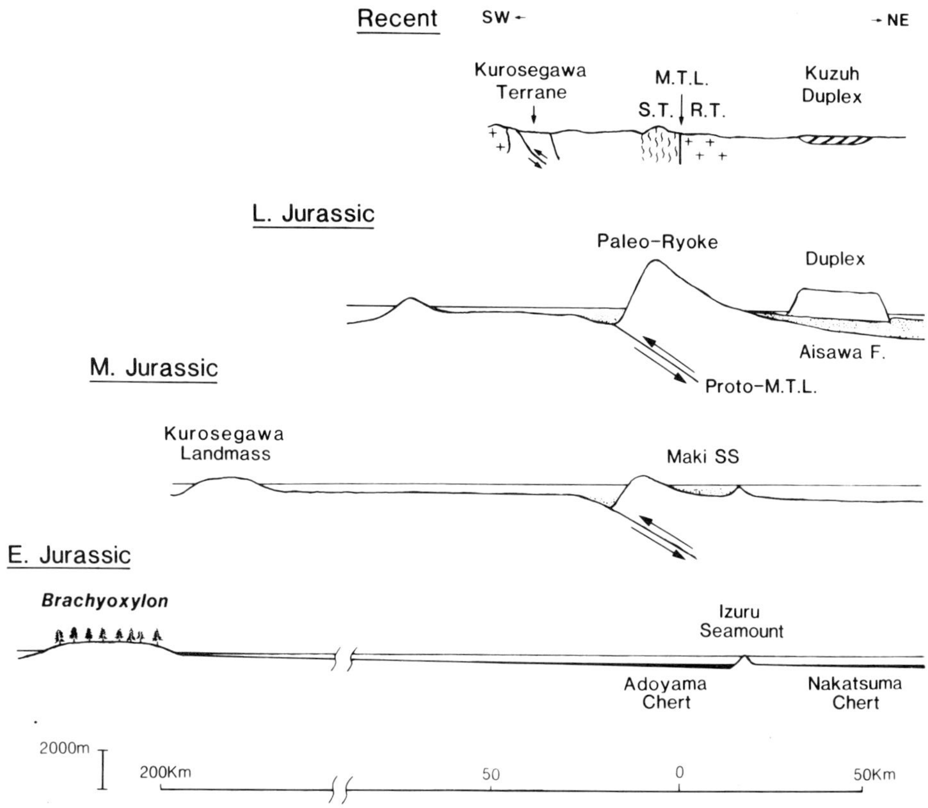

FIGURE 11.12. Schematic NE-SW profiles between the Kuzuh area and Kurosegawa terrane from various geologic ages. M.T.L., Median Tectonic Line; S.T., Sanbagawa metamorphic terrane, R.T., Ryoke metamorphic terrane.

of the marginal sea seems to be the most probable landmass during Early Jurassic time (Fig. 11.12).

The plan and profile of the Kuzuh duplex (Fig. 11.2) are similar to those of duplexes in orogenic belts such as the Appalachians and Caledonides (Boyer and Elliot, 1982) and indicate that the Kuzuh duplex was displaced northeastward by gravitational tectonics. The displacement occurred in Late Jurassic time, because the Kuzuh duplex overrode the shallow-water deposits of the upper part of the Aisawa Formation of Late Jurassic age. The lower part of the Aisawa Formation contains some slide and slump beds composed of Late Triassic bedded chert, chert breccia, and chert-lithic sandstone (Iijima et al., 1978). The systematic changes in thickness and grain-size distribution of the slide and slump deposits as well as the pattern of slump folds (Fig. 11.2) show that the paleoslope of the Late Jurassic Aisawa Formation also faced northeast.

The original position of the strata composing the Kuzuh duplex can be inferred. In Late Jurassic time, the embryonic geanticline of the Eo-Nippon cordillera (Kobayashi, 1941) is considered to have formed in the Paleo-Ryoke terrane on the north side of the proto-Median Tectonic Line (Kimura, 1980). It is therefore reasonable to suggest that the original position of the Kuzuh duplex was on the north side of the embryonic geanticline (Fig. 11.12). If so, the duplex had to be displaced as giant gravitational slides for approximately 40 to 60 km. Similar gravity slide blocks were reported in the Inuyama-Mugi area (Kano, 1979). According to Boyer and Elliot (1982), the lateral position of thrust sheets in such duplexes in orogenic belts is generally in accord with the source area. Consequently, the Adoyama Chert and overlying Maki Sandstone in the southern, chert-dominated thrust sheets must have been to the south of the Izuru Greenstone, Nabeyama Limestone, and Karasawa Formation in the northern, carbonate-dominated thrust sheets. In Triassic and Early Jurassic time, when the Adoyama Chert accumulated, the Nabeyama Limestone capped a basaltic seamount comprising the Izuru Greenstone, which topographically formed a shallow bank. Such basaltic seamounts formed a seamount chain which stretched parallel to the elongated marginal sea, as inferred from the distribution of upper Paleozoic greenstone and shallow-water limestone in the Inner Zone of southwest Japan. To the northeast of the bank, bedded cherts of the Nakatsuma Chert were also widespread. The *Brachyoxylon* wood drifted from the Kurosegawa landmass, settled on the outer shelf or on the upper slope where siliceous organic remains slowly accumulated, and then slumped down the slope into the basin where the Adoyama Chert was accumulating. It is difficult to estimate the distance between the Kurosegawa landmass and the place of deposition of the Adoyama Chert because of the complicated tectonic disturbance in the Outer Zone of the Chichibu geosyncline. In the Kanto district, the present geographic distance is about 30 to 50 km but was likely originally a few hundred kilometers or more.

It is also difficult to estimate the water depth of the Adoyama Chert. Iijima et al. (1978) postulated a relatively shallow sea, compared with an open oceanic basin, on the basis of the bedded cherts intercalated with shallow-water sandstone. The configuration of the marginal sea in Triassic and Early Jurassic time should have been uniform and relatively stable as inferred from the widespread distribution of contemporaneous, monotonous bedded cherts throughout the Chichibu geosyncline. Such an environment corresponds to the volcanically quiet nonspreading stage of the marginal sea that followed the active spreading and associated submarine basic volcanism characteristic of the Permo-Carboniferous time. Hence, the marginal sea would have had its largest lateral extent and its deepest water due to cooling and subsidence of the oceanic crust and a slow rate of sedimentation. Assuming that the slope was 0.5° on the average and the offshore distance 200 to 300 km, the water depth of the Adoyama Chert deposition would have been roughly 2,000 to 3,000 m.

There is another approach which provides a depth estimation that seems more probable. The Middle Jurassic (Bajocian to Bathonian) variegated radiolarian mudrock of the Karasawa Formation overlying the karst surface of the Nabeyama Limestone is interbedded with arkosic sandstone. The provenance of the sandstone is not the Izuru seamount or the Kurosegawa landmass, because bedded chert and siliceous mudstone continued to accumulate in the southwestern basin between the seamount chain and the landmass. Furthermore, the Maki Sandstone possesses the same components and thus the same provenance as the sandstone of the Karasawa Formation. Consequently the vast volume of coarse-grained sandstone of the Maki Sandstone had to be derived from the uplifted Paleo-Ryoke terrane and filled the newly formed small basin that had been the site of deposition of the Adoyama Chert until the arkosic material reached the top of the seamount (Fig. 11.12). The thickness of the Maki Sandstone is around 1,000 m, so the seamount should have stood around 1,000 m above the sea floor. It is, therefore, reasonable to think that the water depth of the Adoyama Chert was around 1,000 m (Fig. 11.12).

Kimura (1980) stated that the marginal sea closed in Late Jurassic time and that the Chichibu geosynclinal basin was divided into northern (inner) and southern (outer) basins by the uplift of its central part along the proto-Median Tectonic Line. In Early Cretaceous time, the topographic high grew to form the Eo-Nippon Cordillera in the Paleo-Ryoke terrane, which was subsequently removed by erosion (Kobayashi, 1941; Kimura, 1980). We generally agree with this geologic evolution. We would like to emphasize, however, that the uplift and appearance of a landmass in the Paleo-Ryoke terrane can be traced back to Middle Jurassic time and that the closing of the marginal sea was probably related to the northwest movement of the Kula plate, as suggested by Larson and Chase (1972) and Cooper et al. (1976).

In Late Jurassic time, the Paleo-Ryoke terrane was further uplifted, as the Kula plate continued to move northwestward. The Adoyama Chert and the overlying Maki Sandstone together with the upper part of the seamount were situated on the north side of the uplifted terrane and slid down the slope as far as the site of deposition of the Aisawa Formation, where they piled up in the form of a duplex (Fig. 11.12). Hence, the inner part of the Chichibu geosynclinal basin came to an end and became the site of Cretaceous magmatism and metamorphism.

There is another possible interpretation of these geologic relations: the radiolarian bedded chert together with oceanic crust migrated from a low-latitude, open-ocean setting via plate motion of the oceanic crust and was attached to the northern Sino-Korean continent (the Hida terrane) as accretionary prisms in a subduction zone, and the Kurosegawa terrane is an accreted continental fragment (e.g., Taira et al., 1981; Hirooka et al., 1983). In deep-sea sedimentary sections below the modern open-ocean floor, however, there are no radiolarian oozes or cherts equivalent to the radiolarian bedded cherts in orogenic belts (Hein and Karl, 1983; Iijima et al., 1985). Kimura (1979) stated that the Early Cretaceous Tetori and Ryoseki flora in the Hida and Kurosegawa terranes comprise quite different assemblages and that the Ryoseki flora indicates a tropical climate. However, Tanai (Hokkaido University, oral communication, 1987) does not recognize a definite distinction between the two floras. Paleobiogeographic discussions based on paleontological data are problematic and should not be used as conclusive evidence for the source of the terranes. *Brachyoxylon* wood has also been reported from Early Cretaceous strata in Shikoku and Sanriku as well as at Choshi and Sanchu in Kanto (Fig. 11.1); *Brachyoxylon nipponicum* Nishida from Choshi is most akin to the *Brachyoxylon* n. sp. from the Adoyama Chert (Nishida, Chiba University, written communication, 1986). So far as is known, all *Brachyoxylon* fossils are considered to have been derived from the Kurosegawa landmass and its environs. Northward migration on continental fragments of *Brachyoxylon* was likely prevented by the wide marginal sea in pre-Cretaceous time and by the Eo-Nippon cordillera in Cretaceous time. The low-latitude hypothesis of deposition of the bedded cherts depends strongly on low-inclination dips of paleomagnetism being retained in the bedded cherts. Nevertheless, inclination dips of fine-grained sedimentary rocks are suspected to be modified by extensive compaction during burial (Hall and Kodama, 1983; Anson and Kodama, 1984), so the inclination dips in such compacted cherts may be unreliable for estimating paleolatitudes (Iijima, 1986). Even if the estimated low latitudes were correct, it is not necessary to presume long-distance migration, because the Sino-Korean continent was situated at low latitudes in Paleozoic time (McElhinny et al., 1981). Bedded chert deposited at the low latitude could have moved northward as part of the Hida and Kurosegawa terranes.

Acknowledgments. We are grateful to M. Nishida of Chiba University for identification of the silicified wood and for valuable comments. Thanks are due all participants in the Third International Conference on Siliceous Deposits held in Dubrovnik—above all, J.R. Hein, U.S. Geological Survey; J. Obradović, University of Belgrade; and R. Siever, Harvard University, for their discussions and comments. T. Tanai of Hokkaido University gave us invaluable comments on the paleobotanical significance of the *Brachyoxylon*. S. Yoshida and R. Tada provided invaluable discussion and read the manuscript. Also, J.R. Hein read the manuscript critically. We offer our gratitude to these people.

References

Anson GL, Kodama KP (1984) The effects of compaction on a synthetic PDRM. EOS 65:862.

Boyer SE, Elliot D (1982) Thrust systems. Bulletin of American Association of Petroleum Geologists 66: 1196–1230.

Cooper AK, Scholl DW, Marlow MS (1976) Plate tectonic model for the evolution of the eastern Bering Sea basin. Bulletin of Geological Society of America 87:1119–1126.

Fujimoto H (1961) Geology of the Tochigi District. With Geological Sheet Map at 1:50,000. Geological Survey of Japan, Tsukuba, pp 1–62.

Hall FR, Kodama, KP (1983) The effects of consolidometer tests on the ARM of a wet magnetic sediment. EOS 64:219.

Hein JR, Karl SM (1983) Comparison between open-ocean and continental margin chert sequences. In: Iijima A, Hein JR, Siever R (eds) Developments in Sedimentology 36, Elsevier, Amsterdam, pp 25–44.

Harland WB, Cox AV, Llewellyn PG, Pickton CAG, Smith AG, Walters R (1982) A Geologic Time Scale. Cambridge University Press, Cambridge, U.K.

Hirooka K, Nakajima T, Sakai H, Date T, Nittamachi K, Hattori I (1983) Accretion tectonics inferred from paleomagnetic measurements of Paleozoic and Mesozoic rocks in central Japan. In: Hashimoto M, Uyeda S (eds) Accretion Tectonics in the Circum-Pacific Regions. Terra Scientific, Tokyo, pp 179–194.

Igo H, Igo H (1977) Upper Permian Fusulinaceans contained in the pebbles of the basal conglomerate of the Adoyama Formation, Kuzu, Tochigi Prefecture, Japan. Transaction of Proceedings of the Palaeontological Society of Japan (N.S.) 106:89–99.

Igo H, Koike T (1983) Conodont biostratigraphy of cherts in the Japanese Islands. In: Iijima A, Hein JR, Siever R (eds) Developments in Sedimentology 36, Elsevier, Amsterdam, pp 65–78.

Iijima A (1986) Genesis of ribbon cherts in orogenic belts. Journal of Geography 93:481–487 (JE).*

Iijima A (1988) Diagenetic transformations of minerals as exemplified by zeolites and silica minerals—a Japanese view. In: Chilingarian GV, Wolf KH (eds) Developments in Sedimentology, Elsevier, Amsterdam (in press).

Iijima A, Utada M (1983) Recent developments in the sedimentology of siliceous deposits in Japan. In: Iijima A, Hein JR, Siever R (eds) Developments in Sedimentology 36, Elsevier, Amsterdam, pp 46–64.

Iijima A, Kakuwa Y, Yamazaki K, Yanagimoto Y (1978) Shallow-sea, organic origin of the Triassic bedded chert in central Japan. Journal of the Faculty of Science, the University of Tokyo, Section II, 19:369–400.

Iijima A, Hein JR, Siever R (1983) Siliceous Deposits in the Pacific Region, Developments in Sedimentology 36, Elsevier, Amsterdam, 472 pp.

Iijima A, Utada M, Matsumoto R, Kakuwa Y, Watanabe Y, Matsuda H (1984) Discovery and its significance of silicified wood from the Adoyama Chert. Abstract Paper of the 91st Annual Meeting of the Geological Society of Japan, Tokyo, p 279 (J).

Iijima A, Matsumoto R, Tada R (1985) Mechanism of sedimentation of rhythmically bedded chert. Sedimentary Geology 41:221–233.

Jenkyns HC, Winterer EL (1982) Paleoceanography of Mesozoic ribbon radiolarites. Earth & Planetary Science Letters 60:351–375.

Kano K (1979) Giant *Deckenpaket* and olistostrome in the eastern Mino district, central Japan. Journal of the Faculty of Science, the University of Tokyo, Section II, 20:31–59.

Kimura T (1979) Late Mesozoic paleofloristic provinces in East Asia. Proceedings of Japan Academy, Series B, 55:425–430.

Kimura T (1980, 1983) Japanese Islands—The Geological History—II-2, III-1. Kokin Shoin, Tokyo, pp 579–1268 (J).

Kobayashi T (1941) The Sakawa orogenic cycle and its bearing on the origin of the Japanese Islands. Journal of the Faculty of Science, Imperial University of Tokyo, Section II, 5:219–578.

Koike T, Igo H, Kinoshita T (1974) Geological significance of the unconformity between the Permian Nabeyama and Triassic Adoyama formations in the vicinity of Kuzuu, Tochigi Prefecture. Journal of the Geological Society of Japan, 80:293–306 (JE).

Larson RL, Chase CG (1972) Late Mesozoic evolution of the western Pacific Ocean. Bulletin of the Geological Society of America 83:3627–3644.

Matsuda H, Iijima A (1985) Origin of the dolostone of the Nabeyama Formation—dolomitization and unconformity. Abstract Paper of the 92nd Annual Meeting of the Geological Society of Japan, Yamaguchi, p 252 (J).

Matsumoto R, Iijima A (1983) Chemical sedimentology of some Permo-Jurassic and Tertiary bedded cherts in central Honshu, Japan. In: Iijima A, Hein JR, Siever R (eds): Developments in Sedimentology 36, Elsevier, Amsterdam, pp 175–192.

Matsumoto R, Urabe T (1980) An automatic analysis of major elements in silicate rocks with X-ray fluorescence spectrometer using fused disc samples. The Journal of the Japanese Association of Mineralogists, Petrologists and Economic Geologists 75:272–278 (JE).

McElhinny MW, Embleton BJJ, Ma XH, Zhang ZK (1981) Fragmentation of Asia in Permian. Nature 293:212–216.

Sashida K, Igo H, Igo H, Takisawa S, Hisada K, Shibata T, Tsukada K, Nishimura H (1982) On the Jurassic radiolarian assemblages in the Kanto district. News of Osaka Micropaleontologists, (special volume) 5:51–66 (J).

Suzuki A, Sato T (1972) Discovery of Jurassic ammonite from Toriashi Mountains. Journal of the Geological Society of Japan 78:213–215 (J).

Tada R, Iijima A (1985) Petrology and diagenetic changes of Neogene siliceous sediments in northern Japan. Journal of Sedimentary Petrology 53:911–930.

Taira A, Saito Y, Hashimoto M (1981) The basic process of the formation of Japanese Islands. Kagaku 51:508–515 (J).

Watanabe T, Mukaiyama H, Kanehira K, Hamada T (1957) 1:50,000 Geological Map of the Ashio Mountains, Tochigi Prefecture, Utsunomiya.

Yanagimoto Y (1973) Stratigraphy and geological structure of the Paleozoic and Mesozoic formations in the vicinity of Kuzuu, Tochigi Prefecture. Journal of the Geological Society of Japan 79:441–451 (JE).

Yao A (1986) Geological age of Jurassic radiolarian zones in Japan and their international correlations. News of Osaka Micropalaeontologists (special volume) 7:63–74 (J).

---

*(J) In Japanese; (JE) in Japanese with English abstract.

12
Paleoenvironmental Implications of Alaskan Siliceous Deposits

Susan M. Karl

Overview

Biogenic siliceous deposits in Alaska originally accumulated in a variety of oceanic and continental margin settings. Oceanic environments are distinguished by the absence of any continental deposits. The continental margin environments are marine, but contain continentally derived sediments. Alaskan siliceous deposits can be grouped in terms of six lithologically defined facies associations that can be used to interpret primary depositional environments. (1) The chert/basalt/ultramafic rock association may represent an oceanic or marginal sea spreading center environment. (2) The chert/basalt association may represent deposition on various types of oceanic or marginal basin volcanic edifices. (3) The chert/basalt/andesite/graywacke association represents a marginal basin in the vicinity of a volcanic arc. (4) The chert/mafic to silicic volcanic rock/clastic rock association may represent a volcanic arc, a marginal rift or marginal transform basin. (5) The chert/carbonate/clastic rock association may represent a continental platform basin or shelf setting. (6) The chert/shale association may represent a restricted basin, or an outer shelf or slope environment.

Specific siliceous facies associations characterize certain periods of geologic time and certain regions in Alaska. These characteristic depositional sequences can be used to compare, and possibly relate, basin histories in tectonically disrupted areas. Lower Paleozoic cherts in Alaska that were deposited in continental margin settings tend to occur north of the Denali fault. On the other hand, Mesozoic volcanic arc-related associations occur throughout the state. The pattern of regional tectonic juxtaposition of siliceous oceanic deposits between contemporaneous continental margin deposits in northern, central, and southeastern Alaska provides insights for unravelling the complex tectonic processes that assembled Alaska.

Introduction

The sedimentology of biogenic siliceous deposits in the Alaska region provides a means of interpreting relations between tectonostratigraphic terranes and identifying major tectonic boundaries. This study synthesizes information from measured sections and the geographic distribution of various types of biogenic siliceous deposits, and discusses the similarities and differences between cherts from different areas. Only biogenically derived cherts are considered. Siliceous deposits will first be described in chronological order, comparing depositional histories of different areas at specific time horizons. Similarities in depositional histories will be used to discriminate between major and minor tectonic boundaries. Finally, the importance of factors contributing to distinctive regional histories will be assessed.

Depositional Environments

All of the siliceous deposits considered in this paper are marine deposits, with radiolarians, sponge spicules, or both, serving as the source of silica. Modern siliceous oozes are deposited beneath zones where nutrients upwell, such as along continental margins, along polar current convergent zones, and along equatorial current divergent zones (Heath, 1974; Kennett, 1982). Similar environments for siliceous deposition in the geologic past are postulated by numerous workers. Radiolarians live throughout the ocean but do not accumulate on abyssal plains, because they are not abundant above these areas, and because they dissolve in the silica-undersaturated seawater before they reach the seafloor (Kennett, 1982). Siliceous oozes also accumulate on volcanic edifices in ocean basins when they are favorably located with respect to nutrient sources and to the carbonate compensation depth (CCD) (Garrison, 1974; Heath, 1974). Above the CCD, siliceous deposition tends to be overwhelmed by prolific carbonate deposition (Garrison, 1974). Also, along continental margins, siliceous deposition is commonly overwhelmed by clastic deposition. When clastic deposition is episodic, such as with turbidites, pelagic ooze has an opportunity to accumulate between clastic depositional events (Garrison, 1974; Kelts and Arthur, 1981).

Ribbon cherts are rhythmic alternations of chert and claystone or shale. Chert layer thickness varies on a centimeter scale, and chert/shale ratios vary from greater than 1 to less than 1. Chert rhythmites result from one or more depositional mechanisms: (1) seasonal fluctuations in plankton productivity and detrital input (ROCC Group, 1986); (2) Milankovitch cycles (ROCC Group, 1986); and (3) redeposition as turbidites or contourites (Heath and Moberly, 1971; Nisbet and Price, 1974; Kelts and Arthur, 1981; Karl, 1982). Diagenetic processes may enhance rhythmic bedding in cherts (Karl, 1984).

Chert is predominantly microcrystalline quartz. There is nothing environmentally diagnostic about microcrystalline quartz. Many permissible depositional environments exist for the biogenic silica that ultimately recrystallizes to chert. However, the diagnostic evidence for these environments lies in the lithologies directly associated with the chert in stratigraphic sequences.

The majority of Alaskan cherts were deposited within the influence of a continental margin, as evidenced by associated hemipelagic and clastic rocks in the stratigraphic sequences that contain the cherts. In this paper, cherts deposited near continents will be referred to as marginal basin deposits, and cherts deposited beyond the limits of continental influence will be referred to as oceanic deposits. Water depth can range from shallow (<200 m) to deep (>200 m) for either type of deposit. Both types of deposits contain pelagic sediments (radiolarian ooze is a pelagic sediment, as are pelagic clays). Marginal basin deposits are distinguished by the presence of hemipelagic sediments in addition to pelagic sediments. Hemipelagic sediments are characterized by the presence of silt and terrigenous material (Kennett, 1982).

Marginal basins are generated in a variety of tectonic settings and include back-arc rift basins, continental rift basins, and wrench fault basins. Marginal basins also occur in more passive settings such as slope basins and interplatform basins. Siliceous deposits are found in all of these types of basins when appropriate nutrients were available. Their deposition is independent of tectonic setting, but when classified according to their associated lithologies, they can be used to infer both depositional environment and tectonic setting.

Classification of Siliceous Deposits

The following types of siliceous deposits are not intended to represent or include every biogenic siliceous environment known. The categories used in Table 12.1 were derived from a synthesis of the deposits found in Alaska and are designed to be as mutually exclusive as possible, although a small amount of overlap is unavoidable. The siliceous deposit types defined below were independently derived but are similar to those in the classification of Murchey et al. (1983). The differences will be explained so that both classifications can be used. Both classifications rely heavily on lithologic associations of siliceous deposits.

Chert with Basalt and Ultramafic Rock

The chert/basalt/ultramafic rock association, termed ophiolitic chert by Murchey et al. (1983), is restricted to cherts that are directly associated with ophiolite sequences in the classical sense of the term (Steinman, 1927). The association includes ultramafic as well as mafic intrusive and extrusive rocks. Oceanic spreading-center ophiolitic cherts are classified as oceanic deposits. These chert sections range up to tens of meters in thickness, and the chert is rhythmically bedded with very thin claystone partings. Manganese and hematite may be interstitial or may form coatings on partings or fractures. Marginal sea spreading-center ophiolitic sequences also include chert rhythmites and commonly include terrigenous and/or arc-derived clastic rocks such as shale, tuff, or graywacke. Oceanic and marginal set-

TABLE 12.1. Classification of Alaskan biogenic siliceous deposits.

Category	Lithologic association	Color	Chert deposit thickness	Fossils	Marine environment	Tectonic environment
1	Chert/basalt/ultramafic rock	Red	Decimeters, meters	Radiolarians (rare sponge spicules)	Oceanic, marginal basin	Mid-ocean spreading center, marginal sea spreading center
2	Chert/basalt	Red, green	Centimeters, decimeters meters	Radiolarians, sponge spicules	Oceanic, marginal basin	Aseismic ridge, oceanic plateau, seamount, oceanic arc, spreading center
3	Chert/basalt/andesite/graywacke	Red, green, gray	Centimeters, meters	Radiolarians, sponge spicules, carbonate fossils	Marginal basin	Volcanic arc
4	Chert/mafic to silicic volcanic rock/clastic rock	Green, gray, black	Centimeters	Radiolarians, sponge spicules, carbonate fossils	Marginal basin	Rift, transform (wrench) basin, volcanic arc
5	Chert/carbonate/clastic rock	Gray, black	Centimeters, meters	Radiolarians, sponge spicules, carbonate fossils	Marginal basin	Shelf edge, shelf basin, platform basin
6	Chert/shale	Gray, black, red, green	Centimeters, meters	Radiolarians, sponge spicules, carbonate fossils, graptolites	Marginal basin	Outer shelf, slope, rise, melange zone, subsidence zone

tings both produce iron-rich chert, which is typically red or brown. Cherts with clastic interbeds may be reddish, greenish, or mottled. Radiolarians strongly dominate the biogenic component. Both types of ophiolite sequences are inferred to represent the creation of oceanic crust.

Chert with Basalt

The chert/basalt association does not include ultramafic rocks, and it corresponds to the alternating pillow basalt/chert association of Murchey et al. (1983). Although the lack of ultramafic rocks in this association may be an artifact of the postdepositional tectonic history of an area, the relationship cannot be demonstrated. The basalt in this association is commonly pillowed, and it alternates with lenses of bedded red, green, gray, or white chert up to several meters thick. Limestone layers or lenses may or may not be present. Exclusively pelagic deposits intercalated with the basalts, such as pelagic limestone and chert, indicate an oceanic environment. Basaltic tuffs may be part of an eruptive sequence in both oceanic and marginal basin environments. Hemipelagic shale or siltstone intercalated with chert and basalt would indicate a marginal basin depositional environment. Radiolarians are the dominant source of silica. Red, oxidized deposits reflect good seawater circulation and exposure to oxygenated water. Green cherts represent less oxidized deposits and typically occur with greater proportions of hemipelagic sediments. The alternating chert/basalt relationship represents a marine volcanic center, such as a seamount, an oceanic plateau, an aseismic ridge, or a spreading ridge. Basement to these deposits is inferred to be oceanic crust.

Chert with Basalt, Andesite, and Graywacke

The chert/basalt/andesite/graywacke association does not correspond to any association of Murchey et al. (1983), and it consists of chert in thin layers, or layered lenses up to several meters thick, intercalated with basalt and graywacke. The chert may be reddish but is usually gray or green. Other associated lithologies include tuff, shale, intermediate volcanic rocks, and melange consisting of blocks of various lithologies in an argillaceous, tuffaceous, or sandy matrix. The graywacke commonly occurs as turbidites, and the melange commonly includes evidence of both soft sediment and mechanical post-lithification deformation, all implying an active tectonic environment. Biogenic silica is deposited continuously in this environment, but beds of siliceous ooze are only able to accumulate in the intervals between volcanic, turbiditic, or slumping events (Kelts and Arthur, 1981). Clayey to silty chert beds may show grading, rhythmic bedding, and other sedimentary structures indicating turbidite deposition. Radiolarians are the dominant source of silica, but sponge spicules or glass shards may also be present.

This lithologic association is derived from a volcanic arc. It is usually impossible to discriminate between fore-arc, interarc, and back-arc deposits. Basement is inferred to be transitional crust when intermediate composition igneous rocks are present, but could include oceanic or continental crust.

Chert with Mafic to Silicic Volcanic Rocks and Clastic Rocks

The association of chert with mafic to silicic volcanic rocks, accompanied by one or more of a variety of lithologies including shale, tuff, exhalite, carbonate, sandstone, and conglomerate, is characterized by thinly laminated chert or silty chert that is typically green, gray, or black. Hydrothermal silica, barite, and sulfide layers may be associated with biogenic siliceous deposits (Lange et al., 1985). Silty chert may show sedimentary structures such as grading, cross-lamination, or flaser bedding. Radiolarians and/or sponge spicules may be present.

Murchey et al. (1983, p 125) attribute the chert/silicic volcanic association to a basin adjacent to a volcanic arc. Since silicic volcanism is also characteristic of rift environments (Williams and McBirney, 1979), the interpretation of this lithologic association is herein extended to include rifted basins along continental margins or wrench basins along transform faults that intersect continental margins. A continental margin setting is implied by the combination of marine sedimentary deposits with silicic volcanism. Basement is inferred to be either transitional or continental crust.

Chert with Carbonate and Clastics

Chert associated with carbonate is characteristically gray or black but may be green or red. Shale may or may not be associated with the chert and carbonate. In replacement deposits, silica may have replaced carbonate, or the reverse may be true. Bed thickness is extremely variable on a centimeter scale, depending on the composition of the primary deposit and nature of the replacement, if replacement has occurred. Radiolarians and sponge spicules provide the source of silica and can be identified in calcareous as well as in siliceous layers. Calcareous fossils and microfossils are commonly present. Associated black siliceous and/or calcareous shales may indicate high productivity and high organic content or may indicate poor circulation in restricted interplatform basins. The ratio of chert to carbonate depends on the relative abundance of calcareous and siliceous organisms in the depositional environment and on the chemistry of the diagenetic environment. This classification applies to carbonates with intercalated chert beds and is not intended to apply to carbonates with diagenetically derived chert nodules. Associated platform carbonate deposits and carbonate-clastic rocks indicate interplatform basins. Deeper-water facies, represented by the fine-grained hemipelagic deposits and occasional slump deposits, may represent an outer shelf, slope, or continental rise environment. Sponge spicules tend to be more abundant than radiolarians in shallow-water deposits, whereas radiolarians dominate sponge spicules in deeper-water slope deposits. Basement to the chert/carbonate/clastic deposits is inferred to be continental or transitional crust. This association does not correspond to any category of Murchey et al. (1983).

Chert with Shale

Chert associated with shale is thin-bedded to lenticular and typically dark gray or black, but it may be variegated. Either chert or shale may dominate a stratigraphic section; relative proportions usually vary through the section. Some chert layers may be argillaceous; some shale layers may be very siliceous. The shales tend to be carbonate-poor, owing to generally suboxic or anoxic depositional conditions (summarized in Kennett, 1982). Radiolarians dominate over sponge spicules. In a subsiding environment, radiolarians tend to increase in abundance with deepening water. Thin layers of carbonate debris, tuff, or fine sand may be present. The carbonate may later be replaced by silica. Some chert/shale sequences are rhythmically bedded and contain sedimentary structures indicative of turbidite deposition (Garrison, 1974; Nisbet and Price, 1974; Kelts and Arthur, 1981). Such redeposition of siliceous ooze has been described along tectonically active continental margins such as in the Gulf of California (summarized in Garrison, 1974; Kelts and Arthur, 1981). Chert/shale deposits may be quite extensive, but they lack marker beds, and original thickness is difficult to determine.

The chert/shale association represents a depositional environment that is very widespread. Chert and shale occur together in the carbonate and volcanic associations described above. The chert/shale association is here distinguished by the lack of carbonate, volcanic, and sandstone beds at an outcrop scale, although centimeter scale lenses of these lithologies (less than 1%) may be present. The chert/shale association represents deeper-water depositional environments along continental margins where upwelling nutrients are abundant (Heath, 1974; summarized in Kennett, 1982). The chert/shale association corresponds to the subsidence association of Murchey et al. (1983), but knowledge of the substrata is required to classify their subsidence association, and not the chert/shale association defined here. The difference is that the subsidence association represents a tectonic environment and the chert/shale association represents a depositional environment. The same lithologic association may result from subsidence or a rise in sea level. Crustal basement for the chert/shale association is inferred to be either transitional or continental.

In summary, lithologic associations define six categories for classification of chert deposits found in Alaska (Table 12.1). All of the cherts occur as centimeter scale beds or lenses. Deposits of chert beds tend to be thicker for oceanic igneous associations and thinner for continental margin igneous, carbonate, or clastic associations. Radiolarians and/or sponge spicules compose the siliceous ooze for all the cherts, but spicules are rare for the chert/basalt/ultramafic association, and radiolarians tend to be rare in the chert/carbonate/clastic association. Only the chert/ophiolite and chert/basalt associations are likely to be oceanic deposits on oceanic crust. All six of the chert association deposit types described may be marginal basin deposits on transitional crust. Only the chert/carbonate/clastic and chert/shale associations are likely to be exclusively marginal basin deposits on continental crust.

Alaskan Siliceous Deposits Through Time

This section will discuss biogenic siliceous deposits in Alaska in chronological order. Certain periods of geologic time tend to be characterized by distinct chert deposit types in Alaska. Also, cherts of certain ages occur in specific geographic regions in the state (Fig. 12.1; Table 12.2).

Precambrian Chert

There are no documented Precambrian biogenic siliceous deposits in Alaska. There are cherts in some lithostratigraphic sequences that are possibly Precambrian, such as the argillite grit and quartzite unit in the east-central part of the state (Foster et al., 1983) and chert with amphibolite and metapelite in the basement rocks in the western Brooks Range (Karl, unpublished data). Light gray, olive, green, black, or white chert is a minor lithology in these units, and the source of silica is not known. For some of these units containing metachert and marble, a chert/carbonate/clastic classification may be applied.

Cambrian Chert

In east-central Alaska, south of the Tintina fault in the Yukon-Tanana terrane (Figs. 12.1a,b), the Wickersham unit (Weber et al., 1985), which contains metachert interbedded with quartzite grit, carbonate, and maroon and green slate, overlies the Precambrian Fairbanks schist unit of Bundtzen (1982), a metapelite. The chert occurs as lenses in shale and is considered to have been deposited on the outer shelf (Weber et al., 1985). In the same area, but north of the Tintina fault there is also chert intercalated with Precambrian or Paleozoic maroon and green slate, grit, and gray marble (Weber et al., 1985).

Ordovician Chert

Ordovician chert is widespread in the interior part of Alaska north of the Denali fault (Fig. 12.1a). In the

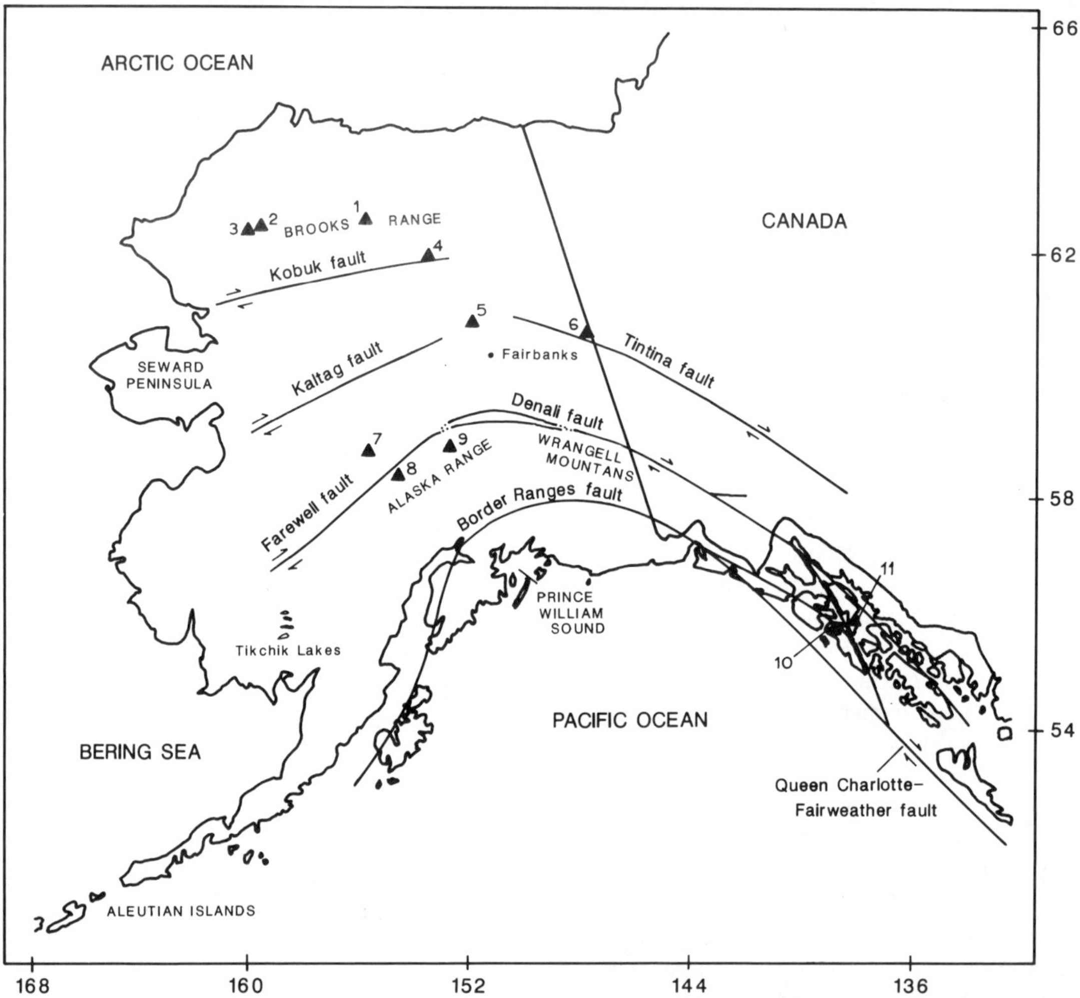

FIGURE 12.1a. Map of Alaska showing key geographic features, faults, and locations of measured sections. (1) Siksikpuk Formation (Fig. 12.10) and Otuk Formation (Fig. 12.11) sections. (2) Siksikpuk Formation section (Fig. 12.10). (3) Kuna Formation section (Fig. 12.8). (4) Angayucham terrane composite section (Fig. 12.9). (5) Livengood Dome Chert (Fig. 12.2) and Amy Creek unit (Fig. 12.4) sections. (6) Road River Formation/McCann Hill Chert section (Fig. 12.3). (7) Medfra section (Fig. 12.12). (8) Shellabarger Pass section (Fig. 12.6). (9) Chulitna section and Broad Pass area (Fig. 12.5). (10) Gambier Bay, Hood Bay, and Cannery Formations sections (Fig. 12.9). (11) Kelp Bay Group section (Fig. 12.13). Note: Adak section (Fig. 12.14) occurs west of this map, in the central Aleutian Islands.

Baldry terrane (Fig. 12.1b), chert is interbedded with early to middle Paleozoic marble, greenstone, and pelitic schist. In the Crazy Mountains (Fig. 12.1b), chert is intercalated with sandstone, grit, slate, and carbonate. In the Innoko, Livengood, and Minchumina terranes (Fig. 12.1b), dark colored to gray or black cherts are interbedded with graptolitic shales (Chapman et al., 1981). The Livengood Dome Chert (Chapman et al., 1980) includes thin silicified carbonate beds and rhythmic graded beds below the graptolite horizon and grades up into silty and tuffaceous beds near the top of the section (Fig. 12.2). There are also tuffaceous beds at the top of the Minchumina section (Chapman et al., 1981). Most of the Minchumina chert beds are dark gray; some are red or tan. Chert comprises about 80% of the Minchumina section, which represents a chert/shale association. The section overlies quartzite, grit, and impure limestone.

North of the Tintina fault, along the International Boundary, the Road River Formation (Churkin and Brabb, 1965) consists of a basal chert conglomerate

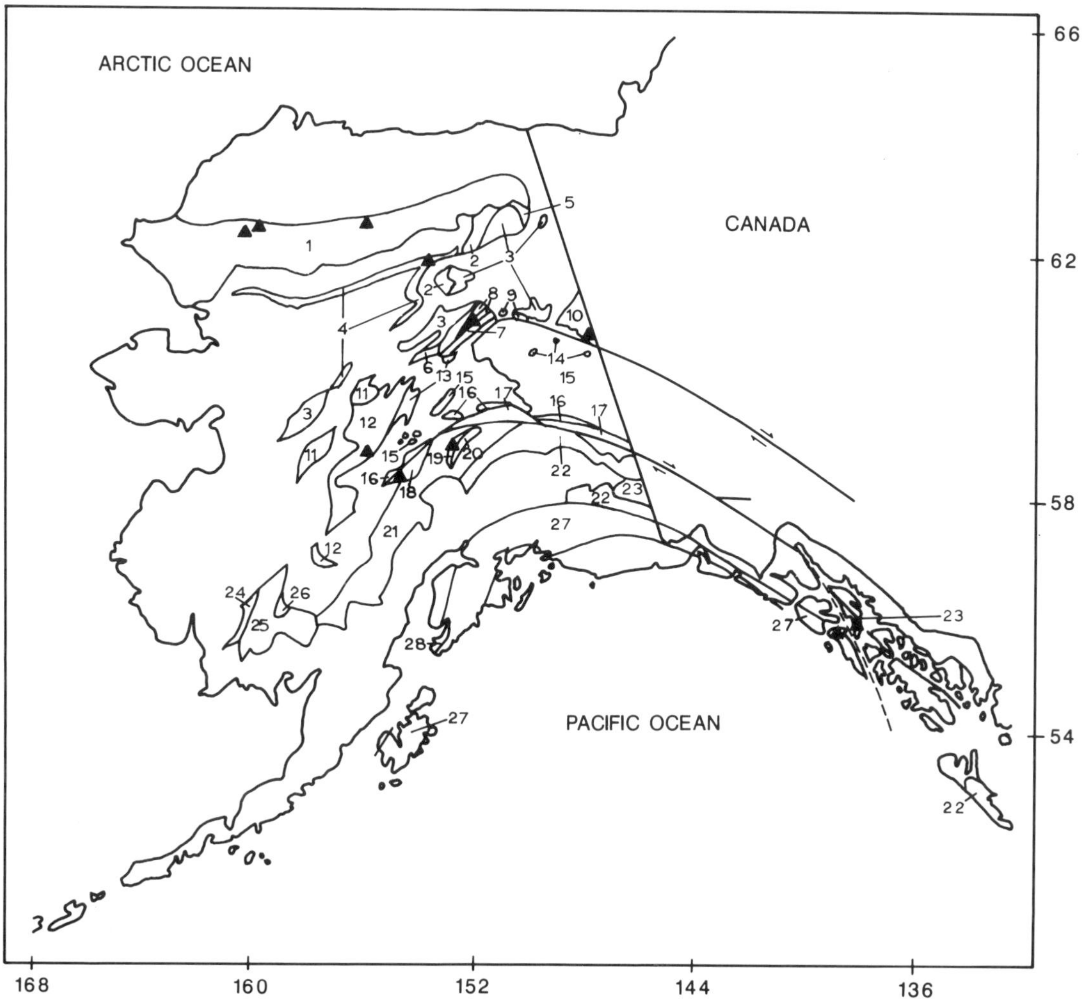

FIGURE 12.1b. Map of Alaska showing some key tectonostratigraphic terranes (after Jones et al., 1984b) and location of measured sections. (1) Endicott, Kagvik, and Delong Mountains terranes. (2) Venetie terrane. (3) Tozitna terrane. (4) Angayucham terrane. (5) Sheenjek terrane. (6) Baldry terrane. (7) Manley terrane. (8) Livengood and White Mountains terranes. (9) Crazy Mountains terrane. (10) North American and Kandik River terranes. (11) Innoko terrane. (12) Nixon Fork terrane. (13) Minchumina terrane. (14) Seventy-mile terrane. (15) Yukon-Tanana terrane. (16) Pingston terrane. (17) McKinley and Windy terranes. (18) Mystic terrane. (19) Chulitna terrane. (20) West Fork and Broad Pass terranes. (21) Kahiltna terrane. (22) Wrangellia terrane. (23) Alexander terrane. (24) Goodnews terrane. (25) Togiak terrane. (26) Tikchik terrane. (27) Chugach terrane. (28) Kachemak terrane.

and black arenite, overlain by 150 to 300 m of dark Lower Ordovician graptolitic shale and grayish black bedded chert with rare dark gray limestone or dolostone lenses (Fig. 12.3), and represents a chert/shale association. The proportion of shale to chert increases upsection. The section is interpreted to represent a progradational environment (Churkin and Brabb, 1965).

In the northeastern Brooks Range, siliceous argillite and argillaceous chert containing radiolarians and Ordovician graptolites have been reported from rocks mapped with the "Neruokpuk Formation" (Thomas E. Moore, U.S. Geological Survey, and Michael Churkin, Atlantic Richfield Company, written communication, 1987). These rocks also represent a chert/shale association.

In southeastern Alaska there are thin-bedded, carbonaceous black cherts interbedded with argillite, graywacke, limestone, and minor volcanics in the Ordovician Hood Bay Formation (Loney, 1964; Carter, 1977). Chert beds are 1 to 6 cm thick and occur in approximately equal amounts with argillite.

TABLE 12.2. Distribution of Alaskan siliceous deposits with respect to time.

Age	Northern Brooks Range	Southern Brooks Range	Central Alaska	Alaska Range	Southeastern Alaska	Southwestern Alaska
Precambrian	Chert/basalt/ clastic		Chert/carbonate/ clastic			
Cambrian			Chert/carbonate/ clastic			
Ordovician	Chert/shale		Chert/shale		Chert/shale	
Silurian			Chert/shale/ (carbonate)			
Devonian		Chert/basalt	Chert/carbonate/ clastic; chert/ basalt/clastic	Chert/carbonate; chert/basalt/ ultramafic	Chert/basalt/ clastic; chert/ carbonate/ clastic	
Mississippian	Chert/carbonate; chert/shale; chert/mafic to silicic vol- canic/clastic	Chert/basalt	Chert/basalt	Chert/basalt/ ultramafic; chert/ carbonate; chert/clastic	Chert/shale; chert/car- bonate/ clastic	
Pennsylvanian	Chert/shale; chert/mafic to silicic vol- canic/clastic	Chert/basalt	Chert/basalt	Chert/carbonate/ clastic	Chert/shale; chert/car- bonate/clastic	
Permian	Chert/shale	Chert/basalt	Chert/basalt/ ultramafic; chert/basalt; chert/ carbonate	Chert/clastic/ carbonate; chert/shale; chert/car- bonate/(basalt)	Chert/carbonate/ clastic	
Triassic	Chert/shale	Chert/basalt/ clastic	Chert/basalt/ andesite/clas- tic; chert/car- bonate/clastic		Chert/basalt/ clastic/ (car- bonate); chert/vol- caniclastic; chert/shale/ (carbonate)	Chert/basalt/ clastic/ carbonate
Jurassic	Chert/shale; chert/shale/ (carbonate)	Chert/basalt/ clastic	Chert/basalt/ clastic	Chert/clastic; chert/shale/ (tuff)/ (carbonate)	Chert/basalt; chert/basalt/ andesite/clas- tic; chert/ shale	
Cretaceous			Chert/ volcaniclastic	Chert/clastic/ (carbonate)	Chert/basalt/ andesite/ clastic; chert/basalt; chert/shale	Chert/basalt/ andesite/ clastic; chert/ basalt; chert/shale
Tertiary					Chert/basalt/ graywacke	Chert/basalt/ andesite/ clastic

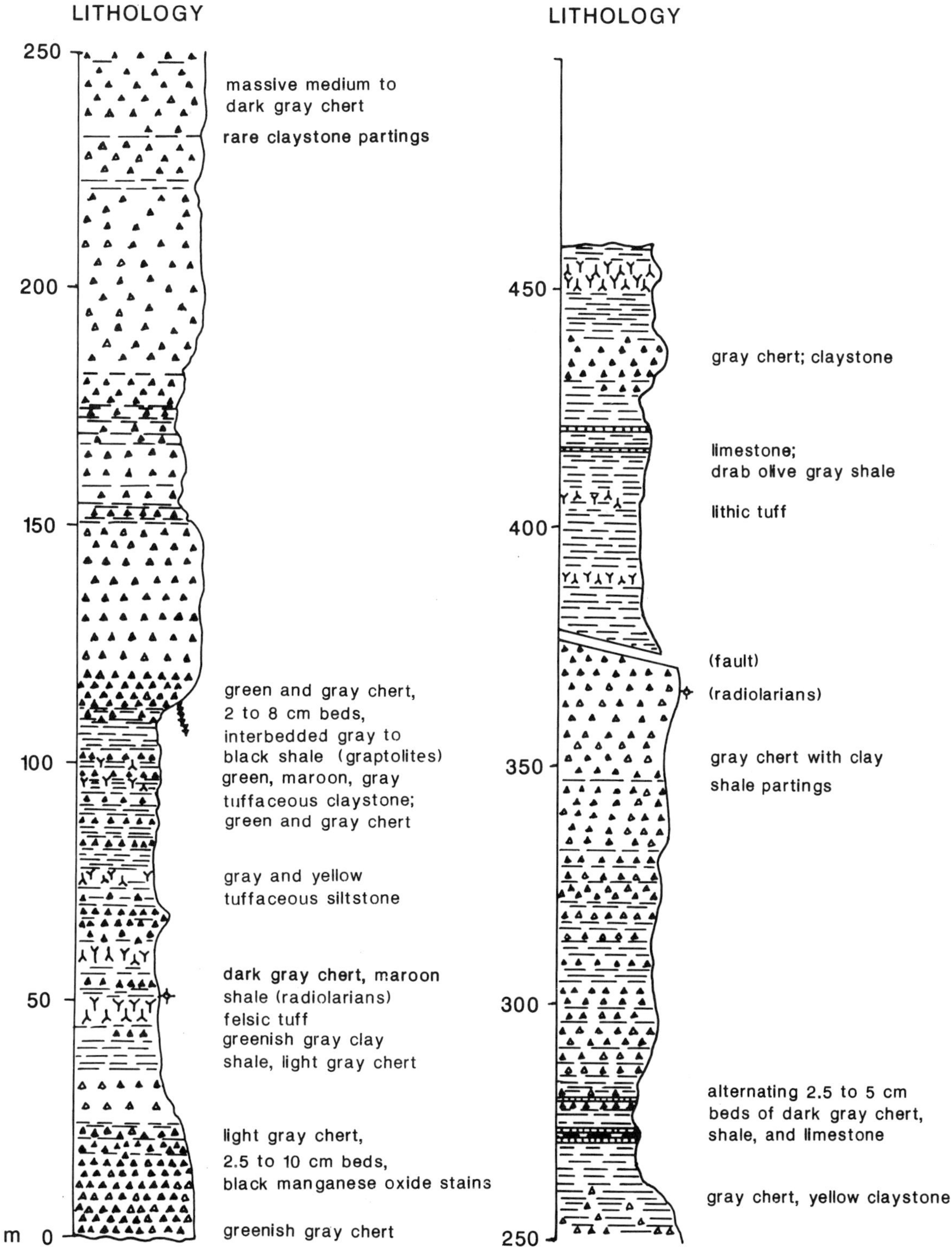

FIGURE 12.2. Stratigraphic section of the Ordovician Livengood Dome Chert, central Alaska, from Chapman et al. (1980). Location 5 on Figure 12.1a. Note: Letters in AGE column in Figures 12.2 to 12.14 represent geologic ages as follows: K, Cretaceous; J, Jurassic; Tr, Triassic; lTr, Late Triassic; eTr, Early Triassic; P, Permian; ℙ, Pennsylvanian; M, Mississippian; D, Devonian; S, Silurian; O, Ordovician.

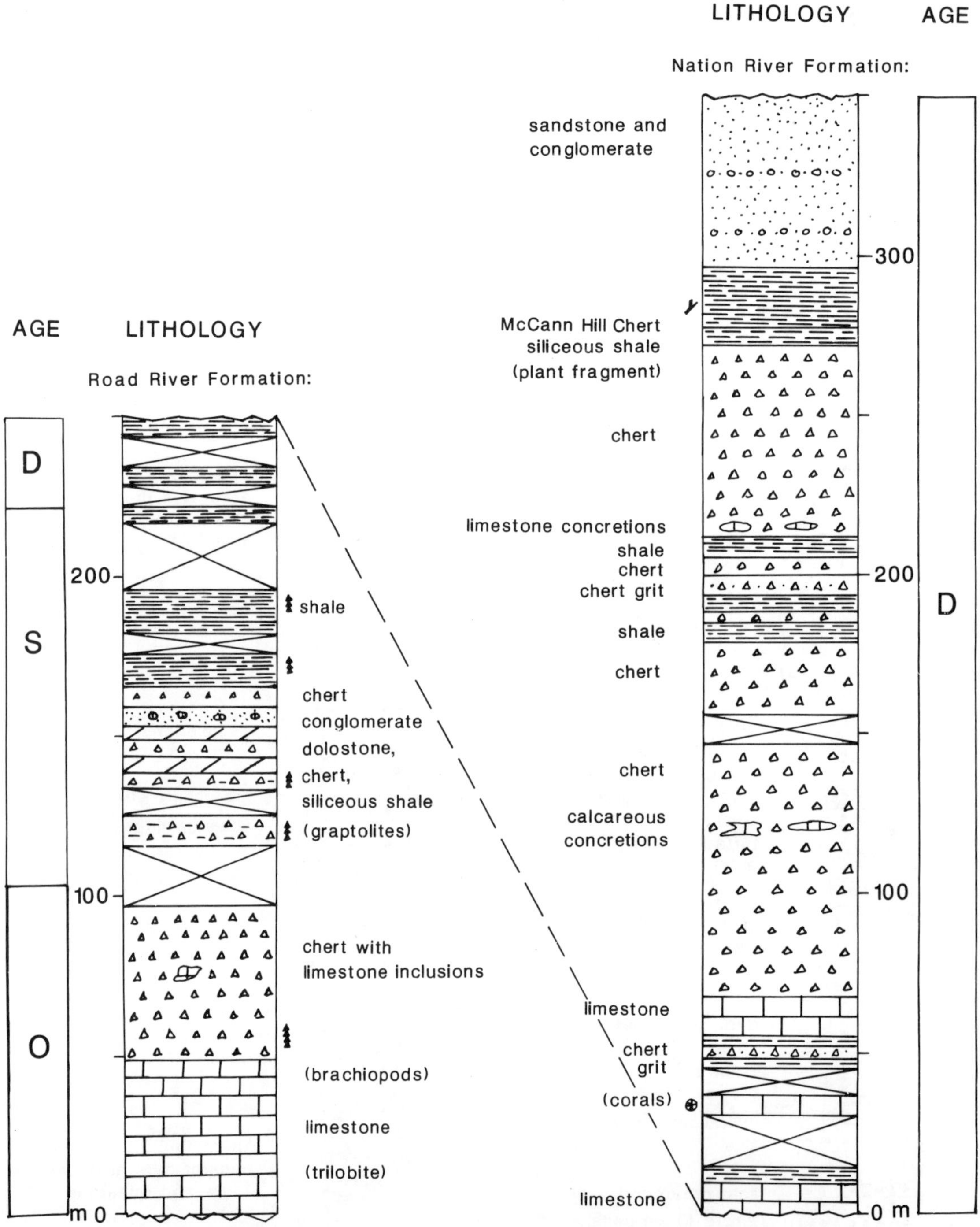

FIGURE 12.3. Composite stratigraphic section of the Ordovician to Devonian Road River Formation and the Devonian McCann Hill Chert, east-central Alaska, after Churkin and Brabb (1965; reprinted by permission of American Association of Petroleum Geologists). Location 6 on Figure12.1a.

Limestone occurs as nodules and as lenses up to 10 m thick, and graywacke occurs as thin turbidite beds within the chert/argillite (Fig. 12.7), indicating a chert/carbonate/clastic association.

Silurian Chert

The upper part of the Road River Formation (Churkin and Brabb, 1965) is Silurian and Devonian and is dominantly shaly. The Ordovician to Devonian rocks in the Beaver, White Mountains, Livengood, and Kandik terranes of east-central Alaska (Fig. 12.1b) are predominantly composed of gray, black, or olive gray siltstone with dolostone lenses less than 1 m thick, dikes, and lenses of gabbro or amygdaloidal basalt, and less than 10% chert (Foster et al., 1983). These cherts are classified with the chert/shale association because of the predominance of shale over other lithologies associated with the chert beds.

Lower Paleozoic Chert

In the Livengood terrane (Fig. 12.1b), the undated Amy Creek unit (Weber et al., 1985) lies in fault contact between Cambrian igneous rocks and the Ordovician Livengood Dome Chert. The inferred base of the Amy Creek unit consists of sooty black siliceous argillite with thin beds of black chert and rare lenses of tuff and carbonate, overlain by a thick section of mafic flows, which is overlain by cross-bedded limestone and silicified dolostone containing irregular masses of black replacement chert (Fig. 12.4). The unit could be as old as Precambrian or as young as Devonian, based on regional correlations, and the chert-bearing part of the section is classified as chert/shale representing a slope environment. The chert in the carbonate part of the section is volumetrically minor and not classified.

On the northeast coast of the Seward Peninsula are gray chert beds associated with dark gray Ordovician graptolitic shale (Michael Churkin Jr., Atlantic Richfield Company, written communication, 1985; Julie Dumoulin, U.S. Geological Survey, oral communication, September 1987). Also, on the eastern Seward Peninsula, pure and impure carbonates containing Ordovician conodonts are associated with quartz-graphite schist containing pre-Devonian radiolarians (Brian Holdsworth, in Till et al., 1986).

In general, Silurian and older cherts in Alaska are associated with shale and are located north of the Denali fault (Fig. 12.16). Lower Paleozoic carbonate rocks in the Brooks Range and central Alaska may have small layers, lenses, or patches of replacement chert.

Devonian Chert

Devonian chert in the central part of Alaska tends to be associated with carbonate and clastic rocks. The McCann Hill Chert (Churkin and Brabb, 1965) overlies the Road River Formation and consists of a basal Lower Devonian fossiliferous limestone followed by thin-bedded light and dark gray chert, siliceous shale, and chert gritstone (Fig. 12.3). Fossils include Devonian plants, conodonts, gastropods, and spores. The sequence is overlain by coarse clastic rocks of the Upper Devonian Nation River Formation. In the same region, the Devonian Woodchopper Volcanics are mafic flows with lenses of shale, chert, sandstone, tuff, and dolostone (Foster et al., 1983). Also in central Alaska, in the Innoko terrane (Fig. 12.1b), chert is interbedded with limestone turbidites of Devonian to Pennsylvania age (Robert Chapman, U.S. Geological Survey, oral communication, December 1985). Elsewhere in the Innoko terrane, Late Devonian to Late Mississippian radiolarians date dark gray, thinly banded cherts with intercalated slaty argillite and felsic tuff (Chapman and Patton, 1979). In the White Mountains and Kandik and Livengood terranes (Fig. 12.1b), a Silurian or Devonian unit includes dolostone and gray to black argillite and siltstone with interlayered limestone, light and dark gray chert, quartzite, and, in some places, amygdaloidal greenstone (Foster et al., 1983). These cherts in central Alaska were probably deposited near carbonate platforms in basins that sustained periodic mafic volcanism (Fig. 12.15).

In the Chulitna terrane (Fig. 12.1b) south of the Denali fault, Famennian to Mississippian conodonts and radiolarians date a thick section of bedded red chert associated with an ophiolite sequence (Fig. 12.5). Uppermost Devonian and Mississippian black chert overlies limestone in the Mystic terranes (Fig. 12.1b) (Jones et al., 1981). A few kilometers to the west, green and gray cherts are interbedded with limestone containing Late Devonian megafossils and with siltstone, quartz sandstone, and chert pebble conglomerate (Fig. 12.6) (Reed and Eberlein, 1972), a chert/carbonate/clastic association.

In southeastern Alaska, in the Alexander terrane, minor amounts of metachert are intercalated with gray phyllite, greenschist, and marble in the Devonian Gambier Bay Formation (Loney, 1964). Also,

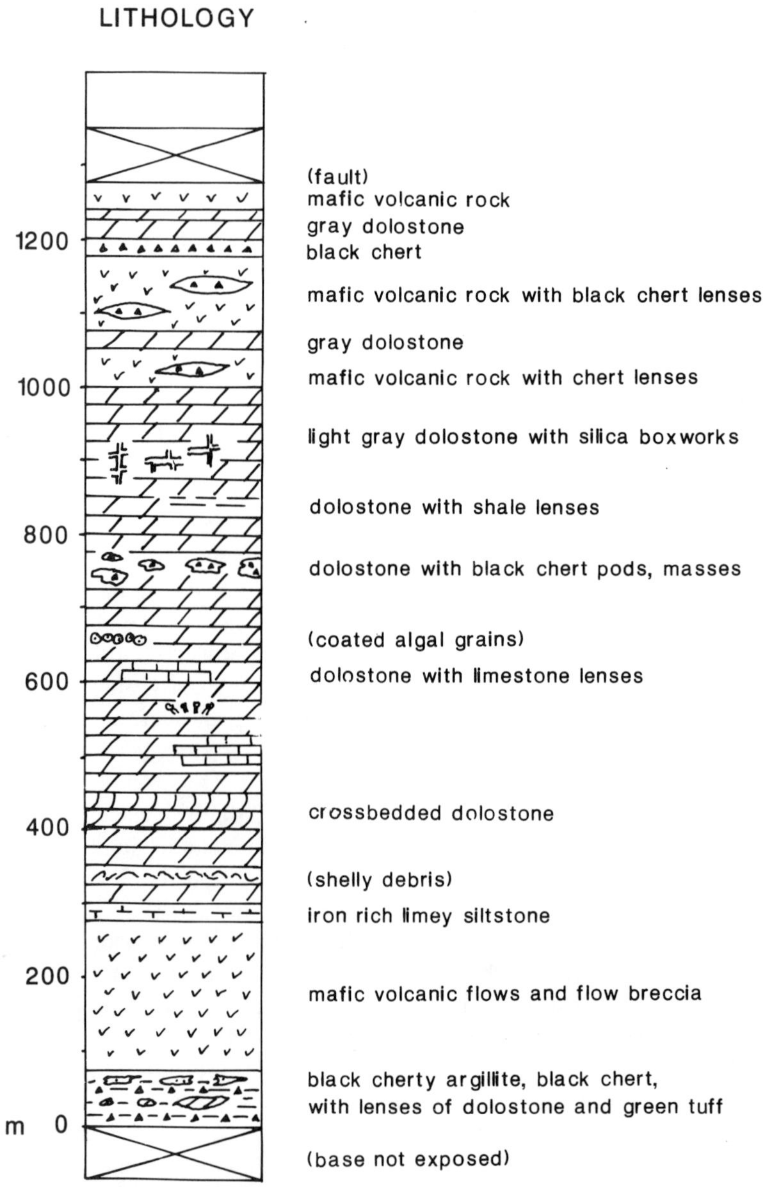

FIGURE 12.4. Stratigraphic section at Amy Creek, Livengood quadrangle, east-central Alaska (Weber, Karl, unpublished field notes). The Amy Creek unit is considered to be early Paleozoic in age (Weber et al., 1985). Location 5 on Figure 12.1a.

uncomformably, or faulted over the Ordovician Hood Bay Formation, lies the Cannery Formation, which contains variegated Devonian to Mississippian radiolarian ribbon cherts interbedded with siliceous argillite and graywacke (Loney, 1964). Mafic rocks and some pillow basalt are included in the Cannery Formation.

In summary, except for the Chulitna ophiolite (Jones et al., 1981), Devonian chert in Alaska is predominantly associated with carbonate and clastic rocks, with minor amounts of mafic volcanics (Fig. 12.15).

Mississippian Chert

In the Brooks Range in northern Alaska, extensive primary and replacement-derived gray to black, thin- to medium-bedded chert occurs in parts of

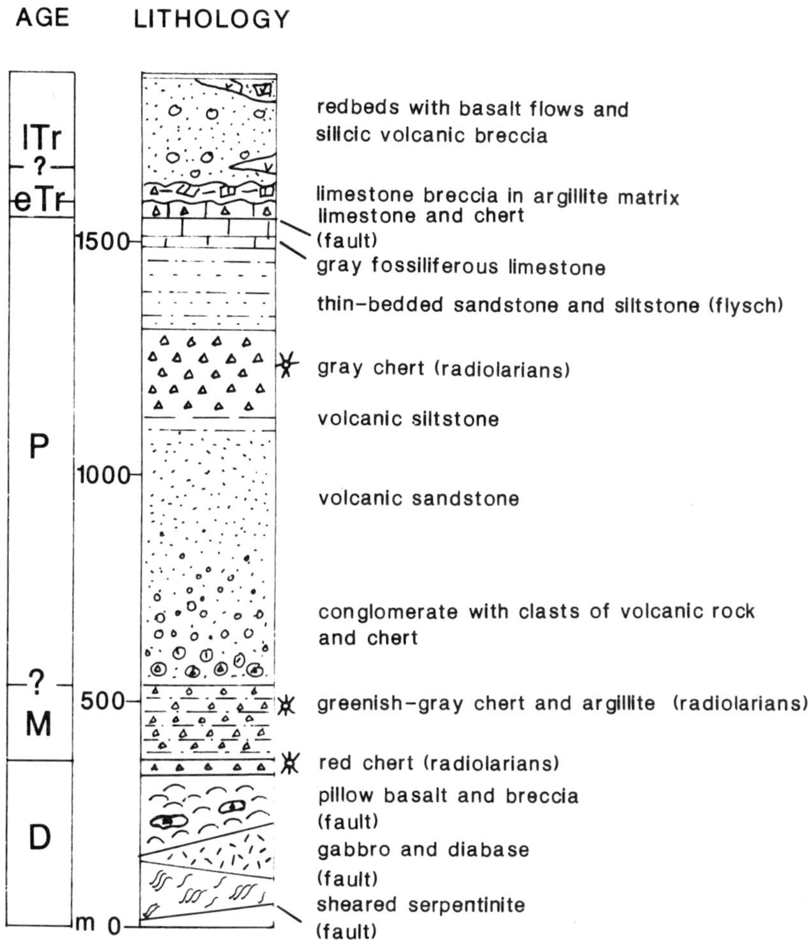

FIGURE 12.5. Composite stratigraphic section for the Chulitna terrane, central Alaska Range, from Jones et al. (1980). Location 9 on Figure 12.1a.

limestones in the Lisburne Group, particularly the Kogruk Formation (Sable and Dutro, 1961). Also in the Lisburne Group is the Kuna Formation (Mull et al., 1982) consisting of black chert and black carbonaceous, siliceous shale (Fig. 12.8). These black cherts represent a restricted basin facies deposited between massive carbonate buildups represented by the Lisburne Group carbonate rocks and may be due to tectonic extensional horst and graben development (Moore et al. 1986). Middle Mississippian to Early Pennsylvanian radiolarians are reported from a barite- and sulfide-bearing siliceous deposit from Red Dog Creek in northwestern Brooks Range (Lange et al., 1985). Mississippian radiolarians from chert and Mississippian K-Ar ages from associated keratophyre and andesite from the same lithostratigraphic unit near Drenchwater Creek in the northwestern Brooks Range (Lange et al., 1985) support a chert/mafic to silicic volcanic/clastic association for these rocks (Fig. 12.18). Unnamed black chert with black siliceous shale of Mississippian to Pennsylvanian age is widely distributed on the north flank of the Brooks Range (Fig. 12.16).

On the south side of the Brooks Range, in the Angayucham terrane, Devonian to Jurassic radiolarians have been extracted from various lenses of ribbon chert alternating with mafic flows and pillow basalts (Fig. 12.9), interpreted as ocean island or oceanic plateau basalts based on their chemical composition (Barker et al., 1988). In some places the basalt and pelagic rocks are faulted against ultramafic rocks, but the protolith age and affinity of the ultramafic rocks are not known. Similar chert and basalt extend southward into interior Alaska as far as the Kaltag fault, in the Angayucham and Tozitna terranes (Figs. 12.1b. 12.17).

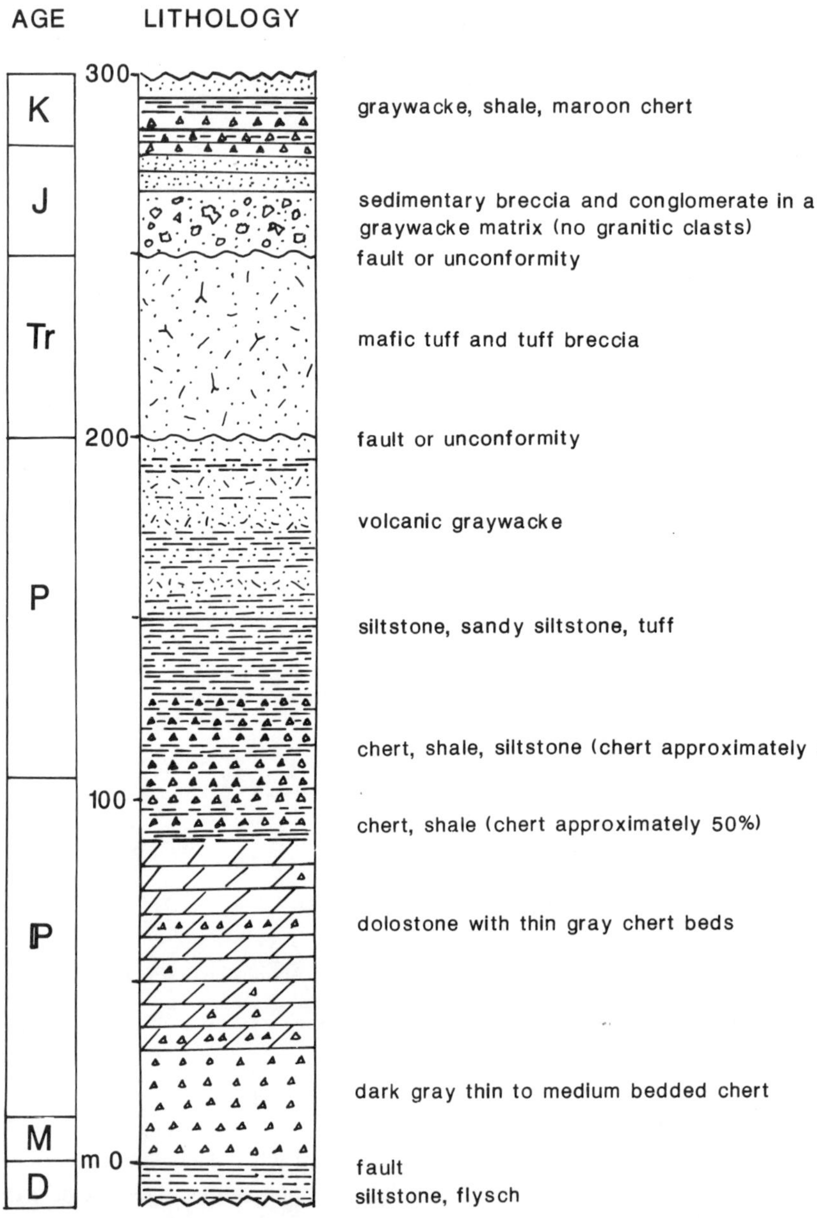

FIGURE 12.6. Stratigraphic section from Shellabarger Pass, northern Alaska Range, from Reed and Eberlein (1972). Location 8 on Figure 12.1a.

In central Alaska, Mississippian to Triassic(?) radiolarian chert occurs with volcanic rocks in the Rampart Group (Chapman et al., 1982) and correlative lithostratigraphic units in the Tozitna terrane. These red and gray cherts are most commonly associated with tuff, in places with limestone, and comprise about 30% of the sedimentary rocks, though only 5% of the Rampart Group overall. In the Livengood area, gray and black, Mississippian and Pennsylvanian Rampart Group cherts comprise 25% to 30% of a shaly section which overlies Cambrian to Devonian metasedimentary and metavolcanic rocks (Weber et al., 1985). In the Medfra area, in the Nixon Fork terrane (Fig. 12.1b), there are two Mississippian to Pennsylvanian units containing radiolarian chert (Chapman and Patton, 1979). One unit

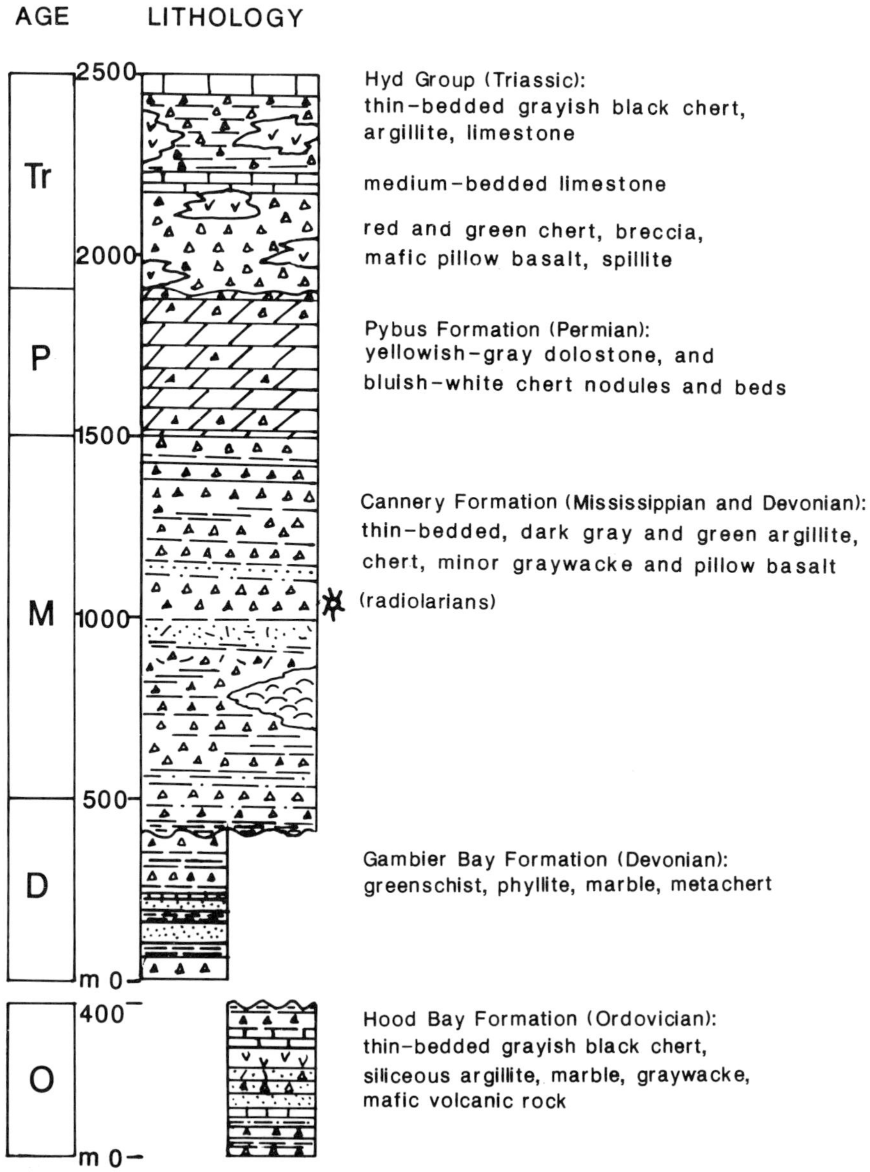

FIGURE 12.7. Composite stratigraphic section from Admiralty Island, southeastern Alaska, after Loney (1964). Location 11 on Figure 12.1a.

is green, gray, black, or red chert in beds 2 to 25 cm thick. The other unit is predominantly carbonate, but includes calcarenite, grit, arkose, and subordinate chert.

South of the Denali fault in the Alaska Range, the chert mapped with the Chulitna ophiolite ranges up into the Mississippian period (Fig. 12.17). Nearby, in the Broad Pass area, Mississippian chert is intercalated with argillite, tuff, and graywacke (Jones et al., 1980). Slightly to the west, at Shellabarger Pass, Mississippian gray bedded chert is faulted over Paleozoic siltstone and grades upward into Pennsylvanian chert and dolostone (Fig. 12.6) (Reed and Eberlein, 1972).

In southeastern Alaska, the Cannery Formation includes Mississippian chert, argillite, and graywacke, with local intercalations of pillow lavas and breccias (Loney, 1964) (Fig. 12.7). The Saginaw Bay Formation includes a volcanic member with pillow breccia, agglomerate, tuff, and limestone with

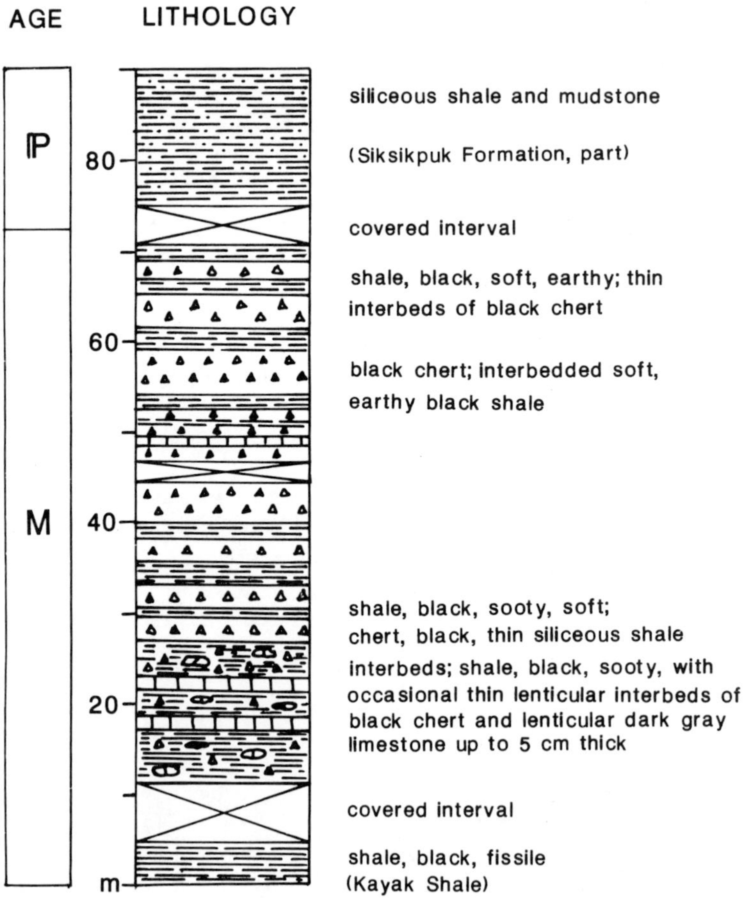

FIGURE 12.8. Stratigraphic section from the Mississippian and Pennsylvanian Kuna Formation at its type locality in the western Brooks Range, Alaska, from Mull and others (1982; reprinted by permission of American Association of Petroleum Geologists). Location 3 on Figure 12.1a.

up to 30% replacement chert lenses and nodules, and a chert and limestone member 150 m thick, with Mississippian to Pennsylvanian thin- to medium-bedded light-brown calcareous chert and subordinate gray fossiliferous cherty or dolomitic limestone (Muffler, 1967). The Saginaw Bay Formation represents a basalt/carbonate/(chert) association, which would be classified with the basalt/chert category (Table 12.1).

Pennsylvanian Chert

The above descriptions of Mississippian through Pennsylvanian cherts will not be repeated here. Alaskan chert deposits of these ages tend to be associated with basalt, shale, or carbonate, with various associations of similar age closely juxtaposed by faulting.

In the northern Brooks Range, the Pennsylvanian to Triassic Siksikpuk Formation (Patton, 1957; Murchey et al., 1981) disconformably overlies limestone, shale, and chert of the Lisburne Group. The lower part of the Siksikpuk Formation in the central Brooks Range includes bioturbated maroon and green siliceous shale and bedded green chert. In some places, the Siksikpuk(?) Formation overlies unnamed Mississippian black cherts and shales. In one section, a debris flow with clasts of oncoids, chert, rip-up clasts, sandstone, shale, and limestone, grades upward into predominantly siliceous siltstone and silty chert containing sponge spicules (Fig. 12.10) (Siok, 1984, 1985b). In the western Brooks Range, the base of the Siksikpuk Formation may extend into the uppermost Mississippian and consists of dark gray chert in beds 5 to 30 cm thick, interbedded with minor amounts of shale (Murchey

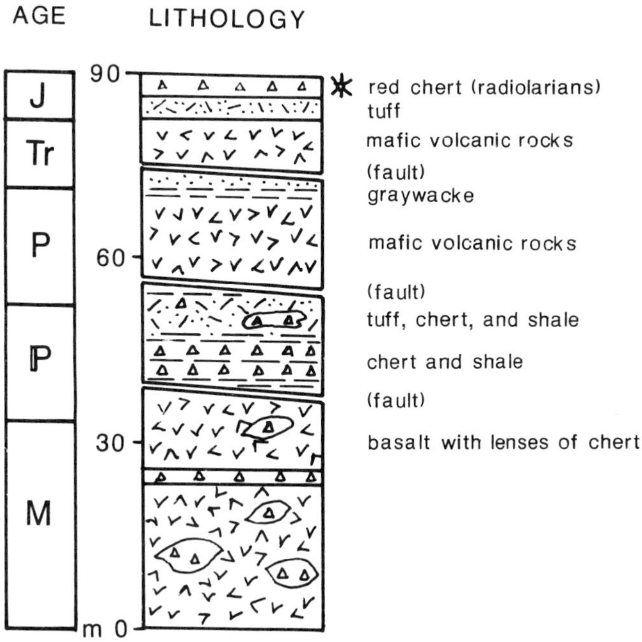

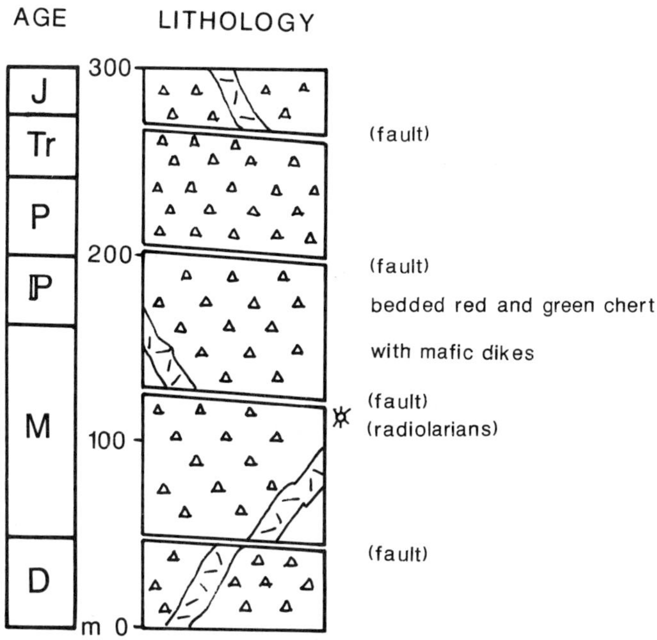

FIGURE 12.9. Generalized stratigraphic sections from the Angayucham terrane, southern Brooks Range, Alaska, after Plafker et al. (1978) and Barker et al. (1988). Location 4 on Figure 12.1a.

et al., 1981). This grades up into 2-cm-thick beds of maroon and green argillite and argillaceous radiolarian chert. The Siksikpuk Formation is a chert/shale association probably representing subsiding basins adjacent to carbonate banks (Murchey et al., 1983).

Pennsylvanian chert occurs in the Angayucham and Tozitna terranes in the interior part of Alaska within the long-ranging basalt/chert and basalt/chert/carbonate/clastic sequences mentioned above. In the Alaska Range, the Pennsylvanian chert and dolostone of Shellabarger Pass grades upsection into a chert/shale section composed of 50% chert (Reed and Eberlein, 1972).

In southeastern Alaska the Saginaw Bay Formation, a basalt/chert association, extends through the Pennsylvanian.

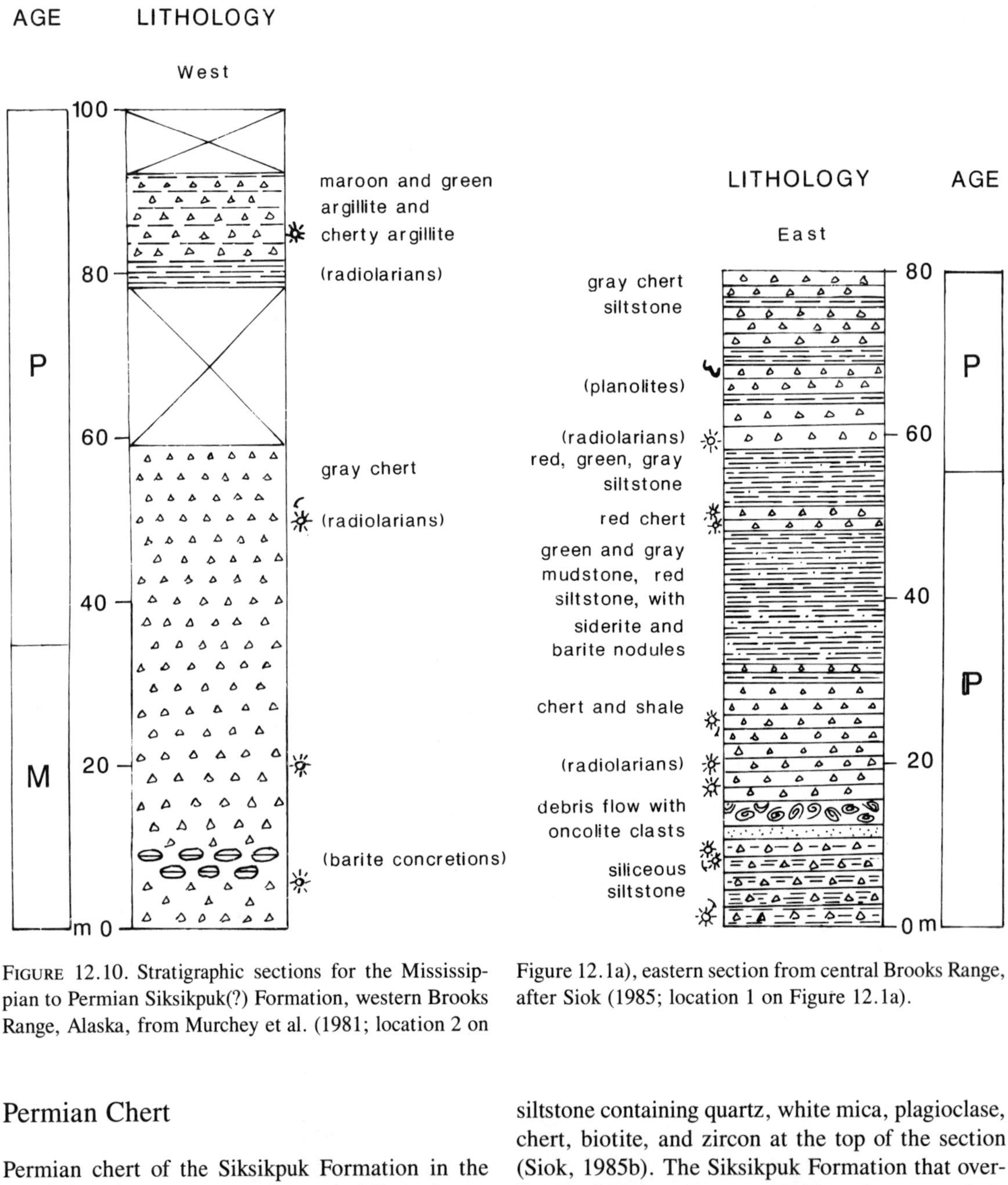

FIGURE 12.10. Stratigraphic sections for the Mississippian to Permian Siksikpuk(?) Formation, western Brooks Range, Alaska, from Murchey et al. (1981; location 2 on Figure 12.1a), eastern section from central Brooks Range, after Siok (1985; location 1 on Figure 12.1a).

Permian Chert

Permian chert of the Siksikpuk Formation in the Brooks Range tends to be silty and shaly and contains Permian brachiopods, corals, gastropods, and radiolarians. Siok (1985a) described a section from the Siksikpuk Formation in the central Brooks Range that contained radiolarians enclosed in glauconite. A generalized section (Fig. 12.10) for the Siksikpuk(?) Formation, which overlies the unnamed Mississippian black cherts and shales, consists of basal green siltstone and claystone, overlain by maroon and green mottled mudstone and siltstone with barite and siderite nodules, overlain by green siliceous mudstone, and ultimately overlain by gray shale and siltstone containing quartz, white mica, plagioclase, chert, biotite, and zircon at the top of the section (Siok, 1985b). The Siksikpuk Formation that overlies the Lisburne Group carbonates has more shale and less chert than the section shown in Figure 12.10. The proportion of chert increases southwestward across the Brooks Range, but the Siksikpuk Formation is dominantly hemipelagic everywhere and the majority of the unit is Early Permian in age (Siok, 1985b). Overall this unit indicates a long period of quiet, relatively deep-water depositional conditions in northern Alaska.

On St. Lawrence Island (located to the southwest of the Seward Peninsula; Fig. 12.1a), Permian radiolarian chert is interbedded with graywacke and mudstone (Benita Murchey, U.S. Geological Survey, personal communication, December 1985).

The chert/basalt sequences of the Angayucham and Tozitna terranes in central Alaska extend through the Permian. In east-central Alaska, Permian cherts overlie mixed sections of basalt, argillite, chert, and limestone, locally faulted against ultramafic rocks near the Tintina fault (Fig. 12.1a). In the 70-mile area (Fig. 12.1b), Permian radiolarian cherts occur in lenses in greenstone associated with ultramafic rocks, and rarely with wacke sandstone. The ultramafic rocks, basalt, and chert are considered to represent klippen of ophiolite (Keith et al., 1981). The clastic rocks in some of these sequences suggest that the ophiolites may be volcanic arc-related.

South of the Denali fault, volcanic and chert pebble conglomerate grading up into volcaniclastic sandstone, overlain by gray bedded radiolarian chert siltstone, and carbonate, lies unconformably over the Chulitna ophiolite section (Fig. 12.5) (Jones et al., 1980). At Shellabarger Pass, Permian chert comprises 30% of a chert/shale/siltstone sequence that gives way upward to sandy siltstone tuff and ultimately volcaniclastic graywacke (Fig. 12.6) (Reed and Eberlein, 1972). Permian chert in the Wrangell Mountains in the eastern Alaska Range are composed mostly of sponge spicules in a shallow-water platform carbonate sequence overlying mafic flows (Murchey, 1985).

In southeastern Alaska, dolostone of the Pybus Formation lies unconformably on the Cannery Formation and includes Cannery chert pebbles at its base. The dolostone is fossiliferous and contains ubiquitous zones of replacement chert (Fig. 12.7). The top 2 m of the Pybus dolostone is the most intensely silicified part of the unit, consisting of 75% chert (Loney, 1964).

In summary, in northern Alaska, most Permian cherts are part of chert/shale associations with interspersed areas of chert/basalt in central Alaska, juxtaposed by faults. Chert in southern Alaska tends to be associated with carbonates and shallower water sequences.

Triassic Chert

In northern Alaska, the silty chert and siliceous shale of the Pennsylvanian to Lower Triassic Siksikpuk Formation grade up into black shale and black and green chert of the Triassic and Jurassic Otuk Formation (Mull et al., 1982). These are rhythmically bedded, often with turbidite sedimentary features and, in places, complete Bouma sequences. The tops of some beds are cream-colored and contain abundant Monotis fossils. Silicified carbonate turbidites are present higher in the section (Fig. 12.11) (Karl, unpublished field notes). The chert/shale deposits represent outer shelf and slope environments, mostly under relatively low-oxygen conditions. There is no evidence of any volcanism associated with the deposition of these units (Siok, 1985b).

On the south side of the Brooks Range, all of the Mesozoic cherts in the Angayucham and Tozitna terranes (Fig. 12.1b) are associated with volcanic rocks in chert/basalt or chert/basalt/andesite/graywacke associations (Fig. 12.18).

In the central part of the state, the Rampart Group and its equivalents in the Tozitna terrane span the Paleozoic-Mesozoic boundary. The radiolarian cherts in these units comprise less than 20% of the chert/basalt/andesite/clastic sequences. The cherts are red, green, or gray, interbedded with argillite, tuffaceous argillite, cherty tuff, lithic tuff, and basaltic to andesitic volcaniclastic rocks (Jones et al., 1984a). Carbonate rocks are rare. Thickness of sequences varies from meters to tens of meters. These sequences are interpreted to represent volcanic arc-related deposition.

In the Nixon Fork terrane (Fig. 12.1b), there is a Triassic radiolarian spiculitic chert with up to 60% sponge spicules (Patton and Dutro, 1979; Patton et al., 1980). This banded, dark gray deposit overlies Norian, Monotis-bearing sandy limestone, conglomerate, and siltstone (Fig. 12.12). This is a small local deposit representing a shallow-water environment. Shallow-water deposits of similar age are rare, small, and widely distributed in the central part of the state.

South of the Denali fault, in the Chulitna terrane (Fig. 12.5), is some sugary white replacement chert associated with Lower Triassic fossiliferous conglomerate (Jones et al., 1980).

In southern Alaska, in the Chugach Mountains and the Kenai Peninsula, there is Middle and Upper Triassic radiolarian chert intercalated with pillow basalt and minor limestone and graywacke in the Kachemak terrane (Fig. 12.1b) (Kelley, 1985). Because of the occurrence of graywacke with the basalts, this association probably represents a volcanic arc environment. In the nearby Chugach terrane, the McHugh Complex contains Upper Triassic radiolarian chert thinly interlaminated with argillite and tuff (Nelson et al., 1987). Such Triassic chert in the McHugh Complex may occur as blocks in a younger

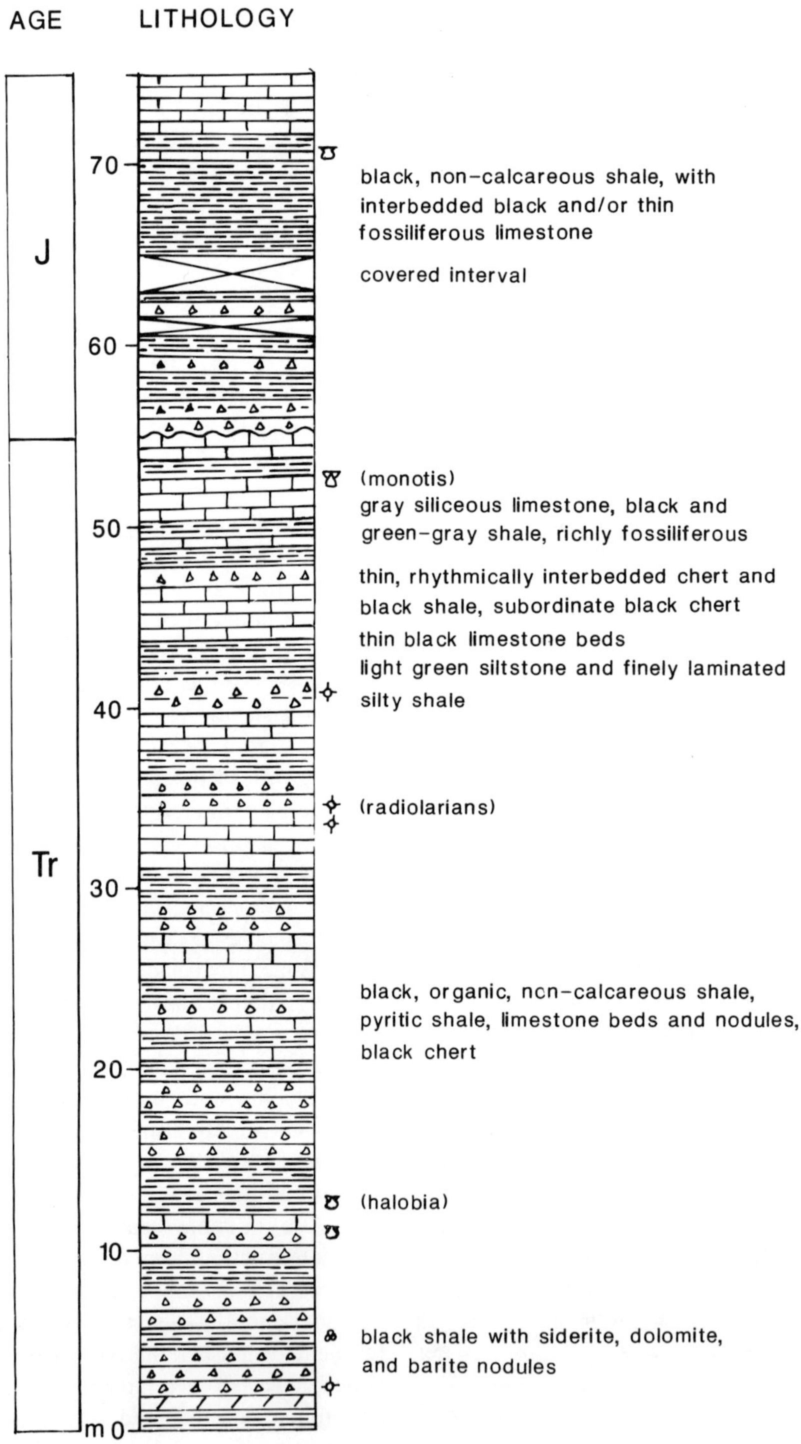

FIGURE 12.11. Composite stratigraphic section for the Triassic and Jurassic Otuk Formation, northern Brooks Range, Alaska, after Bodnar (1984). Location 1 on Figure 12.1a.

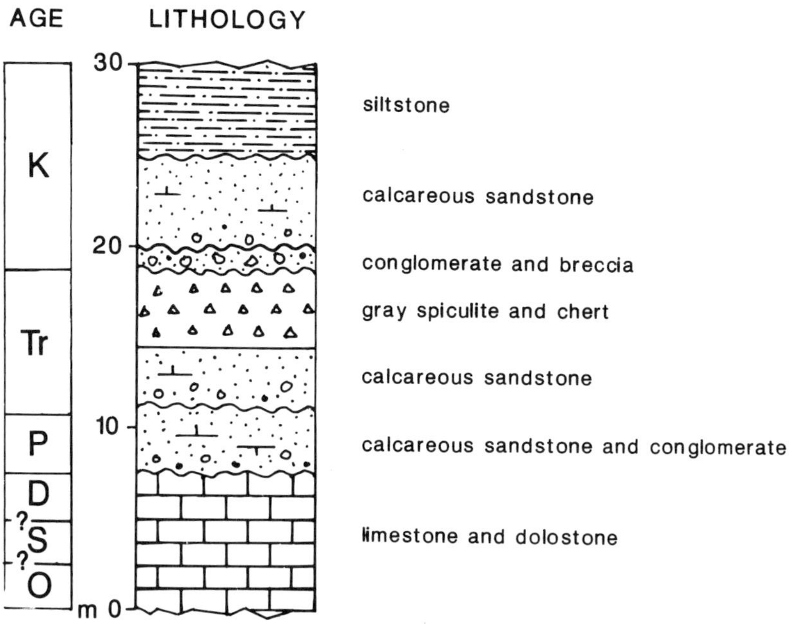

FIGURE 12.12. Stratigraphic section from the northern Kuskokwim Mountains, Medfra quadrangle, central Alaska, from Patton and Dutro (1979). Location 7 on Figure 12.1a.

tuff and argillite matrix melange (which includes blocks of Jurassic and Cretaceous ribbon chert), or it may extend the age of the melange. The McHugh Complex represents an accretionary sequence associated with a volcanic arc.

In the Alexander terrane (Fig. 12.1b) in southeastern Alaska, the Upper Triassic Hyd Group (Loney, 1964) contains 90 m of gray to black argillite and chert, limestone, graywacke and 150 m of pillowed spilitic mafic volcanic rocks. Rapid facies changes suggest volcanic islands. Argillite dominates the chert/argillite; the argillite is gradational to chert beds which are 0.5 to 1.5 cm thick. Locally, the argillite is calcareous or siliceous. Both the argillite and the chert contain radiolarians, and both contain angular, floating clasts of plagioclase, chlorite, sericite, and quartz, suggesting proximity to a continental margin. The black, carbonaceous character of the chert/argillite indicates a low-oxygen environment. Small-scale graywacke turbidites and debris flows also occur in the chert/argillite deposits. Rhyolites of the adjacent Keku Volcanics (Muffler, 1967), combined with the other information above, suggests accumulation of the siliceous deposits in a marginal basin in the vicinity of both a continental margin and a volcanic arc, rift, or wrench basin.

In the Tikchik Lakes and Goodnews Bay area (Fig. 12.1a), in southwestern Alaska, Triassic radiolarian chert lenses occur with pillow basalt, graywacke, and limestone (Hoare and Coonrad, 1978; Jones et al., 1981). These cherts are inferred to be part of a volcanic arc association.

Jurassic Chert

In the northern Brooks Range, chert of the Triassic and Jurassic Otuk Formation is intercalated with a large proportion of shale and silicified calcareous clastic rock, representing slope and outer shelf deposits (Bodnar, 1984). In the Kobuk sequence in the southern Brooks Range, an Upper Triassic to Jurassic part of the Angayucham cherts includes gray, green, and black ribbon chert, black phyllite, tuff, volcaniclastic sandstone, and basalt, which are inferred to represent a volcanic arc association (Plafker et al., 1978).

In central Alaska, the Rampart Group and lithologically equivalent units range into the Jurassic and consist of chert/volcanic rock/graywacke volcanic arc associations.

South of the Denali fault in the Shellabarger Pass section (Fig. 12.6), Jurassic to Cretaceous maroon chert is intercalated with sandstone and conglomerate (with no granitic clasts) unconformably above Triassic mafic tuff and tuff breccia (Reed and Eberlein, 1972). A volcanic arc association is inferred for these rocks. Nearby, in the Lime Hills quadrangle, are Jurassic gray and green radiolarian siltstones (Reed et al., 1985). To the east, the West Fork terrane (Fig. 12.1b) includes a section of Upper Jurassic gray to black rhythmically bedded chert, cherty argillite, siltstone, thin-bedded sandstone, and minor conglomerate, above Lower Jurassic phosphatic calcareous sandstone, sandy limestone, crystal tuff, cherty tuff, and minor conglomerate (Jones et al., 1980). These rocks most likely indicate slope or restricted basin deposition, remotely associated with a volcanic arc, or at least volcanic activity. In the adjacent Chulitna area, Upper Jurassic radiolarian chert is associated with *Inoceramus*- and Buchia-bearing Lower Cretaceous argillite, sandstone, and rare limestone (Jones et al., 1980), indicating a slope facies association.

In southern and southeastern Alaska, the Chugach terrane (Fig. 12.1b) includes the Uyak Complex, the McHugh Complex, the Yakutat Group, and the Kelp Bay Group, which all contain blocks of red radiolarian ribbon chert and blocks with radiolarian chert lenses in basalt, containing radiolarians of Jurassic and Cretaceous age, within argillite and tuff matrix melanges. These blocks are isolated from their original stratigraphic sequences, but various blocks are interpreted to represent oceanic and volcanic arc depositional environments based on the chemical compositions of the chert and basalt (Karl, 1982). There is also chert interbedded with the argillaceous melange matrix, interpreted as slope deposition in the vicinity of a volcanic arc.

Cretaceous Chert

Located between the Kobuk and Kaltag faults (Fig. 12.1a), the Gemuk Group includes volcaniclastic rocks that grade from breccia to sandstone to tuffaceous chert to blue-green chert, in repeating cycles (W.W. Patton, U.S. Geological Survey, oral communication, December 1985). The section is underlain by Lower Cretaceous pillow basalt and capped by coquina and Albian strandline terrigenous clastic rocks. This volcanic facies association is considered to be younger and independent of the volcanic arc represented by the Rampart Group and its lithostratigraphic equivalents. In the Innoko terrane (Fig. 12.1b), slightly to the southwest, there is also green chert interbedded with mafic agglomerate, lapilli tuff, and volcaniclastic shallow-water sandstone containing wood chips and *Inoceramus* of Cretaceous age (Bundtzen and Laird, 1983).

South of the Denali fault, in the West Fork terrane, gray Valanginian chert is intercalated with *Inoceramus*- and Buchia-bearing argillite, sandstone, and limestone—a chert/carbonate/clastic association.

In southern, southwestern, and southeastern Alaska, Lower Cretaceous chert associated with basalt, argillite, volcanic rocks, tuff, and graywacke occurs as blocks in the Chugach terrane melanges (Connelly, 1978; Karl et al., 1979; Nelson et al., 1987; Plafker et al., 1977; Johnson and Karl, 1985). Some of the Cretaceous chert is red radiolarian ribbon chert overlying pillow basalt with chert chemistry and basalt chemistry indicating an oceanic origin (Karl, 1982). Other blocks containing Cretaceous chert consist of interbedded chert and argillite, or chert, argillite, tuff, and graywacke, or chert and graywacke (Fig. 12.13). These chert beds are red or green and commonly show turbidite sedimentary structures. Some chert beds are graded with radiolarian-rich bases and silty tops. Rare granitic clasts have been discovered in associated conglomeratic debris flow deposits. A volcanic arc association is interpreted for these blocks.

Tertiary Chert

Eocene diatoms and rare radiolarians are found in cherts intercalated with siliceous shale, black pyritic shale, tuff, and coarse volcaniclastic rocks in the Andrew Lake Formation on Adak Island (Fig. 12.14) (Scholl et al., 1970; Hein and McLean, 1980). This sequence is interpreted to have been deposited in 200 to 500 m of water in a small marine volcanic arc-associated basin (Hein and McLean, 1980).

Late Cretaceous to middle Eocene radiolarians from chert intercalated with graywacke and basalt of the Orca Group in Prince William Sound is interpreted to represent a cool- or cold-water deposit in a zone of upwelling (Blueford, 1986). Because of the presence of the clastic rocks and the upwelling environment, a slope facies association in the vicinity of a volcanic arc is inferred.

Regional Comparison

Alaska has been divided into tectonostratigraphic terranes by Jones et al. (1981, 1984b). These terranes were defined based on areas with unique stratigraphic sections separated by faults from adjacent areas with different stratigraphic sections. As more is learned about individual terranes, some stratigraphic

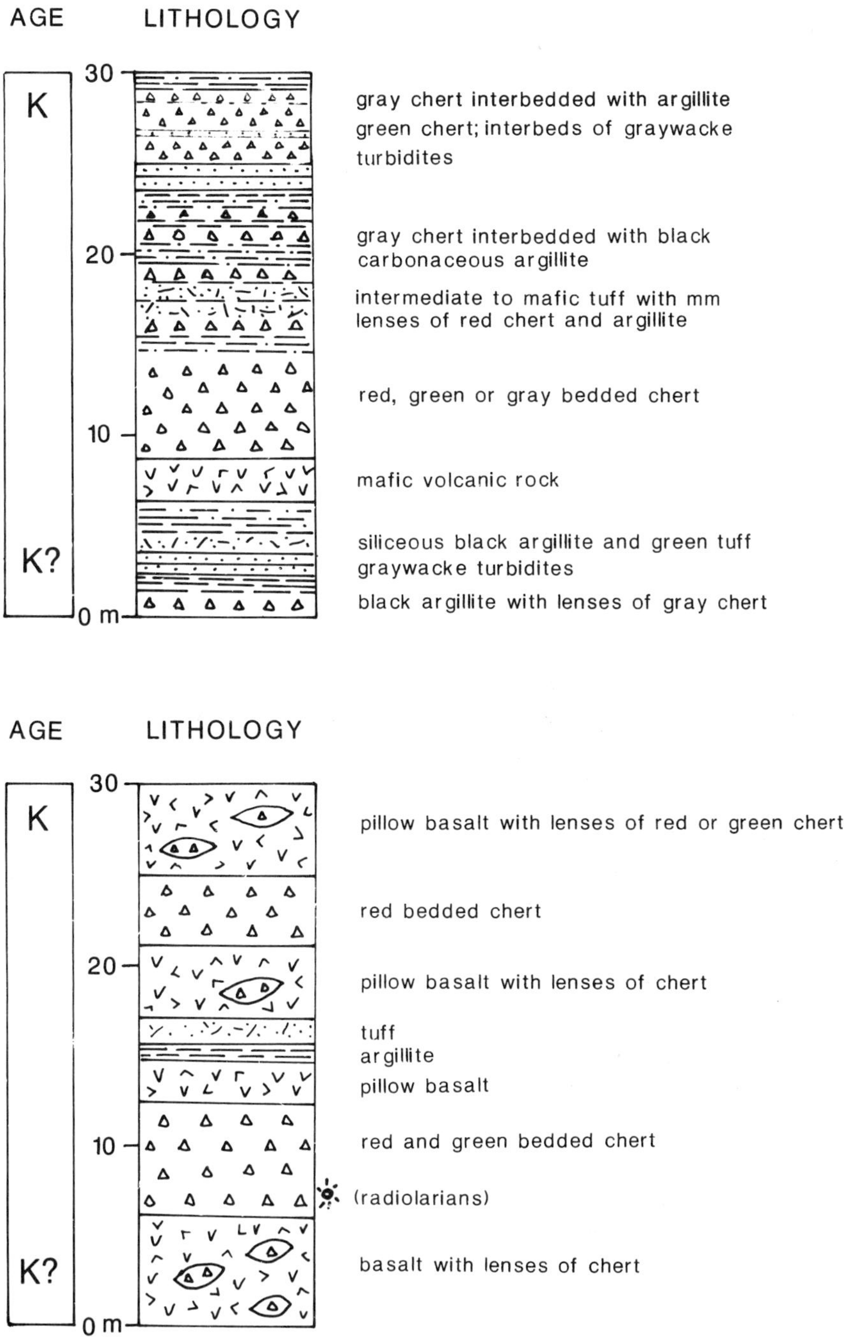

FIGURE 12.13. Generalized stratigraphic section for the Cretaceous and Cretaceous(?) Kelp Bay Group, southeastern Alaska (Johnson and Karl, 1985; Karl, unpublished field notes). Location 10 on Figure 12.1a.

sequences have been related, and some terranes have been combined (Jones et al., 1984b). Based on this regional study, it appears that some terranes in the central part of Alaska can also be combined.

Most of Alaska, in various pieces, was someplace else in Precambrian and Paleozoic time. The state has been assembled mainly since middle Mesozoic time by a complex system of both translational and compressional faults, many of which are still active. Prior to middle Mesozoic time, various parts of Alaska were probably attached to some part of western North America, or they may have been

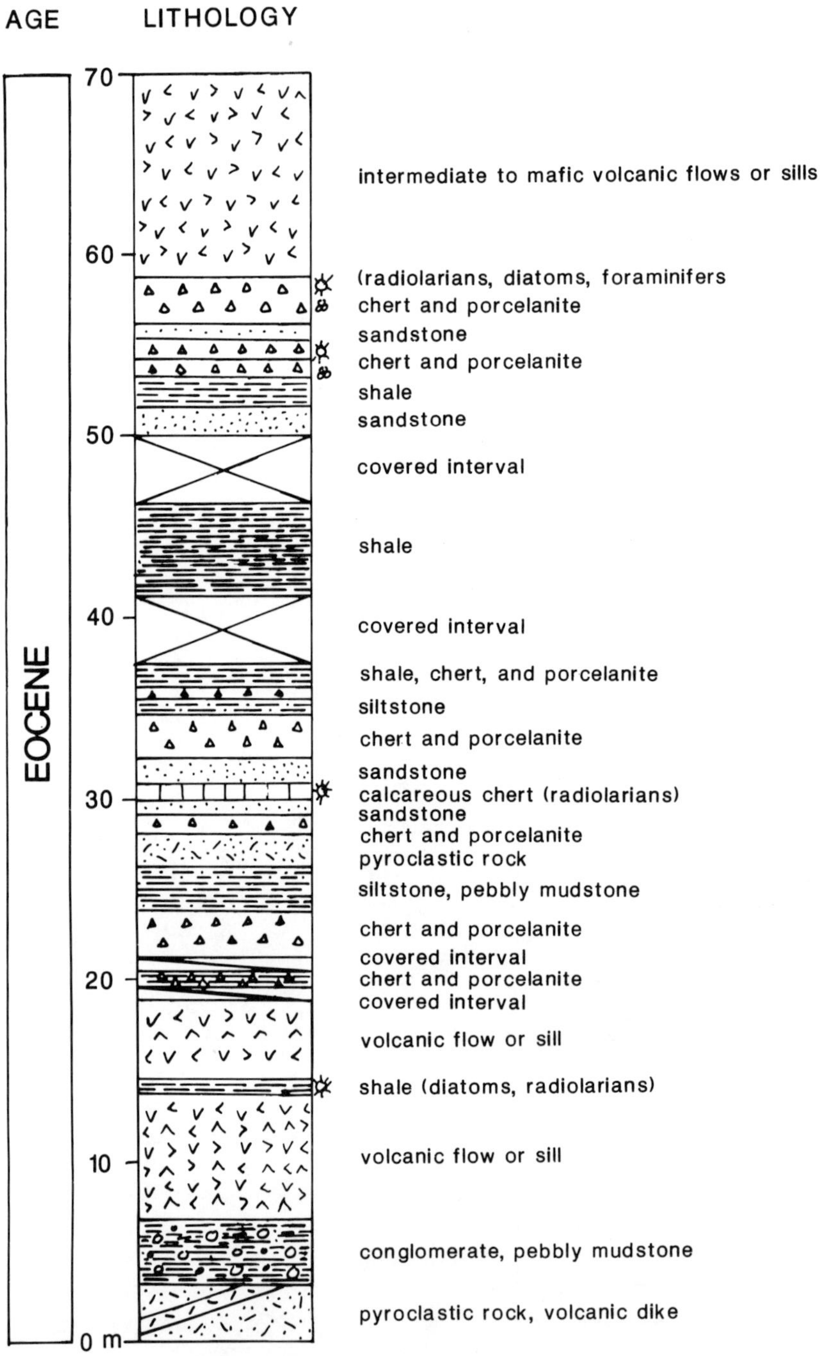

FIGURE 12.14. Stratigraphic section for part of the Eocene Andrew Lake Formation, Adak Island, western Alaska, from Hein and McLean (1980). Adak Island is located in the Aleutian Islands to the west of Figure 12.1a.

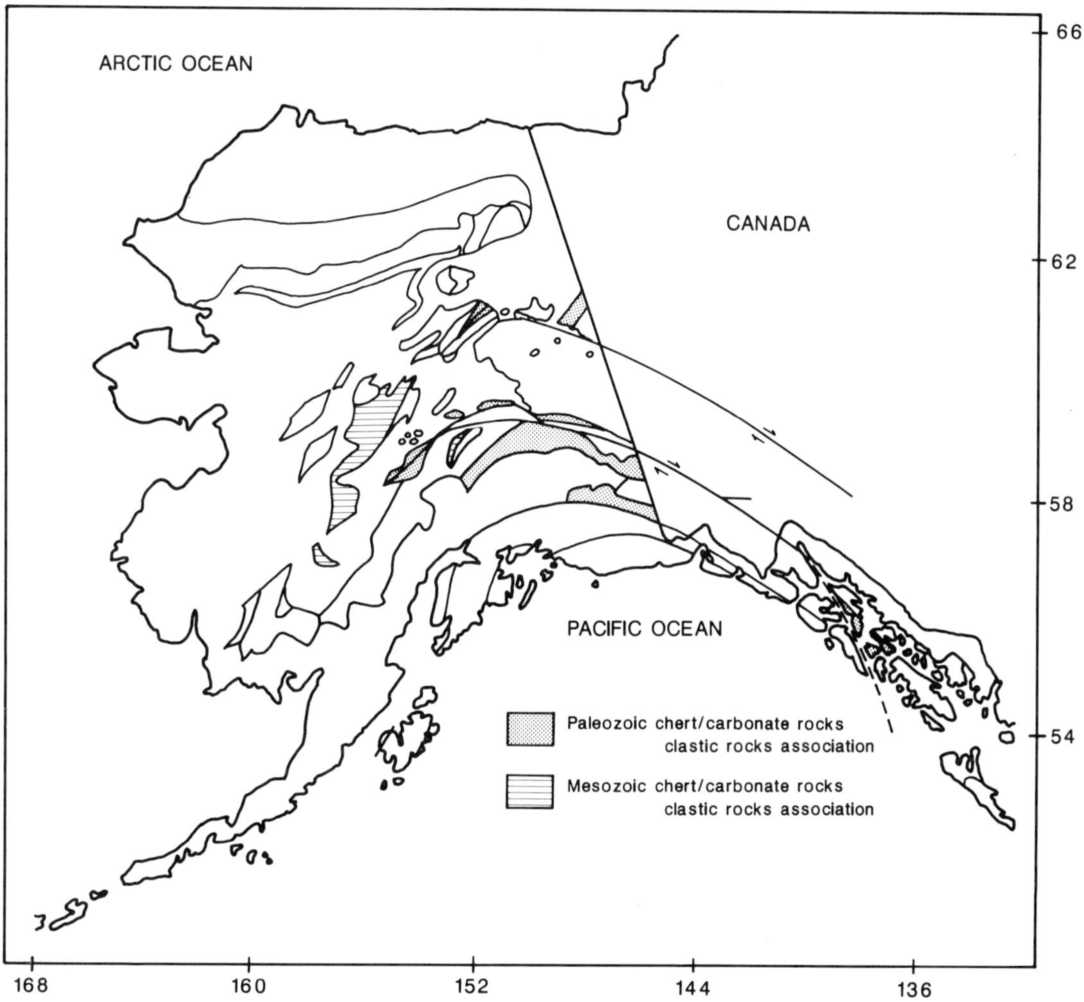

FIGURE 12.15. Map showing the distribution of the chert/carbonate/clastic rock association in Alaska. Outlined areas correspond to terranes in Figure 12.1b.

located somewhere in the proto-Pacific or proto-Arctic Oceans.

Precambrian continental basement rocks have been identified in northern and interior, or central, Alaska, but so far none have been recognized south of the Denali fault. Undated chert in stratigraphic sequences of possible Precambrian age are associated with pelitic and continental clastic rocks, as well as carbonate, various volcanic rocks, and granitic intrusive rocks. These rocks have been interpreted to be shallow-water deposits on continental basement (Weber et al., 1985).

Although stratigraphic relations are not always clear because of faulting, lower Paleozoic chert-bearing sequences occur mainly in northern and central Alaska, where Precambrian continental basement is known or inferred. Some chert beds in central Alaska occur in predominantly carbonate and clastic sequences of Cambrian age. Extensive chert/shale sequences of Ordovician age occur throughout central Alaska. Other Ordovician sequences in central Alaska include volcanic, carbonate, and clastic sequences, representing shallower water depositional environments than the chert/shale sequences. The regional setting for these deposits probably included numerous small basins separated by carbonate buildups, with local volcanic activity. Such basins and platforms apparently covered a broad area over the North American craton. The black chert and shale indicate low-oxygen conditions, commonly found on continental slopes or in restricted basins. Over a broad area, small, relatively shallow basins might not have oceanic connections sufficient to replenish oxygen-depleted water. Shallow-water

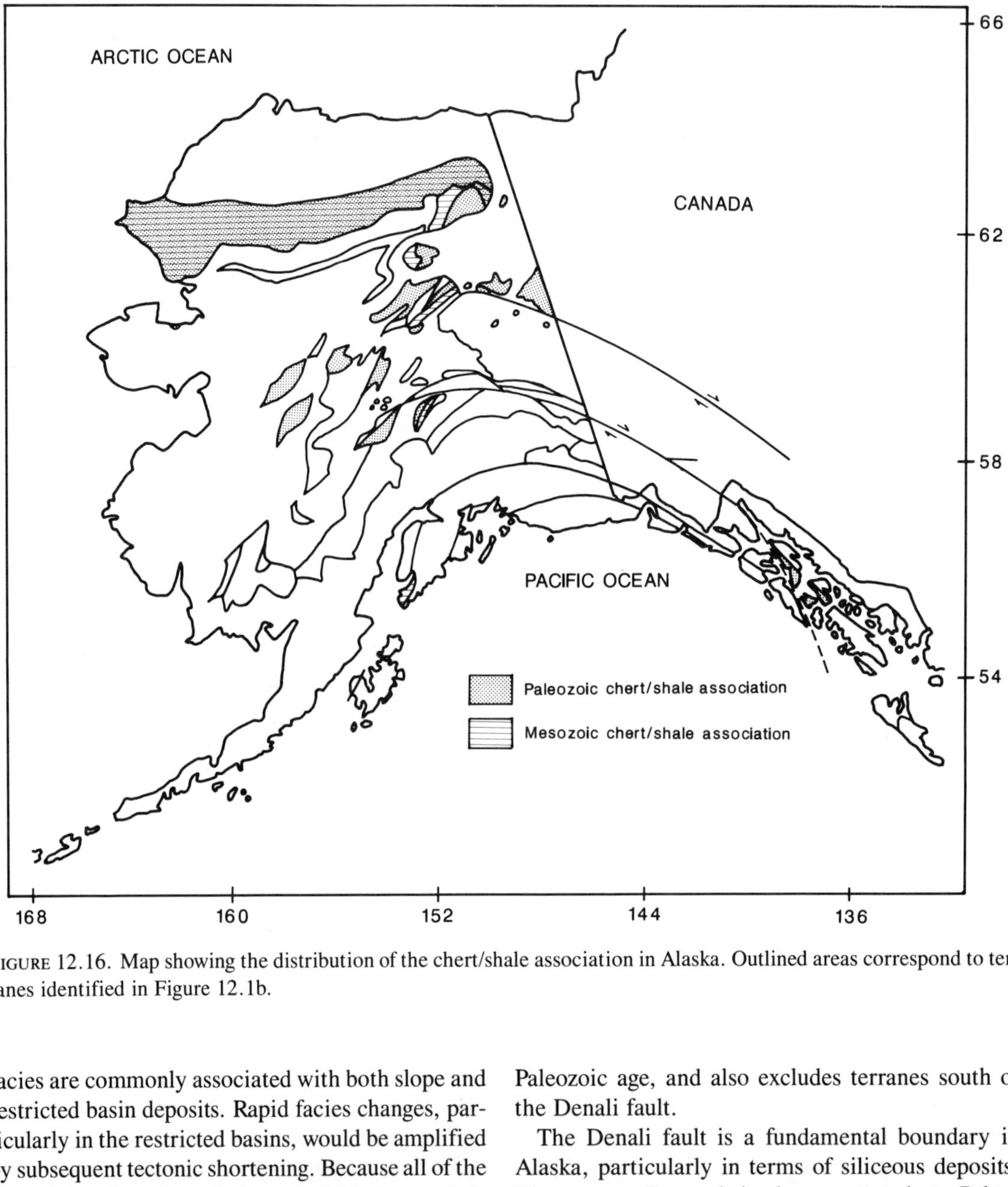

FIGURE 12.16. Map showing the distribution of the chert/shale association in Alaska. Outlined areas correspond to terranes identified in Figure 12.1b.

facies are commonly associated with both slope and restricted basin deposits. Rapid facies changes, particularly in the restricted basins, would be amplified by subsequent tectonic shortening. Because all of the Ordovician chert deposits in central Alaska are shallow-water, platformal, restricted basin, and/or continental slope deposits on continental basement, it seems likely that they were originally part of the same broad region. On this basis, many of the terranes in this area could reasonably be combined. The overlying Paleozoic and Mesozoic chert/clastic sequences over a broad area in central Alaska also consist of facies that need not be considered exotic to the regional depositional environment at any given time horizon. The suggested group of related terranes of central Alaska excludes the Angayucham and Tozitna terranes, which contain oceanic rocks of Paleozoic age, and also excludes terranes south of the Denali fault.

The Denali fault is a fundamental boundary in Alaska, particularly in terms of siliceous deposits. There are no Precambrian basement rocks or Paleozoic continental platform deposits reported from the south of (or outboard of) the Denali fault. There are no siliceous deposits that have been demonstrated or inferred to have continental basement south of the Denali fault. The Denali fault also divides high-paleolatitude radiolarian faunal assemblages typical of the Brooks Range in northern Alaska from low-paleolatitude faunal assemblages characteristic of western cordilleran accreted terranes from southern California to southern Alaska (Blome, 1986). With the exception of the Chulitna ophiolite, and some of the chert blocks in the Chugach terrane, which

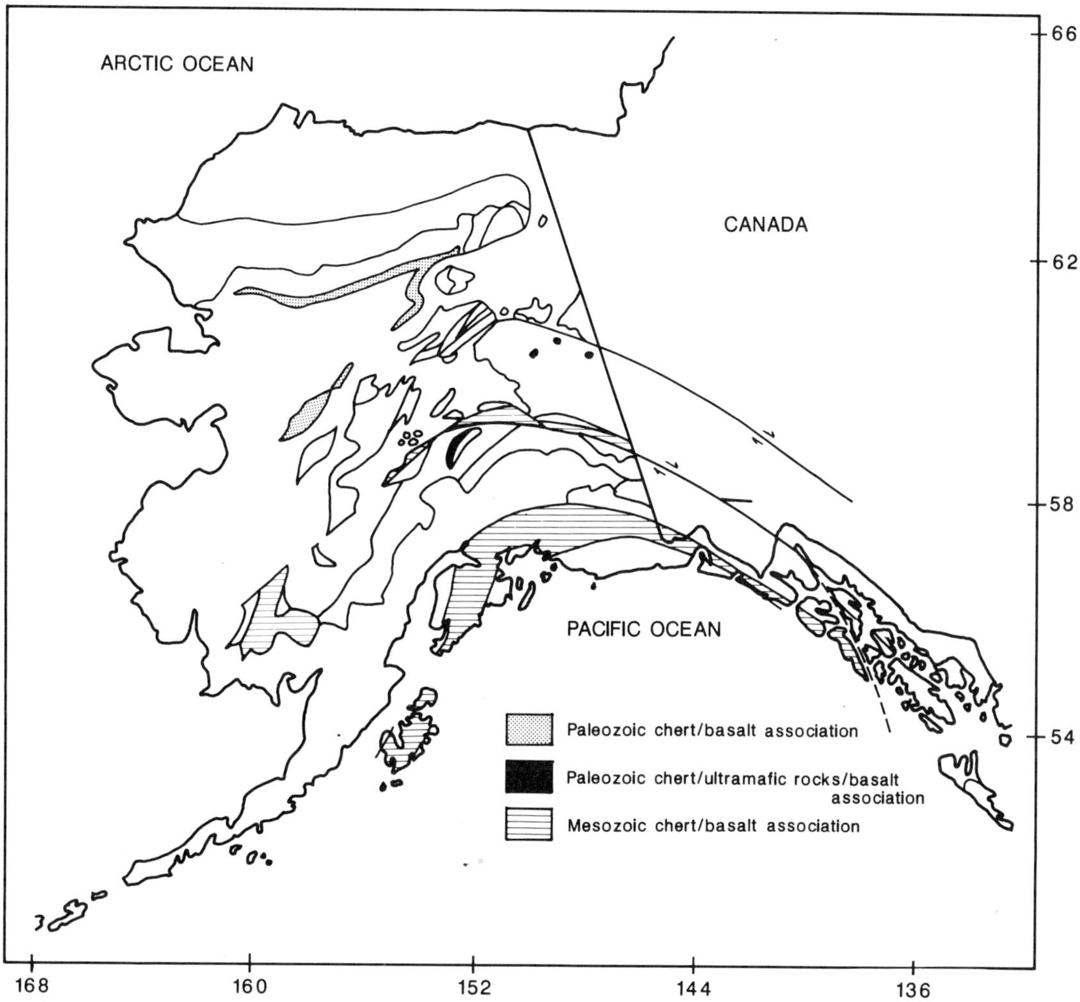

FIGURE 12.17. Map showing the distribution of the chert/basalt and chert/basalt/ultramafic rock associations in Alaska. Outlined areas correspond to terranes identified in Figure 12.1b.

are probably oceanic, all of the siliceous deposits south of the Denali fault were deposited in a volcanic arc or other marginal basin environment, most likely on transitional crust. Stratigraphic sequences of adjacent terranes commonly appear to represent different depositional settings, although there is a common mafic to intermediate volcanic association through most of the siliceous deposits. Differing chemical compositions of igneous rocks and superjacent and subjacent stratigraphy within the terranes suggest that numerous different arcs are represented. Many of these volcanic arcs may have formed somewhere along the western North American margin, but conclusive evidence for such an interpretation is lacking. Some of the arcs may have formed elsewhere in the Pacific basin. Some of the arcs may represent a composite migrating arc system such as Karig et al. (1986) described in the Philippines.

In addition to the major contrast in the nature of Paleozoic and Mesozoic siliceous deposits discussed above, another regional contrast exists in the nature of Mesozoic siliceous deposits. This contrast lies between the cherts of the Pennsylvanian to Jurassic Etivluk Group (Mull et al., 1982) in the Brooks Range and all the Mesozoic cherts to the south of them. The Etivluk Group consists of the Siksikpuk and Otuk Formations described earlier and represents the evolution of a depositional environment from the restricted basins associated with the Mississippian platform carbonates of the Lisburne Group to a progradational shelf-slope margin (Siok, 1985b). The Mesozoic deposits in the Brooks Range are unique sedimentary sequences with respect to

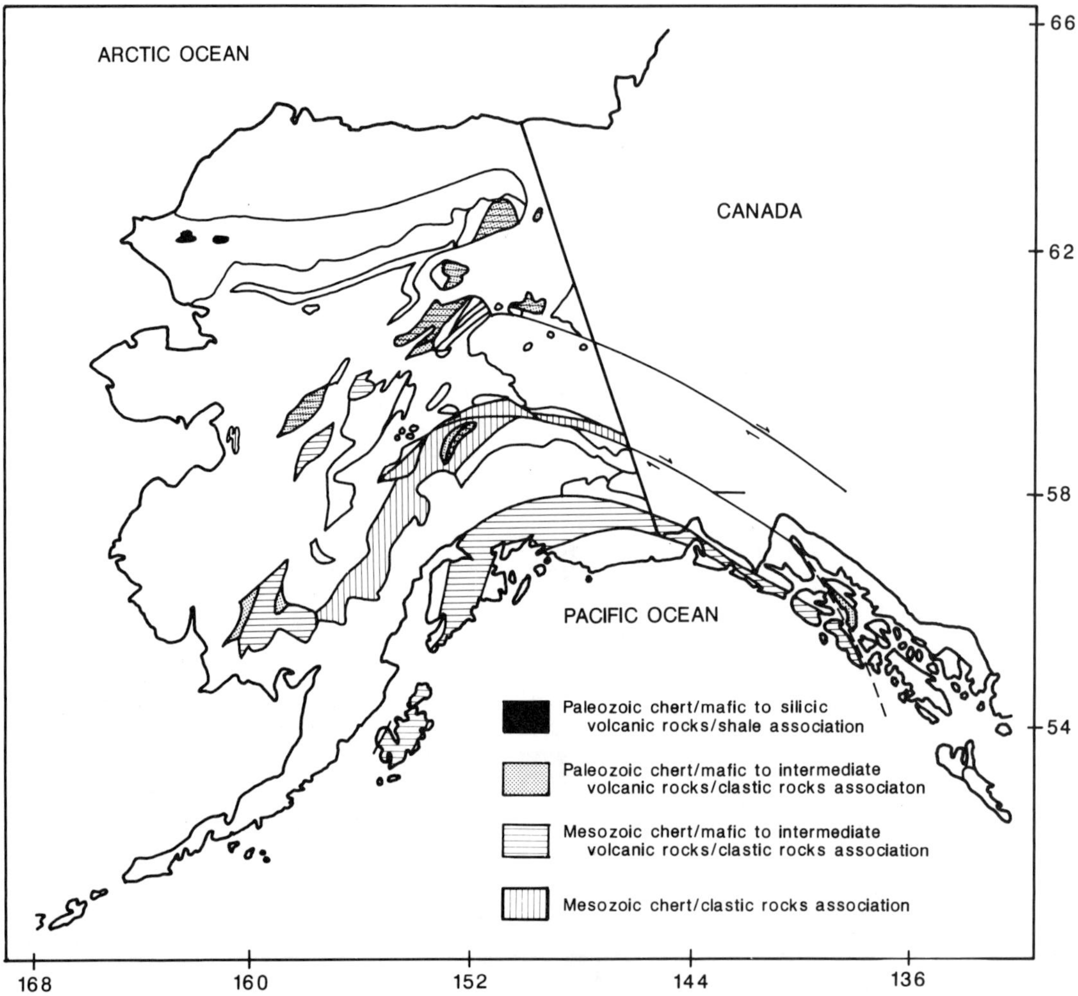

FIGURE 12.18. Map showing the distribution of the chert/basalt/andesite/graywacke and chert/mafic to silicic volcanic rock/clastic rock associations. Outlined areas correspond to terranes identified in Figure 12.1b.

deposits of the same age in the rest of Alaska because of their lack of associated volcanic rocks.

Mesozoic siliceous deposits occurring to the south of the Etivluk Group are associated with varying amounts of volcanic rocks. The inferred volcanic activity apparently overwhelmed most of the sedimentary facies that might have served to distinguish between depositional basins or volcanic arcs, effectively homogenizing middle Mesozoic stratigraphic sequences located to the south of the Brooks Range.

Chert/basalt associations in the Angayucham and Chugach terranes (Fig. 12.1b) may represent Mesozoic (and Paleozoic, in the Angayucham terrane) oceanic depositional environments, as distinguished from volcanic arc or marginal basin environments. However, both the Angayucham and Chugach terranes also contain volcanic arc-related deposits.

Alaska is a complicated patchwork of igneous and sedimentary lithostratigraphic sequences juxtaposed by thrust faults, strike-slip faults, and oblique-slip faults. In a highly simplified tectonic model, the assembly of Alaska may have resulted from a sequential stepping outboard of right transpressive faults, in which bits of western North America and its marginal basins were compressed and/or sliced off and translated northwestward and accreted. As the Pacific and North American plates converged obliquely in postmiddle Mesozoic time, oceanic deposits were accreted and translated northward along the continental plate margin with the fragments of the marginal basin deposits. If a spreading ridge intersected the plate margin at any angle, short sections of ridge with associated long transform faults could have liberated a slice of the continental margin (with its newly accreted material) and translated it northward (somewhat analogously to Baja California). At some irregularity in the plate margin, this slice of crust could have jammed to a halt, forcing a splay in the right lateral system, resulting in the liberation of another slice of continental margin, and translating it northward to a position outboard of the previous slice.

This process could be repeated many times. The steep right lateral faults would separate terranes of continental margin and accreted material. A sedimentary deposit on the continental margin could in this way be transposed from a position south of a related continental sequence to a position west of that same sequence. In the same way, some accreted oceanic deposits (on oceanic crust) could get sandwiched between marginal basin deposits on continental crust. In such a process, pieces of the oceanic Angayucham terrane could end up trapped between contemporaneous continental and margin basin deposits. Additionally, the Denali fault could be analogous to the right lateral splay that stepped outboard of the westernmost Precambrian cratonic crustal basement. Finally, parts of the Chugach terrane could be accreted material that was torn off, translated, and reaccreted.

Conclusion

The sedimentology of siliceous deposits can be used to interpret depositional conditions in the basin where they accumulated. Lithologic associations in chert-bearing stratigraphic sequences can also help discriminate between possible tectonic environments responsible for the development of the depositional basin.

Siliceous deposits in Alaska can be grouped in terms of six lithologic associations that serve to limit the reconstruction of depositional environments. Even in areas of intense disruption, these chert facies associations (Table 12.1) can help interpret basin depth, seawater circulation, relative proximity to a continental margin or volcanic arc, and possible basement rocks. In Alaska, siliceous deposits with continental affinities occur in the northern and central part of the state. Siliceous deposits with volcanic arc affinities occur mainly in the southern half of the state. Most Alaskan siliceous deposits are interpreted as one of a variety of marginal basin deposits. Siliceous deposits with inferred oceanic affinities are trapped as tectonic slivers between these marginal basin deposits, emphasizing the complexities in the evolution of the transpressive tectonic system that assembled Alaska.

Acknowledgments. I would like to thank Benita Murchey of the U.S. Geological Survey for many insights and helpful information. This paper was greatly improved as a result of suggestions from reviewers James R. Hein and Thomas E. Moore. Mark Zayatz provided valuable drafting assistance.

References

Barker F, Jones DL, Budahn JR, Coney P (1988) Ocean plateau-seamount origin of basaltic rocks, Angayucham terrane, central Alaska. Journal of Geology 96.

Blome CD (1986) Late Triassic radiolarian paleogeography for western cordilleran accreted terranes. Geological Society of America, Cordilleran Section Abstracts 3:87.

Blueford JR (1986) New radiolarian data from the Orca Group, Prince William Sound. In: Bartsch-Winkler S, Reed KM (eds) United States Geological Survey in Alaska: Accomplishments During 1985. U.S. Geological Survey Circular 978, pp. 33–34.

Bodnar D (1984) Stratigraphy, age, depositional environments and hydrocarbon source rock evaluation of the Otuk Formation, north-central Brooks Range, Alaska. Unpublished Master's Thesis, University of Alaska, Fairbanks, 232 pp, 26 figures.

Bundtzen TK (1982) Bedrock geology of the Fairbanks Mining District, western sector. Alaska Division of Geological and Geophysical Surveys, Open-File Report 155, 2 pl, scale 1:24,000.

Bundtzen TK, Laird GM (1983) Geologic map of the Iditarod D-1 quadrangle, Alaska. Alaska Division of Geological and Geophysical Surveys Professional Report 78, scale 1:63,360.

Carter C (1977) Age of the Hood Bay Formation, Alaska. In: Sohl NF, Wright WB (eds) Changes in Stratigraphic Nomenclature by the U.S. Geological Survey, 1976. U.S. Geological Survey Bulletin 1435-A, p A117.

Chapman RM, Patton WW Jr (1979) Two upper Paleozoic sedimentary rock units identified in southwestern part of the Ruby quadrangle. In: Johnson KM, Williams JR (eds) United States Geological Survey in Alaska: Accomplishments During 1978. U.S. Geological Survey Circular 804B, pp B59–B61.

Chapman RM, Yeend W (1981) Geologic reconnaissance of the east half of the Kantishna River quadrangle and adjacent areas. In: Albert NRD, Hudson TL (eds) United States Geological Survey in Alaska: Accomplishments During 1979. U.S. Geological Survey Circular 823B, pp B30–B32.

Chapman RM, Weber FR, Churkin M Jr, Carter C (1980) The Livengood Dome chert, a new Ordovician formation in central Alaska, and its relevance to displacement on the Tintina fault. U.S. Geological Survey Professional Paper 1126-F, pp F1–F13.

Chapman RM, Churkin M Jr, Carter C, Trexler JH Jr (1981) Ordovician graptolites and early Paleozoic radiolarians in the Lake Minchumina area date a regional shale and chert belt. In: Albert NRD, Hudson TL (eds) United States Geological Survey in Alaska: Accomplishments During 1979. U.S. Geological Survey Circular 823B, pp B32–B34.

Chapman RM, Patton WW Jr, Moll EJ (1982) Preliminary summary of the geology in the eastern part of the Ophir quadrangle. In: Coonrad WL (ed) United States Geological Survey in Alaska: Accomplishments During 1980: U.S. Geological Survey Circular 844, pp 70–73.

Churkin M Jr, Brabb EE (1965) Ordovician, Silurian, and Devonian biostratigraphy of east-central Alaska. American Association of Petroleum Geologists Bulletin 49(2):172–185.

Connelly W (1978) Uyak Complex, Kodiak Island, Alaska—a Cretaceous subduction complex. Geological Society of America Bulletin 89:755–769.

Foster HL, Laird J, Keith TEC, Cushing GW, Menzie WD (1983) Preliminary geologic map of the Circle quadrangle, Alaska, U.S. Geological Survey Open-File Report 83-170-A, scale 1:250,000.

Garrison RE (1974) Radiolarian cherts, pelagic limestones, and igneous rocks in eugeosynclinal assemblages. In: Hsu KJ, Jenkyns HC (eds) Pelagic Sediments: On Land and Under the Sea. Special Publication of the International Association of Sedimentologists 1:367–399.

Heath GR (1974) Dissolved silica and deep-sea sediments. In: Hay WW (ed) Studies in Paleo-oceanography. Society of Economic Paleontologists and Mineralogists Special Publication 20, pp 77–93.

Heath GR, Moberly R (1971) Cherts from the western Pacific, leg 7, Deep Sea Drilling Project. In: Winterer EL, Riedel WR, et al. (eds) Initial Reports of the Deep Sea Drilling Project. U.S. Government Printing Office, Washington 7:991–1007.

Hein JR, McLean H (1980) Paleogene sedimentary and volcanogenic rocks from Adak Island, central Aleutian Islands, Alaska. In: Shorter Contributions to Stratigraphy and Structural Geology, 1979. U.S. Geological Survey Professional Paper 1126E, pp E1–E16.

Hoare JM, Coonrad WL (1978) Geologic map of Goodnews and Hagemeister Island quadrangles region, southwestern Alaska. U.S. Geological Survey Open-File Report 78-9-B, scale 1:250,000.

Johnson BR, Karl SM (1985) Geologic map of western Chichagof and Yakobi Islands, southeastern Alaska. U.S. Geological Survey Map I-1506, 15 pp. Scale 1:125,000.

Jones DL, Silberling NJ, Csejtey B Jr, Nelson WH, Blome CD (1980) Age and structural significance of ophiolite and adjoining rocks in the upper Chulitna district, south-central Alaska. U.S. Geological Survey Professional Paper 1121-A, pp A1–A21.

Jones DL, Silberling NJ, Berg HC, Plafker G (1981) Tectonostratigraphic terrane map of Alaska. U.S. Geological Survey Open-File Report 81-792, scale 1:500,000.

Jones DL, Silberling NJ, Chapman RM, Coney PJ (1984a) New ages of radiolarian chert from the Rampart district, east-central Alaska. In: Coonrad WJ, Elliott RL (eds) United States Geological Survey in Alaska: Accomplishments During 1981. U.S. Geological Survey Circular 868, pp 39–43.

Jones DL, Silberling NJ, Coney PJ, Plafker G (1984b) Part A—lithotectonic terrane map of Alaska. In: Silberling NJ, Jones DL (eds) Lithotectonic Terrane Maps of the North American Cordillera. U.S. Geological Survey Open-File Report 84-523A, 12 pp, 1 sheet, scale 1:500,000.

Karig DE, Sarewitz DR, Haeck GD (1986) Role of strike-slip faulting in the evolution of allocthanous terranes in the Philippines. Geology 14:852–855.

Karl SM (1982) Geochemical and depositional environments of upper Mesozoic radiolarian cherts form the northeastern Pacific Rim and from Pacific DSDP cores. Unpublished Ph.D. Thesis, Stanford University, Stanford, CA, 365 pp.

Karl SM (1984) Sedimentologic, diagenetic, and geochemical analysis of upper Mesozoic ribbon cherts from the Franciscan Assemblage at the Marin Headlands, California. In: Blake MC Jr (ed) Franciscan Geology of Northern California. Society of Economic Paleontologists and Mineralogists Book 43, pp 71–88.

Karl SM, Decker J, Jones DL (1979) Early Cretaceous radiolarians from the McHugh Complex, southcentral Alaska. In: Johnson KM, Williams JR (eds) United States Geological Survey in Alaska: Accomplishments During 1978. U.S. Geological Survey Circular 804B, pp B88–B90.

Keith TEC, Foster HL, Foster RL, Post EV, Lehmbeck WL (1981) Geology of an Alpine-type pendotite in the Mount Sorenson area, east-central Alaska. U.S. Geological Survey Professional Paper 1170-A, 9 pp.

Kelley JS (1985) Geologic setting of the Kenai Peninsula and Cook Inlet Tertiary basin, south-central Alaska. In: Guide to the Geology of the Kenai Peninsula, Alaska. Alaska Geological Society Guidebook, 19 pp.

Kelts K, Arthur MA (1981) Turbidites after ten years of deep-sea drilling—wringing out the mop? Society of Economic Paleontologists and Mineralogists Special Publication 32, pp 91–127.

Kennett JP (1982) Marine Geology. Prentice-Hall, Englewood Cliffs, NJ, 813 pp.

Lange IM, Nokleberg WJ, Plahuta JT, Krouse HR, Doe BR (1985) Geologic setting, petrology, and geochemistry of stratiform sphalerite-galena-barite deposits, Red Dog Creek and Drenchwater Creek areas, northwestern Brooks Range, Alaska. Economic Geology 80:1896–1926.

Loney RA (1964) Stratigraphy and petrography of the Pybus-Gambier area, Admiralty Island, Alaska. U.S. Geological Survey Bulletin 1178, 103 pp.

Loney RA, Himmelberg GR (1984) Preliminary report on ophiolites in the Yuki River and Mount Hurst area, west-central Alaska. In: Coonrad WL, Elliott RL (eds) United States Geological Survey in Alaska: Accomplishments During 1981. U.S. Geological Survey Circular 868, pp 27–30.

Moore DW, Young LE, Modene JS, Plahuta JT (1986) Geologic setting and genesis of the Red Dog zinc-lead-silver deposit, western Brooks Range, Alaska. Economic Geology 81:1696–1727.

Muffler LJP (1967) Stratigraphy of the Keku Islets and neighboring parts of Kuiu and Kupreanof Islands, southeastern Alaska. U.S. Geological Survey Bulletin 1241-C, pp C1–C52.

Mull CG, Tailleur IL, Mayfield CF, Ellersieck I, Curtis S (1982) New upper Paleozoic and lower Mesozoic stratigraphic units, central and western Brooks Range, Alaska. American Association of Petroleum Geologists Bulletin 66:3:348–362.

Murchey BL (1985) Using radiolarian assemblage variations as a tool for paleogeographic reconstructions. EOS (American Geophysical Union Transactions) 66: 862.

Murchey BL, Swain PB, Curtis S (1981) Late Mississippian to Pennsylvanian radiolarian assemblages in the Siksikpuk(?) Formation at Nigu Bluff, Howard Pass quadrangle, Alaska. In Albert NRD, Hudson TL (eds) United States Geological Survey in Alaska: Accomplishments During 1979. U.S. Geological Survey Circular 823B, pp B17–B19.

Murchey BL, Jones DL, Holdsworth BK (1983) Distribution, age, and depositional environments of radiolarian chert in western North America. In: Iijima A, Hein JR, Siever R (eds) Siliceous Deposits in the Pacific Region: Developments in Sedimentology. Elsevier, Amsterdam, pp 109–131.

Nelson SW, Blome CD, Karl SM (1987) Late Triassic and early Cretaceous fossil ages from the McHugh Complex, southern Alaska. In: Hamilton TD, Galloway JP (eds) Geologic Studies in Alaska by the U.S. Geological Survey During 1986. U.S. Geological Survey Circular 998, pp 96–99.

Nisbet EG, Price I (1974) Siliceous turbidites: Bedded cherts as redeposited ocean ridge-derived sediments. In: Hsu KJ, Jenkyns HC (eds) Pelagic Sediments: On Land and Under the Sea. International Association of Sedimentologists Special Publication 1, pp. 351–366.

Patton WW Jr (1957) A new upper Paleozoic formation, central Brooks Range, Alaska. U.S. Geological Survey Professional Paper 303-B, part 3, pp 41–45.

Patton WW Jr, Dutro JT Jr (1979) Age of the metamorphic complex in the northern Kuskokwim Mountains, west-central Alaska. In: Johnson KM, Williams JR (eds) United States Geological Survey in Alaska: Accomplishments During 1978. U.S. Geological Survey Circular 804B, pp B61–B63.

Patton WW Jr, Moll EJ, Dutro JT Jr, Silberman ML, Chapman RM (1980) Preliminary geologic map of the Medfra quadrangle, Alaska. U.S. Geologic Survey Open-File Report 80-811A, scale 1:250,000.

Plafker G, Jones DL, Pessagno EA Jr (1977) A Cretaceous accretionary flysch and melange terrane along the Gulf of Alaska margin. In: Blean KM (ed) United States Geological Survey in Alaska: Accomplishments During 1976. U.S. Geological Survey Circular 751-B, pp B41–B43.

Plafker G, Hudson TL, Jones DL (1978) Upper Triassic radiolarian chert from the Kobuk volcanic sequence in the southern Brooks Range. In: Johnson KM (ed) United States Geological Survey in Alaska: Accomplishments During 1977. U.S. Geological Survey Circular 722-B, pp B45–B47.

Reed BL, Eberlein GD (1972) Massive sulfide deposits near Shellabarger Pass, southern Alaska Range. U.S. Geological Survey Bulletin 1342, 48 pp.

Reed KM, Blome CD, Gilbert WG, Solie DN (1985) Jurassic radiolaria from the Lime Hills quadrangle. In: Bartsch-Winkler S, Reed KM (eds) United States Geological Survey in Alaska: Accomplishments During 1983. U.S. Geological Survey Circular 945, pp 53–54.

ROCC (Research on Cretaceous Cycles) Group (1986) Rhythmic bedding in upper Cretaceous pelagic carbonate sequences: Varying sedimentary response to climatic forcing. Geology 14:153–156.

Sable EG, Dutro JT Jr (1961) New Devonian and Mississippian formations in the DeLong Mountains, northern Alaska. American Association of Petroleum Geologists Bulletin 45(5):585–593.

Scholl DW, Green HG, Marlow MS (1970) Eocene age of the Adak Paleozoic(?) rocks, Aleutian Islands, Alaska. Geological Society of American Bulletin 81:3583–3592.

Siok JP (1984) A siliceous facies of the Siksikpuk Formation, north-central Brooks Range, Alaska. Geological Society of America, Cordilleran Section, Abstracts, p 333.

Siok JP (1985a) Phosphatic glauconitic sandstone and oncolite deposition at the upper Paleozoic base of Etivluk Group, north-central Brooks Range, Alaska. American Association of Petroleum Geologists, Abstracts, Pacific Section, pp 35–36.

Siok JP (1985b) Geologic history of the Siksikpuk Formation on the Endicott Mountains, and Picnic Creek Allocthons, north-central Brooks Range, Alaska. Unpublished Master's Thesis, University of Alaska, Fairbanks, 253 pp, 13 plates.

Steinman G (1927) Die ophiolithischen Zoren in dem mediterranean Kettengebirge. 14th International Geological Congress, Madrid 2:638–667.

Till AB, Dumoulin JA, Gamble BM, Kaufman DS, Carroll PI (1986) Preliminary geologic maps and fossil data, Solomon, Bendeleben, and southern Kotzebue quadrangles, Seward Peninsula, Alaska. U.S. Geological Survey Open-File Report 86-276, 3 plates, scale 1:250,000, pp A1–18, B1–33.

VanAndel TJH, Komar PD (1969) Ponded sediments of the mid-Atlantic Ridge between 22° and 23° north latitude. Geological Society of America Bulletin 80(7): 1163–1190.

Weber FR, Smith TE, Hall MH, Forbes RB (1985) Geologic Guide to the Fairbanks-Livengood Area, East-Central Alaska. Anchorage, Alaska Geological Society, Guidebook, 44 pp.

Williams H, McBirney AR (1979) Volcanology, Freeman, Cooper, San Francisco, 397 pp.

13
Length-Slow Chalcedony in Sedimentary Rocks of the Mesozoic Allochthonous Terrane in Central Japan and Its Use for Tectonic Synthesis

Isamu Hattori

Overview

Length-slow chalcedony was recently discovered in the Mesozoic Mino Terrane, Central Japan. Lutecite and quartzine, two varieties of length-slow chalcedony, occur in exotic limestone, chert, and dolostone blocks as secondary minerals replacing carbonates or as vein minerals. Clastic fragments of length-slow chalcedony were also found in Mesozoic turbidite sandstone in this terrane. Indigenous rock-formations carrying length-slow chalcedony have not been recognized in Japan. Length-slow chalcedony is generally considered to form under geologic environments different from those for length-fast chalcedony, that is, length-slow chalcedony tends to form in sedimentary rocks deposited on cratonic margins in arid, evaporitic environments. Accordingly, it is likely to consider that the Mino Terrane is composed of tectonic blocks and detritus from cratonic regions containing much length-slow chalcedony formed in evaporitic environments. Recent geologic, paleomagnetic, and paleontologic syntheses of the tectonic evolution of the Japanese Islands have revealed the allochthonous nature of the Mino Terrane. I suggest that length-slow chalcedony can be used as a tracer mineral to identify the provenance of the allochthonous Mino Terrane.

Introduction

I describe the length-slow chalcedony and chalcedonic quartz from Japan and its bearing on the tectonic evolution of Japanese Islands and the circum-Pacific region. Length-slow indicates an optically positive elongation, whereas length-fast denotes an optically negative elongation. Length-slow chalcedony (LSC) includes two varieties, called lutecite and quartzine. Lutecite is characterized by an optically oblique extinction, whereas quartzine is characterized by an optically parallel extinction. Length-fast chalcedony (LFC) includes only one variety, called chalcedonite, which is characterized by an optically parallel extinction (Cayeux, 1916; Frondel, 1962).

The vast majority of chalcedony is LFC, which can be developed in a wide range of geologic environments, geologic ages, and lithologies. In contrast, LSC is not common but is reported rarely from sedimentary rocks in cratonic regions and platforms as is clearly documented in the United States (Fig. 13.1). Virtually all occurrences of LSC in the United States are from the cratonic regions. Much of LSC is said to be formed under arid conditions. This concept dates back to the observation by Pittman and Folk (1971) and Folk and Pittman (1971), who suggested that LSC can be used as a mineralogical indicator of ancient evaporitic climates and vanished evaporitic facies, though the mechanism is far from certain.

Japan is not situated in a cratonic province but forms an orogenic belt, and no reports of LSC have been published. Recently, however, I found the first occurrence of LSC in a Mesozoic allochthonous terrane in Japan (Hattori, 1984, 1985a,b; Hattori and Mizutani, 1986). I suggest that LSC provides an excellent means of detecting the place of origin of allochthonous terranes (Hattori, 1986). In other words, by using LSC as a mineralogical tracer, not only some dispersed continental fragments but also allochthonous tectonostratigraphic terranes may be traced back to their original positions.

In this chapter, modes of occurrence of lutecite and quartzine in sedimentary rocks in a tectonostra-

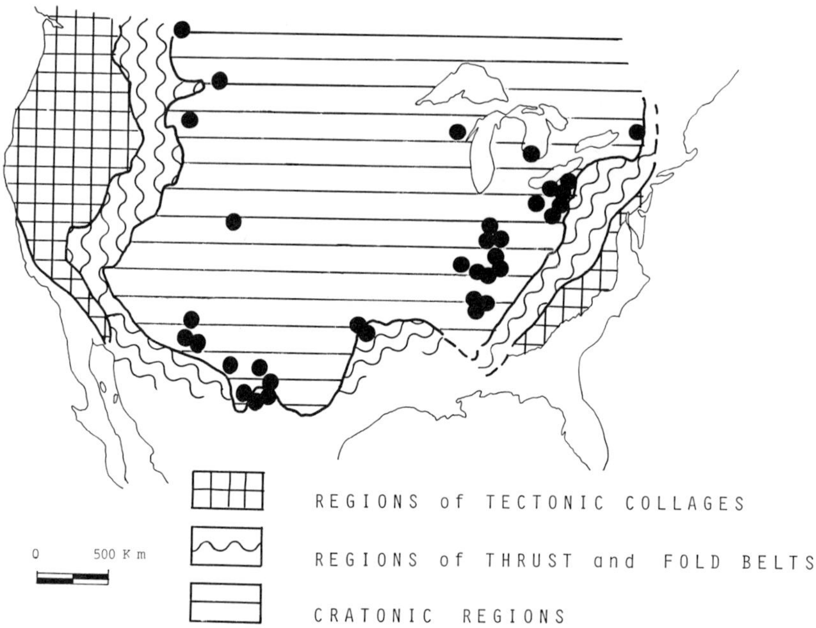

FIGURE 13.1. The distribution of length-slow chalcedony and tectonic divisions in the United States. Length-slow chalcedony is confined to the cratonic region (Chown and Elkins, 1974; Fisher, 1977; Folk, 1973; Folk and McBride, 1976; Folk and Pittman, 1971; Geeslin and Chafetz, 1982; Gill, 1977; Jacka, 1974; Jones and Knauth, 1979; Lougheed, 1983; McBride and Folk, 1977; Meyers, 1977; Meyers and James, 1978; Milner, 1976; Pittman and Folk, 1971; Rapson-McGugan, 1970; Rubin and Friedman, 1977; Schmitt and Boyd, 1981).

tigraphic terrane of central Japan are described, and their origin is discussed from the viewpoint of recent tectonic reconstructions.

Geologic Setting of Length-Slow Chalcedony

The Japanese Islands are interpreted as a tectonic collage composed of nearly 20 allochthonous tectonostratigraphic terranes. Figure 13.2 illustrates the terrane aggregation in southwest Japan. Assembly of these terranes took place in the Mesozoic and Cenozoic along the eastern margin of the Asiatic continent (Hattori, 1982; Saito, 1985; Taira et al., 1983) followed by the separation of the Japanese Islands from Asia in the Miocene (Otofuji and Matsuda, 1983; Otofuji et al., 1985). One well-known terrane is the Mino Terrane (No. 7 in Fig. 13.2), where LSC was first discovered in Japan from exotic rocks.

The Mino Terrane is composed of upper Paleozoic greenstone and limestone and Mesozoic sedimentary rocks (Fig. 13.3) (Hattori, 1982; Mizutani and Hattori, 1983). The Paleozoic rocks occur as blocks in the Mesozoic olistostromes. The typical Mesozoic succession can be characterized by an overall upward-coarsening sequence such as Triassic chert, Lower Jurassic siliceous shale, and Upper Jurassic coarse-grained sedimentary rocks including conglomerate beds. Varied lithologies of coeval ages are also commonplace. Not only did intraterrane recycling and reworking occur, but also incorporation of many extraterrane blocks of greenstone, limestone, dolostone, and chert took place during the Mesozoic processes of terrane transportation, accretion, and collision. Most of LSC has been found in these exotic rocks of central Japan (Fig. 13.3).

Identification of Length-Slow Chalcedony

LSC is identified in thin section. I recommend, however, that before the acceptance of its existence, the chemical composition, crystal structure, and fibrous nature be determined with electron microprobe (EPMA), X-ray diffraction (XRD), and scanning electron microscope (SEM), respectively. Without these determinations, some minerals such

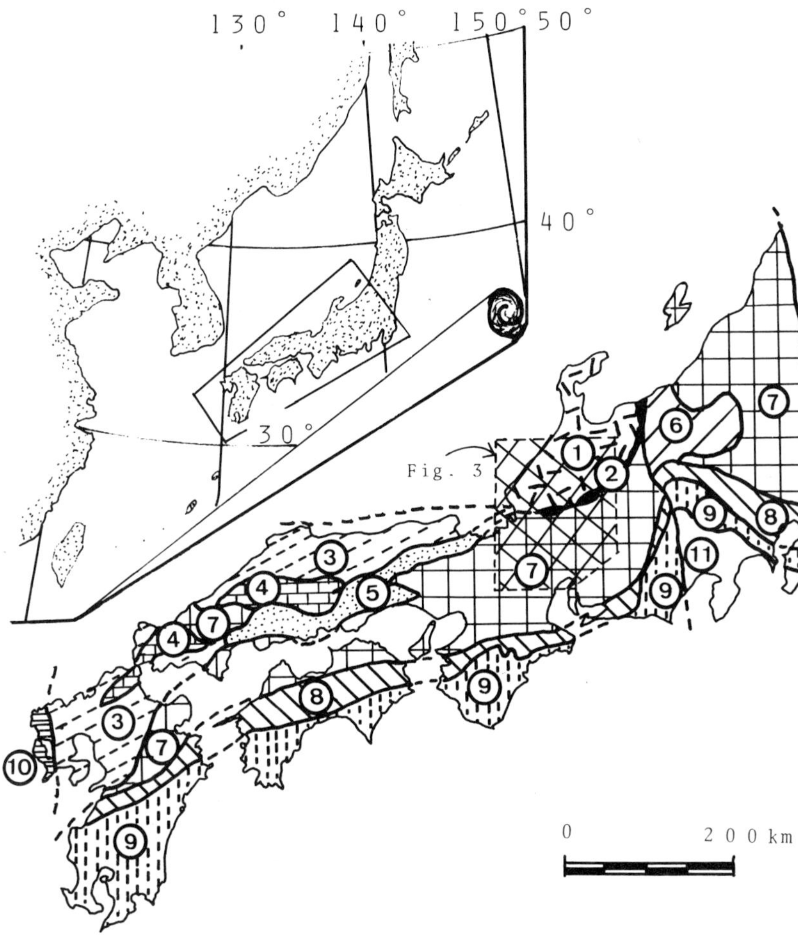

FIGURE 13.2. Tectonostratigraphic map of southwest Japan, modified and simplified after Ozawa et al. (1985). 1, Hida; 2, Circum-Hida; 3, Sangun; 4, Yamaguchi; 5, Maizuru; 6, Joetsu; 7, Tamba-Mino-Ashio; 8, Sambagawa-Chichibu; 9, Shimanto; 10, Nishisonogi; 11, Tanzawa.

as length-slow colorless chlorite may be misidentified as LSC. Flamboyant quartz, microflamboyant quartz, and strained quartz (Folk and Pittman, 1971; Milliken, 1979) may also be misidentified as LSC. Fibrous opal and fibrous cristobalite are also length-slow (Bustillo, 1982). All samples described here are undoubtedly LSC. Figure 13.4 shows a few examples of SEM pictures of LSC and LFC.

Mode of Occurrence of Length-Slow Chalcedony in Japan

More than 20 sites of LSC are known in central Japan (Fig. 13.3). I classify LSC into seven lithologic associations or modes of occurrence: (1) in limestone; (2) in dolostone and associated chert; (3) in bedded chert; (4) in chert cobbles in conglomerate; (5) as clastic grains in turbidite sandstone; (6) in cavities and veins of volcanic rocks; and (7) in veins in manganese nodules (Table 13.1). The first four types of occurrence were selected to be described petrographically. The last three associations are not fully discussed in this paper. In the following description, one or two examples are selected and described as representatives from the individual lithologic associations.

LSC in clastic limestone from site 1, Shirahama (Fig. 13.3) is described (Hattori, 1984, 1985a). LSC occurs as authigenic minerals in clastic limestone and is the first LSC to have been described in Japan. The clastic (allodapic) limestone alternates with muddy red chert containing abundant sponge spicules. The limestone contains fragments of oolitic,

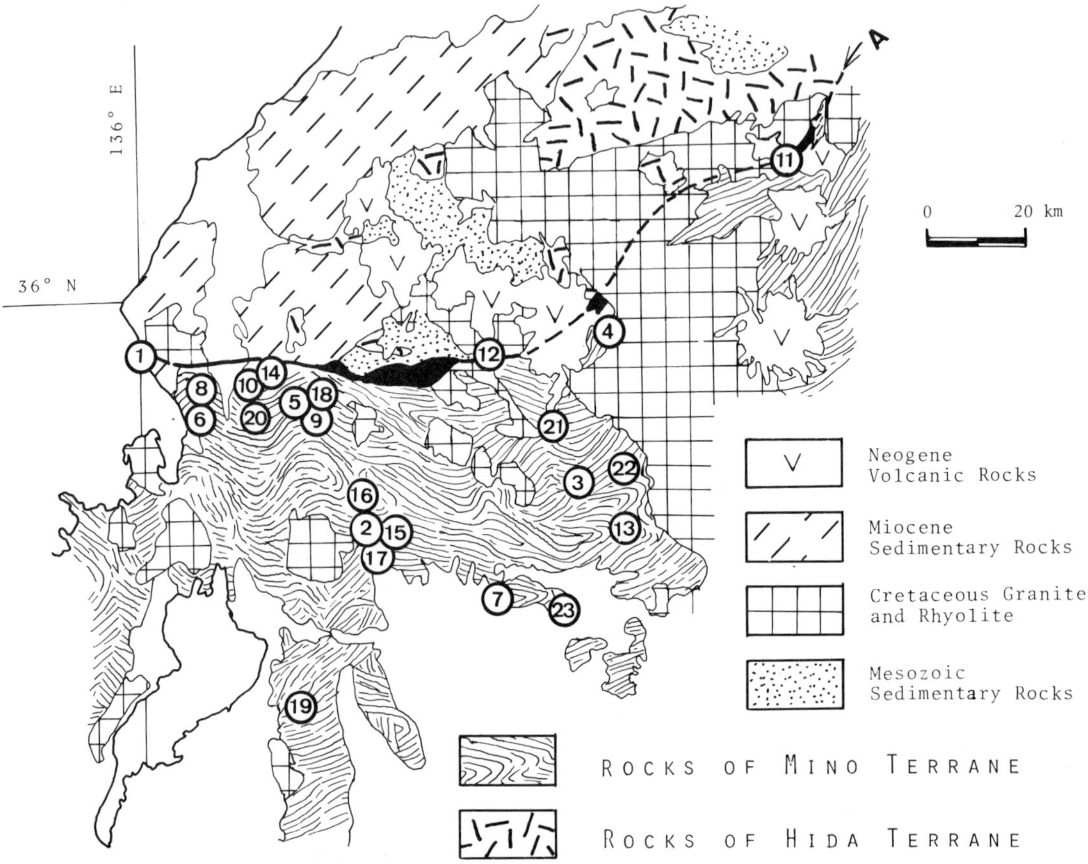

FIGURE 13.3. Occurrence of length-slow chalcedony in central Japan (open circles with numbers). Site names are in Table 13.1. At site 9, LSC-bearing dolostone and conglomerate with LSC-bearing chert clasts occur nearby. A is the circum-Hida Tectonic Zone.

peletal, and crinoidal limestones and abundant fusulinid fossils of Late carboniferous and Early Permian age. Regional geology of central Japan suggests deposition of this unit in the Late Permian. Hattori (1984) concluded that this exotic block was derived in the Early Jurassic from the Paleozoic limestone platform of the Yamaguchi Terrane (No. 4 in Fig. 13.2) and was emplaced as a giant olistolith into a sandstone-shale facies of the Mino Terrane. Limestone is sparsely dolomitized. LSC primarily selectively replaces sparry calcite in the dolomitized limestone, so that the LSC forms spherules in the sparry calcite. Individual LSC spherules were isolated by an acid leach (Fig. 13.5). Where the silicification was pervasive, corroded carbonate fragments float in a matrix of LSC. Diagenetic silica minerals in the limestone are lutecite (Fig. 13.5), but mosaic quartz, chalcedonite, and quartzine are developed in places.

LSC in dolostone is described from site 4, Irasudani (Mizutani et al., 1987); site 5, Fujikura-dani; and site 9, Kanmuri-yama. Dolostone is rare in the Mino Terrane, but three outcrops containing LSC were recently found at sites 4, 5, and 9 in this terrane. The block at site 4 is tens of meters in size and is considered to be a olistolith within the Jurassic sedimentary section. A dolostone layer at site 5 is 10 m thick and more than 100 m long along the strike. This layer occurs between bedded chert of unknown age and Jurassic black shale. A dolostone layer at site 9 is tens of centimeters thick and is intercalated within bedded chert. These dolostones are similar and are described together.

The dolostone is yellow-brown to pale brown on a weathered surface but white to pale green on a fresh surface. At least part of the outcrops appear to be aggregates of rounded cobbles and boulders of dolostone. The dolostone is composed primarily of pro-

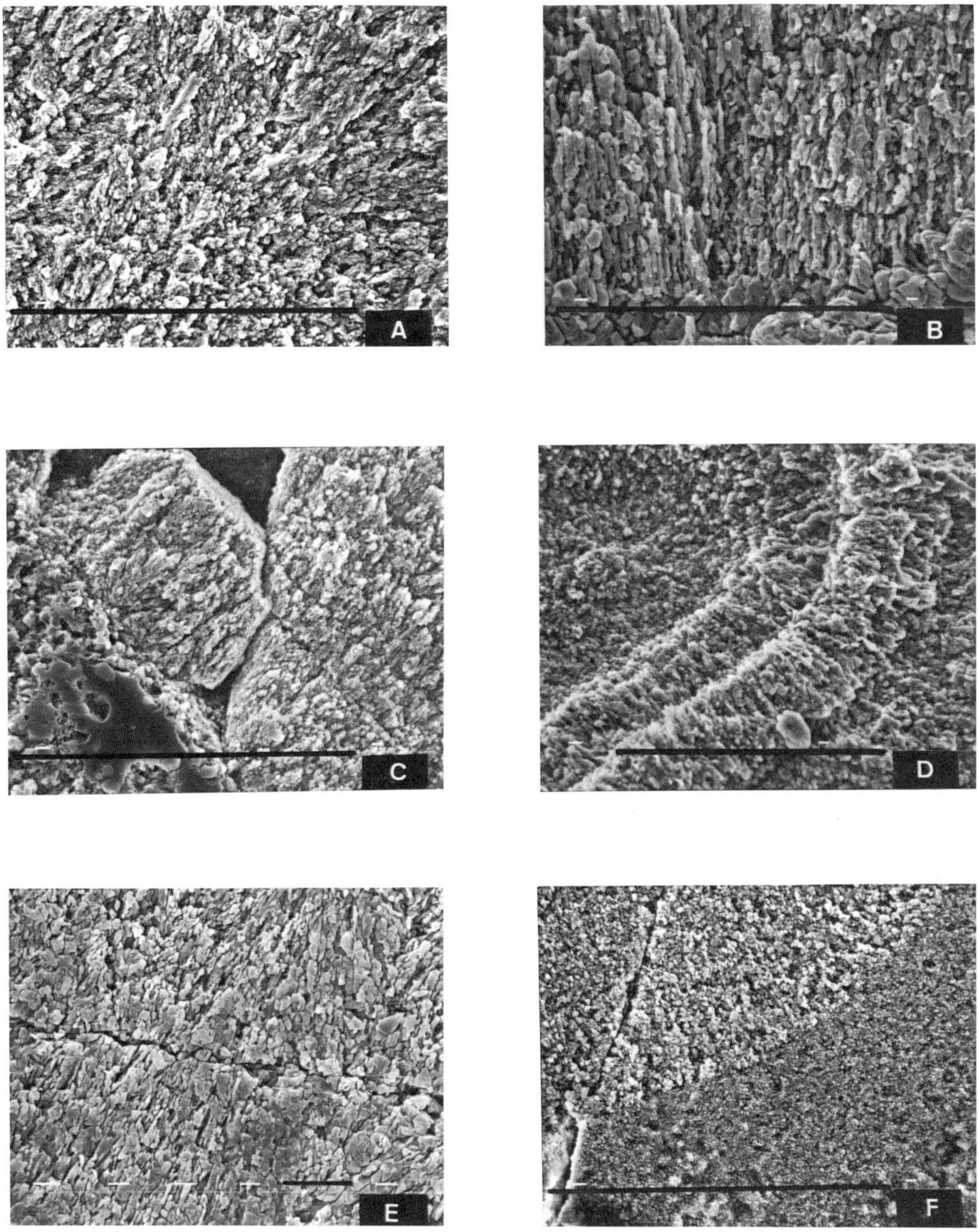

FIGURE 13.4. SEM morphologies of chalcedonic fibers. All bar scales represent 0.1 mm. (A) Lutecite (Shirahama, site 1). (B) Lutecite (Irasu-dani, site 4). (C) Quartzine (Fujikura-dani, site 5). (D) Quartzine occurs to the right of the parallel stripes and chalcedonite to the left (Kanmuri-yama, site 9). (E) Chalcedonite in dolomite (Fujikura-dani, site 5). (F) Coarse and fine chalcedonite (Renkobo-yama).

TABLE 13.1. Modes of occurrence of length-slow chalcedony in the Mino Terrane, central Japan. (Site numbers correspond to those in Fig. 13.3.)

Site	Occurrence, age, and host rocks	Chalcedony type	Reference
A. Length-slow chalcedony in limestone			
1. Shirahama	As spherules in Permian clastic limestone	Lutecite	Hattori (1984, 1985a)
2. Kashiwari-dani	As spherules in Permian limestone breccia	Quartzine	
3. Sobukawa	As spherules in Permian silicified dolomitic limestone	Lutecite	
B. Length-slow chalcedony in dolostone			
4. Irasu-dani	In veins in dolostone and associated chert of unknown age	Lutecite, quartzine	Mizutani et al. (1987)
5. Fujikura-dani	In veins and replacing dolomite in dolostone and associated chert of unknown age	Lutecite, quartzine	
6. Kanmuri-yama	In veins and replacing dolomite in dolostone and associated chert of unknown age	Lutecite, quartzine	
C. Length-slow chalcedony in bedded chert			
7. Yamanaka	In veins in bedded chert of unknown age	Quartzine	
8. Dachiboku-bora	In veins in Triassic bedded chert	Quartzine	Mizutani et al. (1986)
9. Kawachi	In veins in red chert of unknown age	Quartzine	Hattori (1984, 1985a)
D. Length-slow chalcedony in chert cobbles			
10. Kanmuri-yama	As clastic grains and in veins in chert cobbles in Jurassic conglomerate	Quartzine	Hattori (1985) Hattori et al. (1985)
11. Akatani	In veins in chert cobbles in Jurassic conglomerate	Quartzine	Hattori (1985) Hattori et al. (1985)
12. Fukuji	In veins in chert clasts and as clastic grains in Carboniferous conglomerate	Quartzine, lutecite	Mizutani et al. (1986)
13. Tamodani	In veins in chert clasts in Cretaceous conglomerate	Quartzine	Saida (1987)
E. Length-slow chalcedony as clastic grains			
14. Kowa-dani	As clastic grains in Jurassic oligomictic sandstone	Quartzine	
15. Higashimata	As clastic grains in Jurassic sandstone	Quartzine	
16. Takachi-gawa	As clastic grains in Permian basic tuffaceous shale	Quartzine	
F. Length-slow chalcedony in volcanic rocks			
17. Sugihara	Replacing calcite in veins and cavities of Permian greenstone	Quartzine, lutecite	
18. Ozu	In veins in Permian greenstone	Quartzine	
19. Tashiro	With calcite in cavities of Permian greenstone	Quartzine	
20. Nabeziri-yama	With chlorite and quartz in cavities of Permian greenstone	Quartzine	
G. Length-slow chalcedony in manganese nodules			
21. Ogura-dani	With barite in veins in Jurassic manganese nodules	Quartzine	Hattori (1985a)
22. Hachiman	In veins in Jurassic manganese nodules	Quartzine	Adachi (1986)
23. Okukanayama	With barite in veins in Jurassic manganese nodules	Quartzine	Adachi (1986)
24. Unuma	Replacing barite in veins in Jurassic manganese nodules	Quartzine	Adachi (1986)

todolomite with considerable amounts of iron (Mizutani et al., 1987, and my unpublished data), although calcite and stoichiometric dolomite are also present. Microscopically, various types of crystals and crystal aggregations of dolomite are distinguished: micritic dolomite, zoned dolomite, slender crystals of dolomite, xenotopic dolomite, hypidiotopic dolomite, and idiotopic dolomite. The zoned dolomite is generally formed with central opaque domains, probably of iron oxides, surrounded with clear euhedral dolomite. Dolomite pseudomorphs presumably after gypsum are often recognized (Fig. 13.6).

Secondary silica minerals occur in several ways, such as layered chert concretions, small white spherules (Fig. 13.7), and irregular veins. There also commonly occur angular fragments of chert. Some

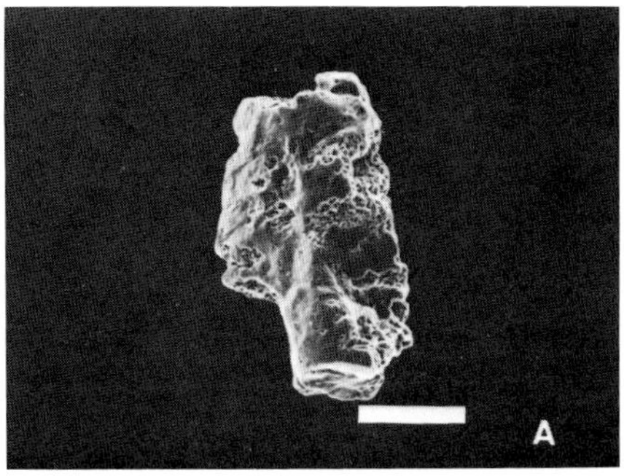

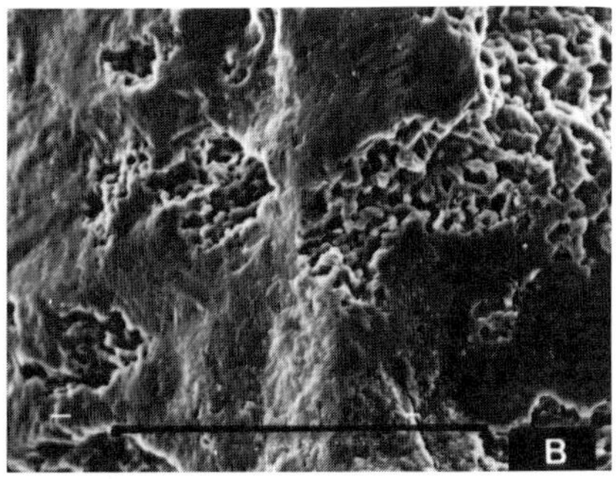

FIGURE 13.5. (A) SEM photographs of a lutecite spherule separated from clastic limestone at Shirahama (site 1). Bar scale represents 1 mm. (B) Closeup of A. Bar scale represents 0.1 mm. (C) Thin-section photograph of lutecite spherule replacing sparry calcite in clastic limestone at Shirahama (site 1). Cross polars. Bar scale represents 0.1 mm.

of these fragments may have been derived from nearby chert layers, and others may have originated from secondary silicification. The silica minerals are mosaic quartz, but chalcedony commonly occurs as vein minerals and as a replacement product. Pseudocubic quartz is also recognized. The chalcedony veins are irregularly curved, deflected, and truncated. The chalcedony species are lutecite, quartzine, and chalcedonite (Figs. 13.6, 13.8). When they co-occur in the tripartite veins, the paragenesis is chalcedonite, LSC, and mosaic quartz (Mizutani et al., 1987).

At sites 4 and 5, irregular layers of chert and nodular chert are developed between dolostone and bedded chert. Bedded chert underlies dolostone, but the exact stratigraphic relation is unknown. At site 9, the boundary between dolostone and chert is disturbed by a small fault, but the contact looks relatively sharp. Very rarely, conodontlike remains are encountered in the bedded chert, but its age is unknown. Small amounts of lutecite and quartzine are recognized in these bedded cherts.

LSC in bedded chert is described from site 6, Yamanaka. The bedded chert in the Mino Terrane generally contains Permian and Triassic radiolarians and is considered to be of pelagic origin. The chalcedony found in this type of chert is dominantly LFC. Exceptionally, white bedded chert, devoid of radiolarian fossils, contains LSC at site 6. The bedded chert section, some tens of meters thick, is truncated at the base by a fault and is underlain by black shale. The chert is also overlain by black shale followed by a turbidite sequence. The boundary between the chert and the overlying black shale is disturbed but seems to be depositional. The lower and upper black shales both bear Jurassic radiolar-

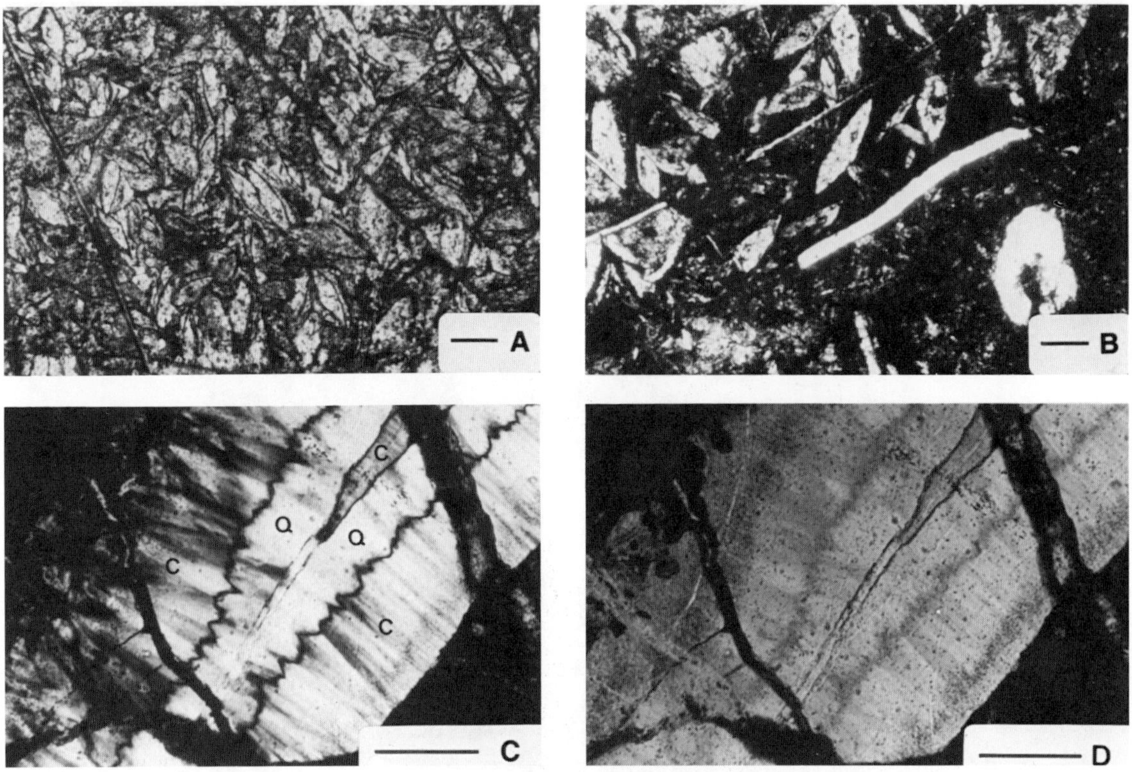

FIGURE 13.6. (A) Dolomite pseudomorphs probably after gypsum (Kanmuri-yama, site 9), parallel polars. Bar scale represents 0.1 mm. (B) Dolomite pseudomorphs probably after gypsum (Kanmuri-yama, site 9). A white stripe at center and a white mass at right bottom are mosaic quartz. Cross polars. Bar scale represents 0.1 mm. (C,D) Chalcedonite (C) and quartzine (Q) in a vein (Kanmuri-yama, site 9). (C) Cross polars; (D) parallel polars. Bar scales represent 0.1 mm.

ians. The chert is considered to be an olistolith in the Jurassic Mino Terrane.

The stratification of the chert is highly deformed and commonly indistinct. In outcrop, the chert looks more crystalline than most radiolarian chert. Microscopically, the chert is composed of microcrystalline quartz with sporadic dolomite. Part of the chert looks like aggregates of angular fragments of chert. LSC as well as LFC occurs in the chert as irregular veins and patches (Fig. 13.8).

LSC in the chert cobbles is described from site 9, Kanmuri-yama (Hattori, 1985b). Intraformational conglomerate beds are exposed at some localities in the northern Mino Terrane (Hattori et al., 1985). The conglomerate contains rounded boulders of leucocratic granite, and angular to subrounded cobbles of chert and limestone. The chert is white or black and contains much LSC. The origin of the chert cobbles is enigmatic. Part of the chert originates from silicification of carbonate rocks, as indicated by corroded dolomite and calcite grains scattered in the chert, and from some carbonate crystals in optical continuity with nearby corroded carbonate crystals. Barite, celestite, and pseudocubic quartz also occur in the chert. From these observations, Hattori (1985b) concluded that the chert formed by silicification of carbonate rocks deposited on a cratonic margin during the middle or late Paleozoic. Quartzine was found in veins and as angular fragments in chert cobbles (Fig. 13.8).

Discussion and Summary

LSC is found in many parts of the world in rocks of many ages (Fig. 13.9). Hattori (1986) recognized five lithologic associations or types of LSC: (1) LSC that occurs in evaporitic rocks such as gypsum beds

FIGURE 13.7. Dolostone-chert association at Fujikuradani (site 5). (A) Fractured surface of dolostone and nodular chert. Pen for scale is about 15 cm long. (B) Polished surface of dolostone with abundant white dots of chalcedony and mosaic quartz. Bar scale represents 10 cm. (C) Polished surface of dolostone (dark) with nodular chert (light). Bar scale represents 10 cm.

(Gill, 1977; Milliken, 1979; Milner, 1976; Nickel, 1982; Siedlecka, 1975, 1976; West, 1964, 1973; West et al., 1968); (2) LSC that occurs in sedimentary rocks, presumably deposited under evaporitic conditions (Carver, 1980; Lougheed, 1983; Meyers and James, 1978); (3) LSC that occurs in silcrete, calcrete, and geodes which form under arid conditions (Chown and Elkins, 1974; Elorza and Rodrigues-Lazaro, 1984a,b; Ross and Chiarenzelli, 1985; Smale, 1973, 1978; Thiry and Millot, 1987; Watts, 1978, 1980); (4) LSC of diagenetic origin in which the relationship to evaporitic facies has not been established (Jacka, 1974; Meyers, 1977; Wilson, 1966); (5) enigmatic LSC of speculative origin (Folk and McBride, 1976, 1978; McBride and Folk, 1979).

It is probable that some occurrences of the fourth type of LSC were formed diagenetically in an evaporitic environment. The Devonian Caballos novaculite (McBride and Thomson, 1970), which bears LSC of the fifth type (Folk and McBride, 1976; McBride and Folk, 1977), occurs as sedimentary rocks probably deposited on the stable cratonic margin and has a close but equivocal relationship to evaporitic facies.

There are many worldwide occurrences of LSC comparable to those in Japan. Folk and Siedlecka (1974) and Siedlecka (1972) reported lutecite and quartzine from Carboniferous fusulina limestone. Jacka (1974) described spherules of LSC in Permian limestone of the Delaware Basin. Bellanca et al. (1986) suggested that LSC-bearing Messinian alternations of diatomite and carbonate were deposited under evaporating conditions. These examples are very similar to LSC occurrence in clastic limestone

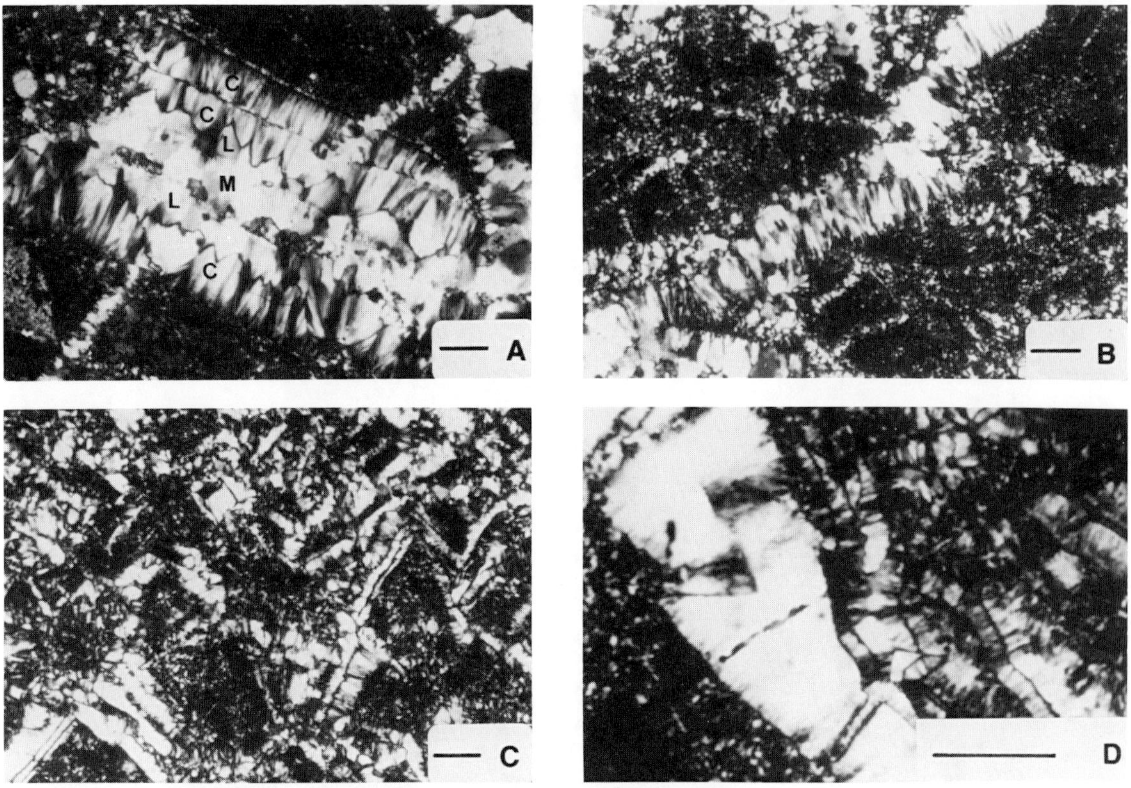

FIGURE 13.8. Photographs of length-slow chalcedony. All bar scales represent 0.1 mm. (A) Tripartite vein with chalcedonite (C), lutecite (L), and mosaic quartz (M) paragenesis, Fujikura-dani (site 5). This vein ends abruptly to the left. (B) Quartzine vein in bedded chert, Yamanaka (site 6). (C) Irregular quartzine and chalcedonite veins in chert clast, Kanmuri-yama (site 9). (D) Quartzine vein (thick white vein) in chert clast, Kanmuri-yama (site 9).

in Japan (sites 1, 2, and 3). LSC occurrences in the Caballos novaculite are comparable to those of chert associated with dolostone (sites 4, 5, and 9), bedded chert (sites 6, 7, and 8), and probably chert cobbles (sites 9 and 10) in Japan.

LSC is commonly found in evaporitic dolostone (Chown and Elkins, 1974; Eriksson and Truswell, 1974; Tucker, 1976a,b). However, few examples of LSC are recorded in dolostone associated with bedded chert. Mizutani et al. (1987) regarded the dolostone-chert association at site 4 as of probable sedimentary origin, formed on a sabkha; some bedded chert (Gutstadt, 1968; Steinitz, 1977) and some dolostone (Bein and Land, 1983; Butler, 1969; Friedman, 1966; Zenger, 1972) can be deposited under evaporitic conditions.

Paleoclimatic interpretation based on the presence of LSC may be valuable where other evidence of environmental conditions is lacking. However, as noted by Adachi (1986), Folk (1975), Keene (1975, 1983), and Simonson (1987) in rock samples and by Kastner (1980, 1981), Oehler (1976), and White and Corwin (1961) in hydrothermal experiments, there are exceptions to formation of LSC in an evaporitic environment and objections against the use of LSC as an indication of such environments. For example, Keene (1975, 1983) recognized authigenic LSC in abyssal Pacific sediment and sea-floor basalts (Fig. 13.8). LSC in manganese nodules and greenstone from Japan possibly formed in a similar environment.

Owing to these apparently exceptional occurrences, West (1979) recommended applying more than one of the 11 mineralogical and textural criteria before concluding formation of LSC under ancient evaporitic conditions. Some of his selected mineralogical indicators are occurrence of pseudomorphs after gypsum, pseudomorphs after anhydrite, quart-

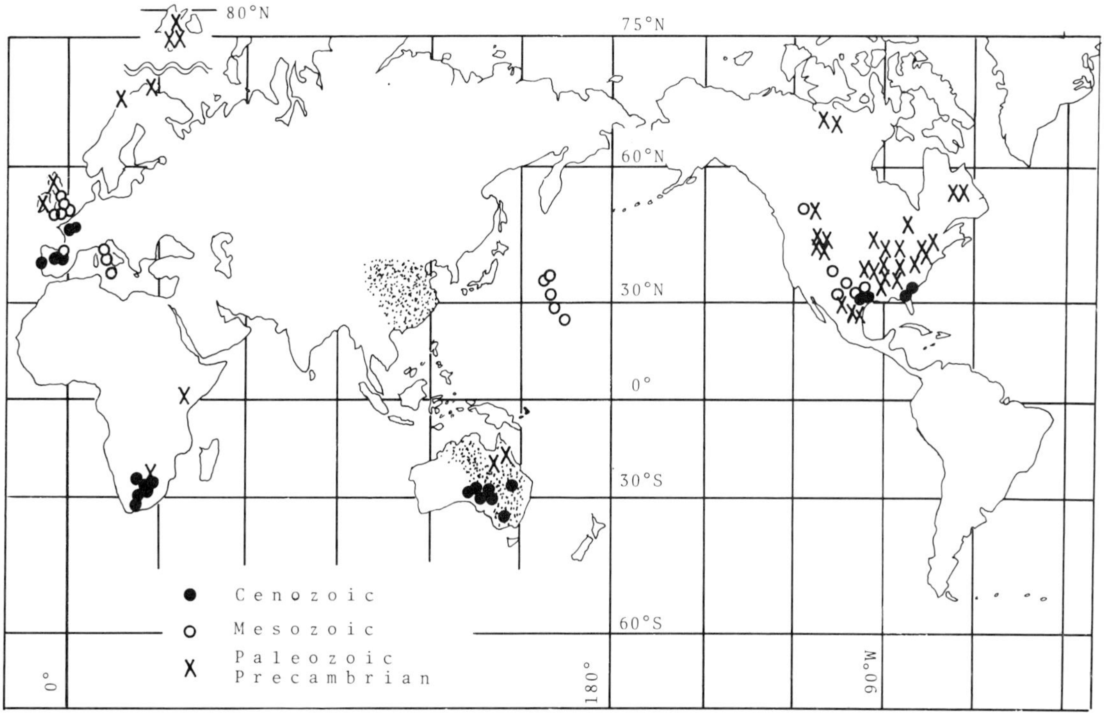

FIGURE 13.9. Worldwide occurrence of Precambrian to Recent length-slow chalcedony. The stippled regions show the distribution of Paleozoic evaporitic and carbonate facies in China and Australia, where the rocks containing the Japanese length-slow chalcedony probably originated. Localities of Japanese length-slow chalcedony are shown in Figure 13.3. Data sources are listed in Hattori (1986).

zine, lutecite, celestite, and barite. McBride and Folk (1977), Milliken (1979), and Rapson-McGugan (1970) suggested that pseudocubic quartz may be pseudomorphs after halite. LSC and the lithologic associations in the Mino Terrane meet many of these criteria.

As shown in Figure 13.1, almost all LSC localities in the United States are restricted to the cratonic regions irrespective of the origin. This is also true for the rest of the world (Fig. 13.9). Based on this distribution, I interpret that LSC-bearing exotic rocks in the Mino Terrane originated at the margins of cratonic provinces or platforms and were probably related to an evaporitic and carbonate environment. Igo (1956, 1961) studied the Fukuji area (site 11 in Fig. 13.3) where exotic Carboniferous aluminous shale is associated with LSC-bearing chert cobbles and independently postulated that the sedimentary rocks had been deposited in an arid environment. The clastic grains of LSC in sedimentary rocks (sites 11, 13, 14, and 15, Fig. 13.3) in Permian tuffaceous shale and in Jurassic sandstone cannot be discussed from the viewpoint of lithologic association, because the clastic grains in turbidite sandstone may also be derived from vanished cratonic provinces with evaporitic facies. A similar occurrence was noted by Schreiber (1974) in submarine sediment of the Balearic Sea, where clastic grains of LSC were interpreted to have been derived from a nearby evaporitic facies.

The Japanese Islands are not situated in a cratonic region, and nowhere do they contain evaporitic sedimentary sequences. Thus, the origin of the LSC may appear enigmatic; however, evidence based on paleomagnetism, paleontology, and regional geology indicates that the Mino Terrane was situated far to the south in early Mesozoic time (Hattori, 1982; Hattori and Hirooka, 1979; Sasajima, 1981). Recent tectonic syntheses suggest that not only China and Australia but also many other cratonic areas in Southeast Asia formed a single continent in the early Paleozoic (Burrett and Stait, 1985; Ji Xiong and Coney, 1985; Lin et al., 1985; Zhang et al., 1984; Zonenshain et al., 1985). It has been suggested that the continent was fragmented in the late Paleozoic to early Mesozoic, and the many fragments dispersed throughout the circum-Pacific region (Hallam, 1986; Howell et al., 1985; McElhinny et al., 1981; Nur and Ben-Avraham, 1983). The stippled regions in Figure 13.9 show the distribution of early to middle Paleozoic evaporitic and carbonate facies in China and Australia (Habicht, 1979; Kemezys, 1974; Veevers, 1984; Yang et al., 1986). Presumably, abundant LSC will be found in these regions. I suggest that the LSC-bearing sedimentary rocks that occur as exotic blocks in the Mino terrane were derived from these cratonic areas. I would also like to stress the usefulness of LSC in tracing the origin of allochthonous tectonostratigraphic terranes found in orogenic belts.

Acknowledgments. I would like to thank S. Mizutani, Nagoya University, for his critical review of the manuscript and suggestions for its improvement. Thanks are extended to J.R. Hein, U.S. Geological Survey, who kindly reviewed and improved the English of the manuscript. Financial support for this study was partly met by grant 59460044 for scientific research from the Ministry of Education, Science, and Culture of Japan.

References

Adachi M (1986) Rock-forming minerals of Mn-carbonate nodules and layers in Jurassic shales of the Mino Terrane: A preliminary report. News Osaka Micropaleontologists (Special Volume 7:275–286 (in Japanese with English abstract).

Bein A, Land LS (1983) Carbonate sedimentation and diagenesis associated with Mg-Ca-chloride brines: The Permian San Andres Formation in the Texas Panhandle. Journal of Sedimentary Petrology 53:243–260.

Bellanca A, Calderone S, Neri R (1986) Isotope geochemistry, petrology and depositional environments of diatomite-dominated Tripoli Formation (lower Messinian), Sicily. Sedimentology 23:729–743.

Burrett C, Stait B (1985) South East Asia as a part of an Ordovician Gondwanaland—a paleobiogeographic test of a tectonic hypothesis. Earth and Planetary Science Letters 75:184–190.

Bustillo MA (1982) Ageing features in inorganic continental opals. Estudios Geológicos 38:335–344.

Butler GP (1969) Modern evaporite deposition and geochemistry of coexisting brines, the sabkha, Trucial coast, Arabian Gulf. Journal of Sedimentary Petrology 39:70–89.

Carver RE (1980) Petrology of Paleocene-Eocene and Miocene opaline sediments, southeastern Atlantic Coastal Plain. Journal of Sedimentary Petrology 50: 569–582.

Cayeux ML (1916) Introduction à L'étude pétrographique des roches sédimentaires. Mémoires pour Servir à L'explication de la Carte Géologique Détaillée de la France. Imprimerie Nationale, Paris, 529 pp.

Chown TM, Elkins JE (1974) The origin of quartz geodes and cauliflower cherts through the silicification of anhydrite nodules. Journal of Sedimentary Petrology 44: 885–903.

Elorza JJ, Rodrigues-Lazaro J (1984a) Late Cretaceous quartz geodes after anhydrite from Burgos, Spain. Geological Magazine 121:107–113.

Elorza J, Rodrigues-Lazaro J (1984b) Existencia de structuras nodulosas de celestina afectadas por silicificacion en el Valle de Losa (N. de Burgos). Estudios Geológicos 40:41–48.

Eriksson KA, Truswell JF (1974) Tidal flat associations from a lower Proterozoic carbonte sequence in South Africa. Sedimentology 21:293–309.

Fisher IS (1977) Distribution of Mississippian geodes and geodal minerals in Kentucky. Economic Geology 72: 864–869.

Folk RL (1973) Evidence for peritidal deposition of Devonian Caballos novaculite, Marathon Basin, Texas. American Association of Petroleum Geologists Bulletin 57:702–725.

Folk RL (1975) Third-party reply to Hatfield: Discussion of Jacka AD 1974, "Fossils by length-slow chalcedony and associated dolomitization; Journal of Sedimentary Petrology 44:421–427." Journal of Sedimentary Petrology 45:952.

Folk RL, McBride EF (1976) The Caballos novaculite revisited. I. Origin of novaculite members. Journal of Sedimentary Petrology 46:659–669.

Folk RL, McBride EF (1978) Radiolarites and their relation to subjacent "oceanic crust" in Liguria, Italy. Journal of Sedimentary Petrology 48:1069–1102.

Folk RL, Pittman JS Jr (1971) Length-slow chalcedony: a new testament for vanished evaporites. Journal of Sedimentary Petrology 41:1045–1058.

Folk RL, Siedlecka A (1974) The "schizohaline" environment: Its sedimentary and diagenetic fabrics as exemplified by late Paleozoic rocks of Bear Island, Svalbard. Sedimentary Geology 11:1–15.

Friedman GM (1966) Occurrence and origin of Quaternary dolomite of salt flat, west Texas. Journal of Sedimentary Petrology 36:263–267.

Frondel C (1962) The System of Mineralogy, Vol 3: Silica Minerals. John Wiley, New York, 334 pp.

Geeslin JH, Chafetz HS (1982) Ordovician Aleman ribbon cherts, an example of silicification prior to carbonate lithification. Journal of Sedimentary Petrology 52:1283–1293.

Gill D (1977) Salina A-1 sabkha cycles and the late Silurian paleogeography of the Michigan Basin. Journal of Sedimentary Petrology 47:979–1017.

Gutstadt AM (1968) Petrology and depositional environments of the Beck Spring dolomite (Precambrian), Kingston Range, California. Journal of Sedimentary Petrology 38:1280–1289.

Habicht JKA (1979) Paleoclimate, Paleomagnetism, and Continental Drift. American Association of Petroleum Geologists Studies in Geology 9, Tulsa, OK, 31 pp + 11 foldouts.

Hallam A (1986) Evidence of displaced terranes from Permian to Jurassic faunas around the Pacific margins. Journal of Geological Society of London 143: 209–216.

Hattori I (1982) The Mesozoic evolution of the Mino Terrane, central Japan: A geologic and paleomagnetic synthesis. Tectonophysics 85:313–340.

Hattori I (1984) Alternating clastic limestone and red chert as olistolith in the Mino Terrane, central Japan. Journal of Geological Society of Japan 90:43–54.

Hattori I (1985a) Length-slow chalcedony in the Paleozoic-Mesozoic strata of the Mino Terrane, Fukui Prefecture, central Japan, and its geological significance. Journal of Geological Society of Japan 91:453–461 (in Japanese with English abstract).

Hattori I (1985b) Length-slow chalcedony in the chert clasts of the Jurassic Kanmuriyama Conglomerates in the northwestern Mino Terrane, central Japan, indicates a pre-Jurassic evaporitic climate. Memoir of Faculty of Education Fukui University, Series III, 35:49–65.

Hattori I (1986) Geologic significance of length-slow chalcedony—review and example. News of Osaka Micropaleontologists (Special Vol) 7:265–273 (in Japanese with English abstract).

Hattori I, Hirooka K (1979) Paleomagnetic results from Permian greenstones in central Japan and their geologic significance. Tectonophysics 57:211–235.

Hattori I, Mizutani S (1986) Length-slow chalcedony from the Mesozoic Mino Terrane, central Japan, and its geologic implications. Third International Congress on Siliceous Deposits IGCP Project 187, J. Obradović (ed) Field Guide and Abstract Volume, Yugoslavia, p 105.

Hattori I, Hattori A, Ueyama K (1985) Kanmuriyama conglomerates—Mesozoic conglomerates in the northwestern Mino Terrane: A comparative study of the Mesozoic conglomerates in the Mino and the Hida Terranes in Fukui Prefecture, central Japan. Memoir of Faculty of Education Fukui University, Series III, 35:33–47 (in Japanese with English abstract).

Howell DG, Jones DL, Schermer ER (1985) Tectonostratigraphic terranes of the circum-Pacific Region. In: Howell DG (ed) Tectonostratigraphic Terranes of the Circum-Pacific Region. Circum-Pacific Council for Energy and Mineral Resources Earth Science Series No. 1, Houston, pp 3–30.

Igo H (1956) On the Carboniferous and Permian of the Fukuji district, Hida Massif, with special reference to the fusulinid zones of the Ichinotani Group. Journal of Geological Society of Japan 62:217–240 (in Japanese with English abstract).

Igo H (1961) On the disconformity and aluminous shales of the Carboniferous Ichinotani Formation, Hida Massif. Journal of Geological Society of Japan 67:261–273 (in Japanese with English abstract).

Jacka AD (1974) Replacement of fossils by length-slow chalcedony and associated dolomitization. Journal of Sedimentary Petrology 44:421–427.

Ji Xiong, Coney PJ (1985) Accreted terranes of China. In: Howell DG (ed) Tectonostratigraphic Terranes of the Circum-Pacific Region. Circum-Pacific Council for Energy and Mineral Resources Earth Science Series No. 1, Houston, pp 349–361.

Jones DL, Knauth LP (1979) Oxygen isotopic and petrographic evidence relevant to the origin of the Arkansas novaculite. Journal of Sedimentary Petrology 49:581–598.

Kastner M (1980) Length-slow chalcedony: The end of the new testament. Transactions, American Geophysical Union (EOS) 61:399.

Kastner M (1981) Authigenic silicates in deep-sea sediments: Formation and diagenesis. In: Emiliani C (ed) The Oceanic Lithosphere, Vol 7: The Sea. John Wiley, New York, pp 915–980.

Keene JB (1975) Cherts and Porcellanites from the North Pacific, DSDP Leg 32. Initial Reports of the Deep Sea Drilling Project 32. U.S. Government Printing Office, Washington, pp 429–507.

Keene JB (1983) Chalcedonic quartz and occurrence of quartzine (length-slow chalcedony) in pelagic sediments. Sedimentology 30:449–454.

Kemezys KJ (1974) Climatology and metallogenesis in the Tasman geosyncline. In: Denmead AK, Tweedal GW, Wilson AF (eds) The Tasman Geosyncline, A Symposium in Honour of Prof D Hill. Geological Society of Australia, Incorporated Queensland Division, pp 259–267.

Lin J, Fuller M, Zhang W (1985) Preliminary Phanerozoic polar wander paths for the north and south China blocks. Nature 313:444–449.

Lougheed MS (1983) Origin of Precambrian iron-formations in the Lake Superior region. Geological Society of America Bulletin 94:325–340.

McBride EF, Folk RL (1977) The Caballos novaculite revisited. II. Chert and shale members and synthesis. Journal of Sedimentary Petrology 47:1261–1286.

McBride EF, Folk RL (1979) Features and origin of Italian Jurassic radiolarites deposited on continental crust. Journal of Sedimentary Petrology 49:837–868.

McBride EF, Thomson A (1970) The Caballos novaculite, Marathon Region, Texas. Geological Society of America Special Paper 122, 129 pp.

McElhinny MW, Embleton BJJ, Ma XH, Zhang ZK (1981) Fragmentation of Asia in the Permian. Nature 293:212–216.

Meyers WJ (1977) Chertification in the Mississippian Lake Valley Formation, Sacramento Mountains, New Mexico. Sedimentology 24:75–105.

Meyers WJ, James AT (1978) Stable isotopes of cherts and carbonate cements in the Lake Valley Formation (Mississippian), Sacramento Mts, New Mexico. Sedimentology 25:105–124.

Milliken KL (1979) The silicified evaporite syndrome—two aspects of silicification history of former evaporite nodules from southern Kentucky and northern Tennessee. Journal of Sedimentary Petrology 49:245–256.

Milner S (1976) Carbonate petrology and syndepositional facies of the lower San Andres Formation (middle Permian), Lincoln County, New Mexico. Journal of Sedimentary Petrology 46:463–482.

Mizutani S, Hattori I (1983) Hida and Mino: Tectonostratigraphic terranes in central Japan. In: Hashimoto M, Uyeda S (eds) Accretion Tectonics in the Circum-Pacific Regions. Terra Scientific, Tokyo, pp 169–178.

Mizutani S, Hattori I, Tsukamoto H (1986) Clastic lutecite, wherefrom? International Symposium on Pre-Jurassic East Asia, IGCP Project 224, Japan, Report and Abstract, pp 130–131.

Mizutani S, Isomi H, Tsukamoto H (1987) Length-slow chalcedony from the Kiyomi area, Gifu Prefecture, central Japan. Bulletin of Geological Survey of Japan 38:113–130 (in Japanese with English abstract).

Nickel E (1982) Alluvial-fan-carbonate facies with evaporites, Eocene Guarga Formation, southern Pyrenees, Spain. Sedimentology 29:761–796.

Nur A, Ben-Avraham Z (1983) Break-up and accretion tectonics. In: Hashimoto M, Uyeda S (eds) Accretion Tectonics in the Circum-Pacific Regions. Terra Scientific, Tokyo, pp 3–18.

Oehler JH (1976) Hydrothermal crystallization of silica gel. Geological Society of America Bulletin 87:1143–1152.

Otofuji Y, Matsuda T (1983) Paleomagnetic evidence for the clockwise rotation of southwest Japan. Earth and Planetary Science Letters 62:349–359.

Otofuji Y, Matsuda T, Nohda S (1985) Paleomagnetic evidence for the Miocene counter-clockwise rotation of Northeast Japan—rifting process of the Japan Arc. Earth and Planetary Science Letters 75:265–277.

Ozawa T, Taira A, Kobayashi F (1985) The tectonic processes with regards to zonal arrangement of terranes in southwest Japan. Kagaku 55:4–13 (in Japanese).

Pittman JS Jr, Folk RL (1971) Length-slow chalcedony after sulfate evaporite minerals in sedimentary rocks. Nature Physical Sciences 230:64–65.

Randazzo AF, Hickey EW (1978) Dolomitization in the Floridan aquifer. American Journal of Science 278: 1177–1184.

Rapson-McGugan JE (1970) The diagenesis and depositional environment of the Permian Ranger Canyon and Mowitch Formations, Ishbel Group, from the southern Canadian Rocky Mountains. Sedimentology 15:363–417.

Ross GM, Chiarenzelli JR (1985) Paleoclimatic significance of widespread Proterozoic silcretes in the Bear and Churchill Provinces of the northwestern Canadian Shield. Journal of Sedimentary Petrology 55:196–204.

Rubin DM, Friedman GM (1977) Intermittently emergent shelf carbonates: An example from the Cambro-Ordovician of eastern New York State. Sedimentary Geology 19:81–106.

Saida T (1987) Triassic and Jurassic radiolarians in chert clasts of the Tetori Group in Tamodani area of Izumi Village, Fukui Prefecture, central Japan. Journal of Geological Society of Japan 93:57–59 (in Japanese).

Saito Y (1985) Jurassic geologic framework in the Japanese Islands. In: Howell DG (ed) Tectonostratigraphic Terranes of the Circum-Pacific Region. Circum-Pacific Council for Energy and Mineral Resources. Earth Science Series No. 1, Houston, pp 401–407.

Sasajima S (1981) Pre-Neogene paleomagnetism of Japanese islands (and vicinities). In: McElhinny MW, Valencio DA (eds) Paleoreconstruction of the Continents. Geodynamic Series 2. American Geophysical Union, Washington, D.C., pp 115–128.

Schmitt JG, Boyd DW (1981) Patterns of silicification in Permian pelecypods and brachiopods from Wyoming. Journal of Sedimentary Petrology 51:1297–1308.

Schreiber BC (1974) Vanished evaporites revisited. Sedimentology 21:329–331.

Siedlecka A (1972) Length-slow chalcedony and relicts of sulphates—evidences of evaporitic environments in the upper Carboniferous and Permian beds of Bear Island, Svalbard. Journal of Sedimentary Petrology 42:812–816.

Siedlecka A (1975) The petrology of some Carboniferous and Permian rocks from Bjørnøya, Svalbard. Norsk Polarinstitutt Årbok 1973:53–72.

Siedlecka A (1976) Silicified Precambrian evaporite nodules from northern Norway: A preliminary report. Sedimentary Geology 16:161–175.

Simonson BM (1987) Early silica cementation and subsequent diagenesis in arenites from four early Proterozoic iron formations of North America. Journal of Sedimentary Petrology 57:494–511.

Smale D (1973) Silcretes and associated silica diagenesis in southern Africa and Australia. Journal of Sedimentary Petrology 43:1077–1089.

Smale D (1978) Silcretes and associated silica diagenesis in southern Africa and Australia. In: Langford-Smith T (ed) Silcrete in Australia. University of New England Press, Armidale, N.S.W., pp 261–279.

Steinitz G (1977) Evaporite-chert associations in Senonian bedded cherts, Israel. Israel Journal of Earth Sciences 26:55–63.

Taira A, Saito Y, Hashimoto M (1983) The role of oblique subduction and strike-slip tectonics in the evolution of Japan. In: Hilde TWC, Uyeda S (eds) Geodynamics of the Western Pacific-Indonesian Region, Geodynamic Series 11. American Geophysical Union, Washington, D.C., pp 303–316.

Thiry M, Millot G (1987) Mineralogical forms of silica and their sequence of formation in silcrete. Journal of Sedimentary Petrology 57:343–352.

Tucker ME (1976a) Replaced evaporites form the late Precambrian of Finnmark, Arctic Norway. Sedimentary Geology 16:193–204.

Tucker ME (1976b) Quartz replaced anhydrite nodules ("Bristol diamond") from the Triassic of the Bristol District. Geological Magazine 113:569–574.

Veevers JJ (ed) (1984) Phanerozoic Earth History of Australia. Oxford Geological Sciences Series No. 2, Clarendon Press, Oxford, U.K., 418 pp.

Watts SH (1978) A petrographic study of silcrete from inland Australia. Journal of Sedimentary Petrology 48: 987–994.

Watts NL (1980) Quaternary pedogenic calcretes from the Kalahari (southern Africa): Mineralogy, genesis, and diagenesis. Sedimentology 27:661–686.

West IM (1964) Evaporite diagenesis in the lower Purbeck beds of Dorset. Proceedings of Yorkshire Geological Society 34:315–326.

West I (1973) Vanished evaporites – significance of strontium minerals. Journal of Sedimentary Petrology 43: 278–279.

West I (1979) Review of evaporite diagenesis in the Purbeck Formation of southern England. In: Symposium Sedimentation Jurassique W. Européen, ASF Special Publication 1:407–416.

West IM, Brandon A, Smith M (1968) A tidal flat evaporite facies in the Visean of Ireland. Journal of Sedimentary Petrology 38:1079–1093.

White JF, Corwin JF (1961) Synthesis and origin of chalcedony. American Mineralogist 46:112–119.

Wilson RCL (1966) Silica diagenesis in upper Jurassic limestones of southern England. Journal of Sedimentary Petrology 36:1036–1049.

Yang Z, Cheng Y, Wang H (1986) The geology of China. Oxford Monographs of Geology and Geophysics, No. 3. Clarendon Press, Oxford, U.K., 303 pp.

Zenger DR (1972) Significance of supratidal dolomitization in the geologic record. Geological Society of American Bulletin 83:1–12.

Zhang ZhM, Liou JG, Coleman RG (1984) An outline of the plate tectonics of China. Geological Society of America Bulletin 95:295–312.

Zonenshain LP, Kuzmin MI, Kononov MV (1985) Absolute reconstructions of the Paleozoic oceans. Earth and Planetary Science Letters 74:103–116.

14

Thermal Alteration of Chert in the Ophiolite Basement of Southern Central America

Hans-Jürgen Gursky and Monika M. Gursky

Overview

Radiolarian chert and volcaniclastic-rich siliceous rocks of the Nicoya Complex were studied by optical mineralogy, X-ray diffractometry, and SEM. Various phenomena of thermal alteration occur due to post-depositional mafic magmatism. Progressive destruction of sedimentary features is classified into four stages from weak to complete recrystallization. Increases in quartz crystallinity reflecting diagenetic to high-grade thermometamorphic conditions parallel the recrystallization noted in thin sections. Mineral assemblages especially in volcaniclastic-rich rocks help estimate the metamorphic grade. Chert with smectite and illite-smectite mixed-layer minerals or heulandite and clinoptilolite, respectively, indicates low-temperature conditions of less than 200°C. Piemontite, garnet, and diopside reflect low to medium or high (?) grade, very low-pressure contact metamorphism. Mineral assemblages with scapolite were found close to igneous contacts and indicate medium to high grade metamorphism. Secondary minerals in veins with dominantly barite and zeolites commonly formed under low temperature conditions of less than 150°C. These different petrologic changes are discussed as to their usefulness in defining thermal grades.

Introduction

Basement rocks of southern Central America are Jurassic to lowermost Tertiary ophiolite units called the Nicoya Complex (Dengo, 1962) in Costa Rica and western Panama (Fig. 14.1). The Nicoya complex is overlain by uppermost Cretaceous to Tertiary dominantly marine sedimentary sequences. Cenozoic volcanic complexes form the prominent geomorphic structure of the present isthmus.

The most complete ophiolitic sequence is exposed in northwestern Costa Rica. Except for the Santa Elena ultramafic body, the Nicoya Complex is composed of tholeiitic basalt (mostly flows and sills), some mafic plutonites, volcaniclastic breccia, and minor sedimentary intercalations (Kuijpers, 1980; Wildberg et al., 1981; Hein et al, 1983; Gursky and Schmidt-Effing, 1983; Gursky et al., 1984; Gursky, 1985). The Lower Nicoya Complex (Gursky et al., 1982) represents ridge-derived oceanic crust of Jurassic age, and the Upper Nicoya Complex consists of the early Upper Cretaceous Oceanic Series and the late Upper Cretaceous Primitive Island-Arc Series (Wildberg, 1984).

Several types of sedimentary intercalations are present (Gursky et al., 1982). Most abundant are noncalcareous radiolarian-chert sequences of the Punta Conchal Formation (Gursky and Schmidt-Effing, 1983), which are up to 45 m thick and mainly intercalated between the Lower and Upper Nicoya Complex. Lenses of fine-grained siliceous, calcareous, and/or volcaniclastic sedimentary rocks occur in places, mainly in the Upper Nicoya Complex. Basalt may intrude and underlie or overlie the sedimentary rocks and incorporate them as xenoliths; thus, thermal alteration of various types is widespread (Wildberg et al., 1981; Hein et al., 1983). Thermal alteration was studied by field geology, petrography of thin sections and polished rock slabs, X-ray diffractometry, and scanning electron microscopy.

No general, regional, or burial metamorphism exists in the Nicoya Complex of Costa Rica (Gursky, 1986). However, the basalts and plutonites of the Lower Nicoya Complex show effects of an early, low- to medium-temperature hydrothermal altera-

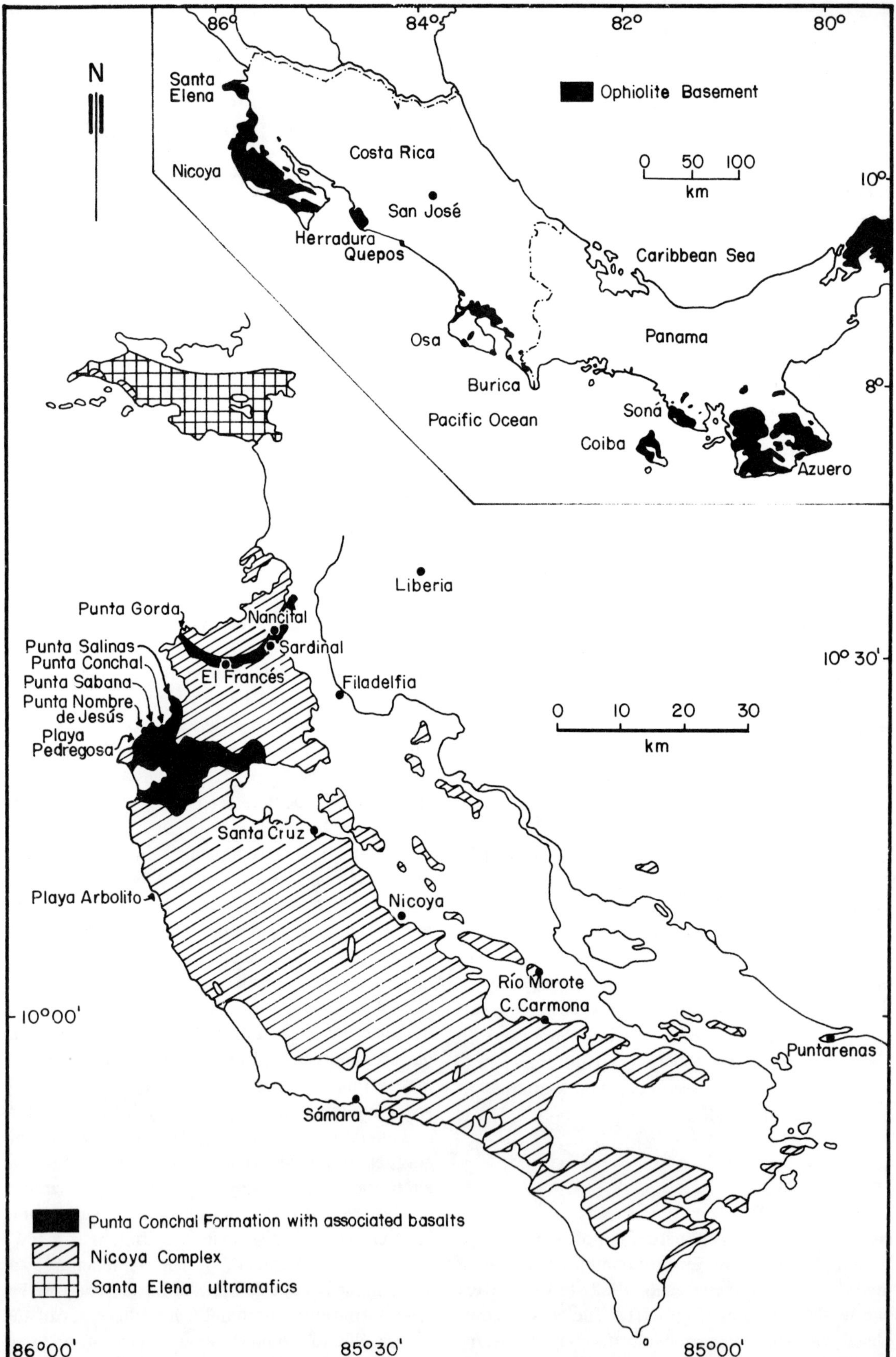

FIGURE 14.1. Outcrops of ophiolite basement in southern Central America; northwestern Costa Rica is enlarged. Localities cited in the text are indicated.

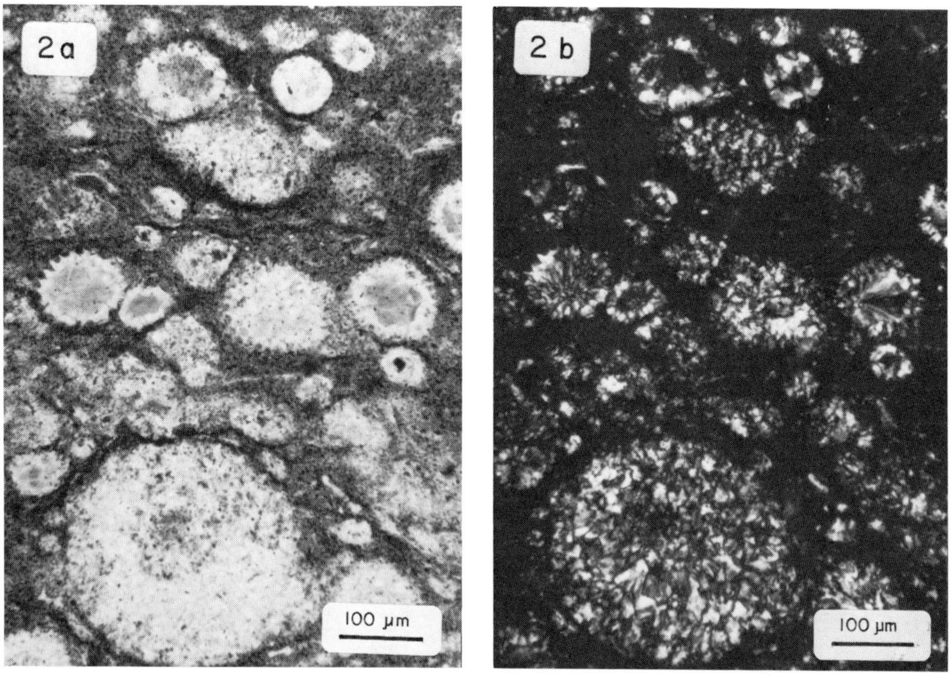

FIGURE 14.2. Weakly recrystallized radiolarian chert (stage 1) with good radiolarian preservation. Sample 2087, Playa Pedregosa. (a) Plane light; (b) crossed nicols.

tion comparable to the ocean ridge metamorphism (sensu Miyashiro et al., 1971; Liou and Ernst, 1979) and a later zeolite-facies alteration comparable to the ocean floor metamorphism (sensu Cann, 1979; Liou and Ernst, 1979). The Punta Conchal Formation is generally not metamorphic, but it is characterized by low-temperature mineralization. Near some postdepositional basalts, the radiolarites are thermally metamorphosed and show secondary mineralization of higher-temperature zeolites (Gursky, 1986). The igneous rocks of the Upper Nicoya Complex show zeolite-facies alteration at slightly lower temperatures than in the Lower Nicoya Complex. Thin sedimentary intercalations within the Upper Nicoya Complex have generally been thermally metamorphosed and are crosscut by veins containing minerals formed at various temperatures (Gursky, 1986).

The sedimentary rocks overlying the Nicoya Complex show sedimentary and diagenetic mineral assemblages (e.g., opal-CT and zeolites). Under diagenetic conditions, volcaniclastic minerals were partly transformed into clinoptilolite, heulandite, and stilbite, which also occur as cement and fracture fillings. Only the few meters of sedimentary rock directly overlying basalt have been influenced by local, very low temperature mineralizing fluids that emanated from the volcanic basement; the fluids produced heulandite, stilbite, and analcite.

Recrystallization Noted in Thin Section

The most typical petrographic characteristic of many cherts of the Nicoya Complex is recrystallization, which affects radiolarian tests and matrix and gradually leads to a quartzite texture. Grain growth textures are noted for rocks that underwent low-temperature diagenetic conditions (in places in chert of the overlying sequences; Gursky, 1984) but are more typical of rocks that underwent advanced diagenetic and thermometamorphic conditions (in ophiolite chert). Recrystallization textures can be similar, so the geologic field relations must be considered for genetic interpretations.

Red radiolarian chert with well-preserved primary sedimentary structures can be progressively converted into strongly recrystallized hornfelslike rocks characterized by a pale color (see Thurston, 1972). Sedimentary structures are destroyed, and radiolarians and matrix are recrystallized. The dominant manifestation is the progressive coarsening (grain

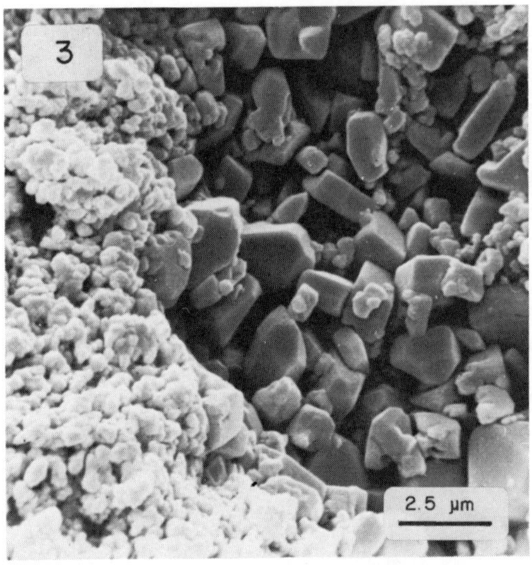

FIGURE 14.3. SEM of the boundary between a radiolarian test (left; fine-grained quartz) and its filling (right; coarse-grained quartz). Sample 2087, Playa Pedregosa; surface HF etched.

growth) of the quartz grains (cf. Iijima et al., 1978; Thurston, 1972) documented in detail by Keller et al. (1985) using SEM.

Four recrystallization stages are distinguished in thin section for the Nicoya Complex radiolarites; rocks rich in volcaniclastic detritus are commonly too poor in quartz for classification of the recrystallization stages to apply.

Stage 1

Weakly recrystallized chert is characterized by generally well to excellently preserved radiolarians showing internal skeletal elements, spines, and pores. Species with delicate tests are also well preserved. Radiolarians show sharp contacts with the matrix and with infillings. Infillings include spherulitic chalcedony and very fine grained mixtures of quartz, iron oxide, and clay minerals, similar to the matrix. Quartz is generally submicroscopic (< 10 µm) in the matrix, and microcrystalline (up to 30 µm and more) in the tests. Cryptocrystalline hematite pigment is abundant and uniformly disseminated, causing intensive reddish brown colors (Fig. 14.2a,b, 14.3).

Stage 2

In moderately recrystallized chert, the best-preserved radiolarians have lost their sharp contacts with the matrix and infillings. Test ornamentation is rarely preserved. Matrix quartz is commonly 10 to 20 µm (Fig. 14.4a,b).

Stage 3

Strongly recrystallized chert is considerably deformed. Radiolarians show diffuse, partly transitional contacts with the matrix. Tests and infillings recrystallized, forming quartz spherules; test ornamentation is completely lost. Test infillings of quartz, iron oxide, and clay minerals are rarely preserved. Matrix and test quartz have similar grain sizes (15 to more than 30 µm), and microquartzite textures develop. The iron oxide gradually loses its coloring effect and tends to concentrate in discrete minute specks; chert becomes paler (Fig. 14.5a,b).

Stage 4

Completely recrystallized chert is typical at contacts with igneous sills and dikes and in xenoliths. Radiolarians are rarely identified; metamorphic quartzite (metachert) formed. Quartz grain sizes range between 20 and 70 µm with a maximum of 150 µm. Equigranular to heterogranular, polygonal textures developed starting from isolated areas (Spry, 1969); grains show even, simple faces with triple-point contacts (Keller et al., 1985). Quartz extinction is no longer undulatory. In places, chlorite crystallizes along quartz grain contacts or may replace some radiolarian infillings. Pigment grains have coarsened (up to more than 5 µm) and have become distinct, angular, and less abundant. Reddish hematite is progressively transformed into opaque magnetite, causing pale gray to whitish rock colors (Figs. 14.6a,b, 14. 7–14.9).

From SEM studies of Japanese cherts, Kakuwa (1984) proposed a classification for the preservation of radiolarian individuals based on pore structure and cementation grade. Using his classification, Nicoya Complex cherts mainly show radiolarians of the classes PPF (partly preserved, filled) and NPF (not preserved, filled), but some WPF (well preserved, filled) also occur. Iijima et al. (1978) classified the preservation of radiolarians in non-metamorphic Japanese cherts, differentiating grades a to d for individuals and grades A to D for faunal assemblages; our stages 1 and 2 comprise the entire range of their grades A to D. Individual test preservation is variable within each of our stages in the Nicoya Complex cherts. However, in contrast to the classification of Iijima et al. (1978), our classification is based on the average preservation optimum

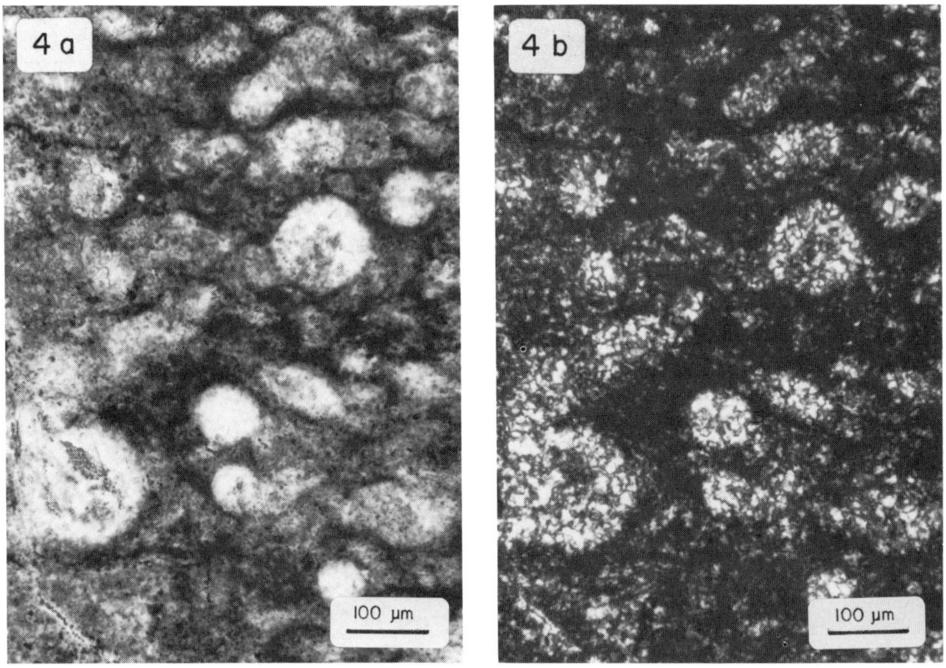

FIGURE 14.4. Moderately recrystallized radiolarian chert (stage 2). Sample 2307, Nancital. (a) Plane light; (b) crossed nicols.

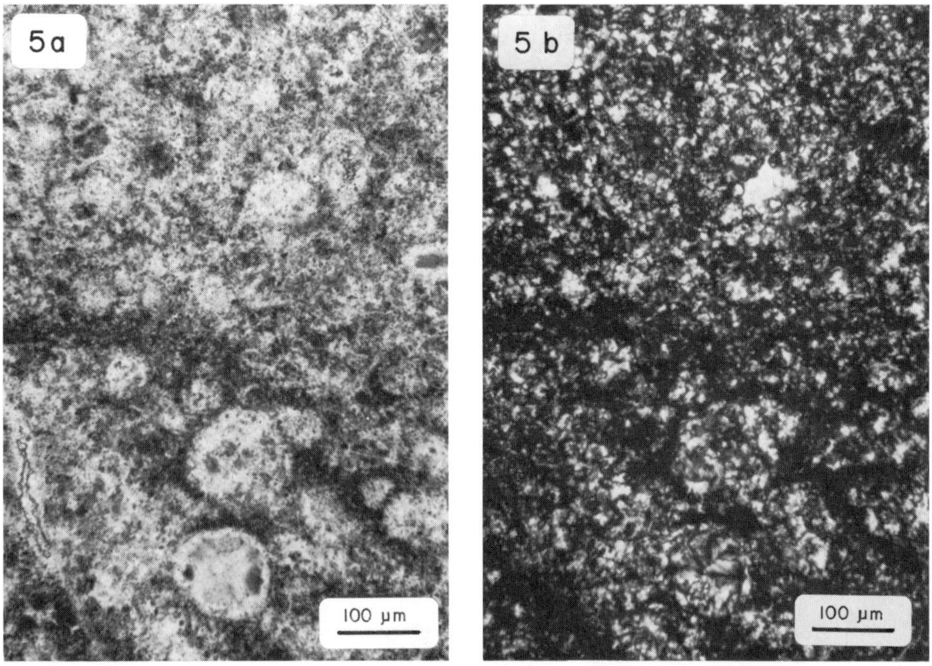

FIGURE 14.5. Strongly recrystallized radiolarian chert (stage 3) with poor radiolarian preservation. Sample 2292, Sardinal. (a) Plane light; (b) crossed nicols.

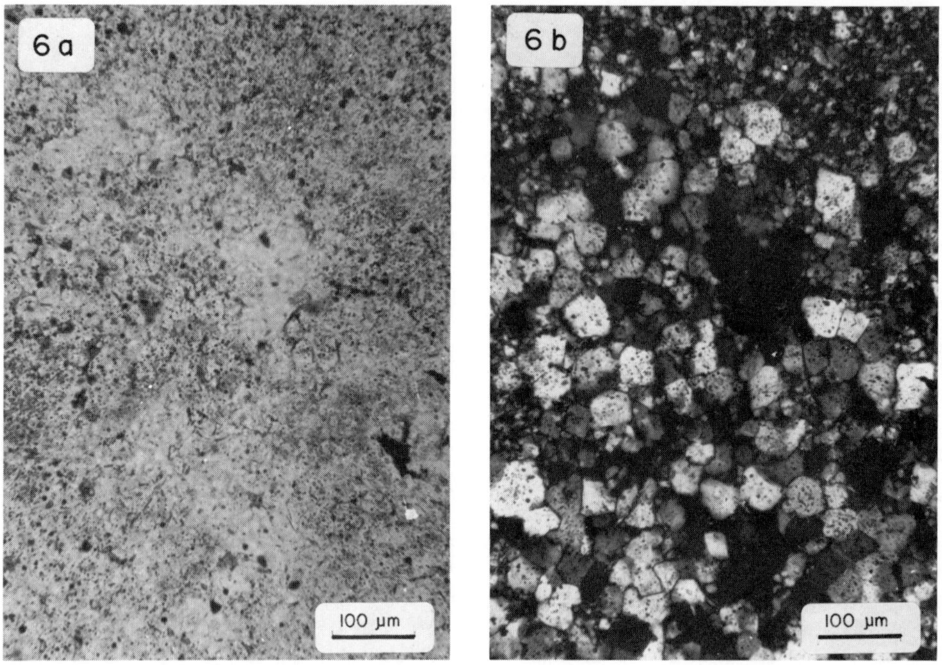

FIGURE 14.6. Completely recrystallized radiolarian chert (stage 4). Sample 2215, Punta Salinas. (a) Plane light; (b) crossed nicols.

of the radiolarian assemblages and includes the matrix texture.

The presence of stage 4 chert at contacts with post-depositionally erupted basalt, in addition to the formation of metamorphic minerals, demonstrates the thermometamorphic origin of these micro-quartzites. With increasing distance from the contact (generally several meters), the rocks grade to stage 3 chert, indicating decreasing temperature. A rapid transition to stage 2 or 1 chert within a few centimeters occurs in places, reflecting varying cooling histories probably due to different thicknesses and temperatures of intrusion.

Tullis and Yund (1982) experimentally simulated the gradual transformation of fine-grained sedimentary chert to coarse-grained quartzite. Granoblastic to homeoblastic textures similar to those described above formed on heating the samples. Tullis and

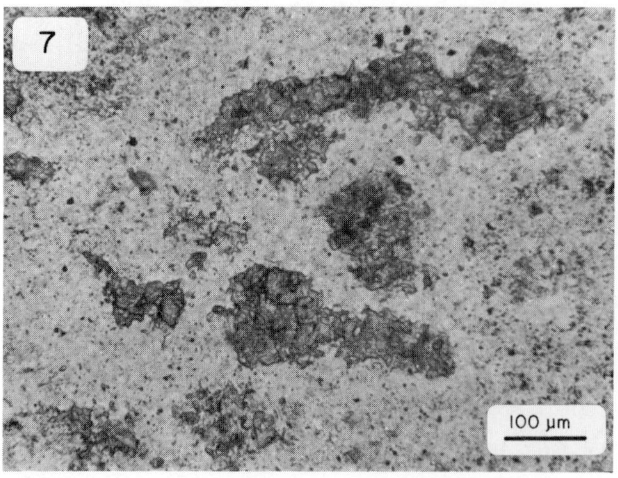

FIGURE 14.7. Completely recrystallized radiolarian chert (stage 4). The gray chlorite specks mark areas of former radiolarian infillings. Sample 2045, Punta Gorda; plane light.

FIGURE 14.8. SEM of a recrystallized radiolarian from Figure 14.7. The test consists of coarse-grained, triple-point quartz (some damaged by etching). The chlorite aggregate in the center is a metamorphic relic of the infilling. Sample 2045, Punta Gorda; surface HF etched.

FIGURE 14.9. SEM of a chlorite aggregate within a radiolarian test (enlarged detail from Figure 14.8). Sample 2045, Punta Gorda; surface HF etched.

Yund stress the importance of coexisting water ($<$ 2%) to enhance solubility along grain contacts and the grain-to-grain transport of silica, by increasing the silica-acid concentration in the fluid. The presence of water accelerates the recrystallization. Some sections of stage 4 chert show that grain growth starts at isolated centers and then extends through the rock by coalescence of coarse patches. Spry (1969) pointed out that final grain size and temperature and duration of heating are not directly correlated. Thurston (1972) found that impurities like iron oxide, manganese oxide, or clay minerals may hinder the migration of pore solutions and slow down the recrystallization process. Our data support this, and we find low amounts of impurities in rocks having very coarse grains, whereas clay-rich chert shows small grain sizes after recrystallization. Alternatively, impurity-rich, intensely colored material may be bleached by grain growth and phase transformations, as described above.

Quartz Crystallinity

Through diagenetic and metamorphic processes, quartz undergoes crystallographic changes, which can be measured in terms of quartz crystallinity (e.g., Murata and Norman, 1976). In selected chert sections of the Nicoya Complex, measurements of the quartz crystallinity were made by X-ray diffractometry, and the Crystallinity Index (CI; Murata and Norman, 1976) was calculated based on the degree of resolution of the d(212) peak (Fig. 14.10; additional measurements are in Gursky, 1984).

In the Punta Gorda and Punta Sabana sections, radiolarite is underlain and overlain at baked contacts by postdepositional basalt (Fig. 14.11). At Punta Salinas, postdepositional basalt only crops out at the baked base of the section and includes a metachert xenolith (sample 2244). At Nancital and Punta Conchal, the basalts are not exposed; the cherts are stratigraphically incomplete owing to recent erosion. In sections with radiolarite in contact with basalt, CIs decrease statistically, but not in individual samples, away from the contacts. A symmetrical distribution is noted at Punta Sabana (Fig. 14.11). This strongly suggests that the CI increases result from the heat input. The high CI values at the strongly recrystallized tops of Nancital and Punta Salinas sections (samples 2553, 2315, 2316, 2419) are probably due to heating by postdepositional basalts that were eroded. Murata and Norman (1976) concluded that CIs of up to 3 represent diagenetic quartz with a disordered lattice and values of 6 to more than 10, igneous and metamorphic, well-ordered quartz. CI distributions mapped here confirm these observations (Fig. 14.11).

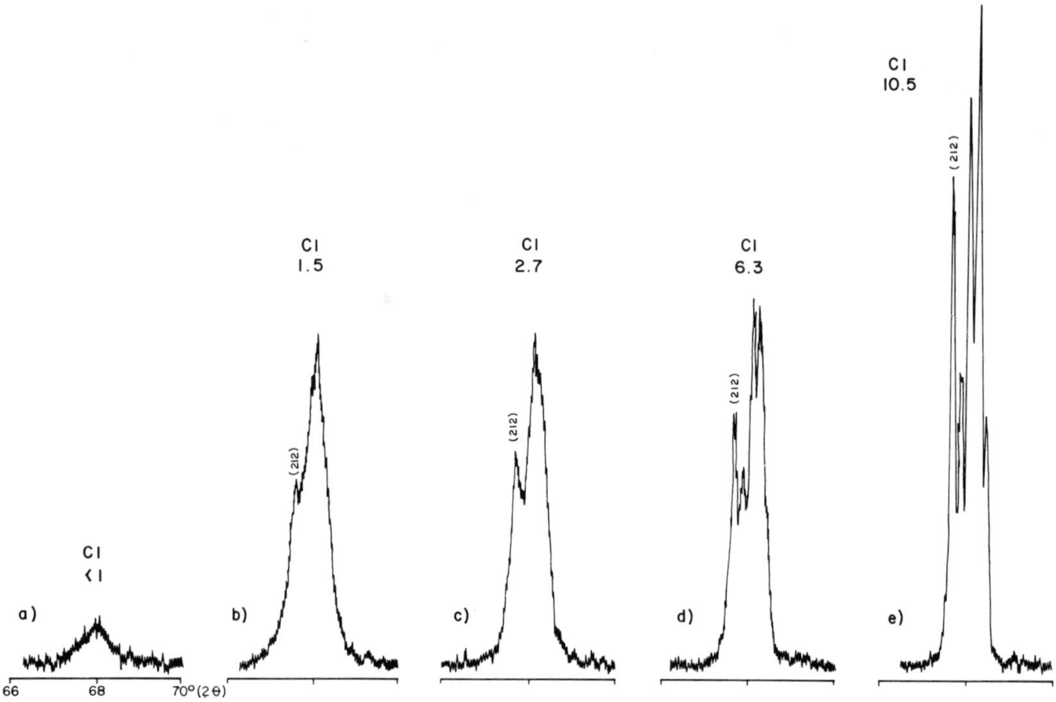

FIGURE 14.10. Examples of quartz crystallinity indices (CI) in siliceous rocks from the Nicoya Peninsula; interval measured by X-ray diffractometry: 2 θ = 66–70°. (a) Sample 2112, Paleogene chert from the sedimentary cover of the Nicoya Complex, Sámara. (b) Sample 2365, Paleogene chert from the sedimentary cover of the Nicoya Complex, 6 km N of Santa Cruz. (c) Sample 2314b, Punta Conchal Formation, Nancital. (d) Sample 2075b, Punta Conchal Formation, Punta Sabana. (e) Sample 2074, Punta Conchal Formation, from the contact margin of a sill, Punta Sabana.

Values of approximately 1.5 to 3.5 occur in chert thermally unaffected by basalt. Samples 2548 and 2413 with CIs of 6.4 and 5.9, respectively, are reddish radiolarites with gray, irregular laminae extraordinarily rich in secondary, concretional(?), silica. CIs of Paleogene siliceous rocks overlying the Nicoya Complex are higher in silica-rich concretions than in typical bedded/chert (Gursky, 1984). This suggests that, besides heat input, other factors may influence the quartz crystallinity (Elorza et al., 1985; Bustillo and Iglesia, 1979).

CIs and recrystallization stages or grain sizes are not directly correlated. Coarse-grained, recrystallized chert indeed shows high CIs (e.g., Punta Salinas), but stage 2 and 3 chert shows variable low (Nancital and Punta Conchal) or relatively high CIs (Punta Sabana). We infer that the ordering of the quartz lattice precedes the step-by-step grain growth.

Metamorphic Minerals

General Aspects

Most of the Nicoya Complex cherts are not metamorphosed but rather are characterized by sedimentary and diagenetic zeolites and clay mineral assemblages including smectite, illite-smectite mixed-layer minerals ("ISML"), and illite, in places with chlorite. Illite is detrital or a minor weathering product of volcaniclastic components. At 70 to 90°C, smectite transforms into illite through an intermediate mixed-layer phase if sodium and potassium are available; if magnesium is available, smectite transforms into chlorite also through an intermediate mixed-layer phase (Dunoyer de Segonzac, 1970). Smectite and mixed-layer minerals become unstable at 150° to 200°C, the lower boundary of metamorphism (Dunoyer de Segonzac, 1970;

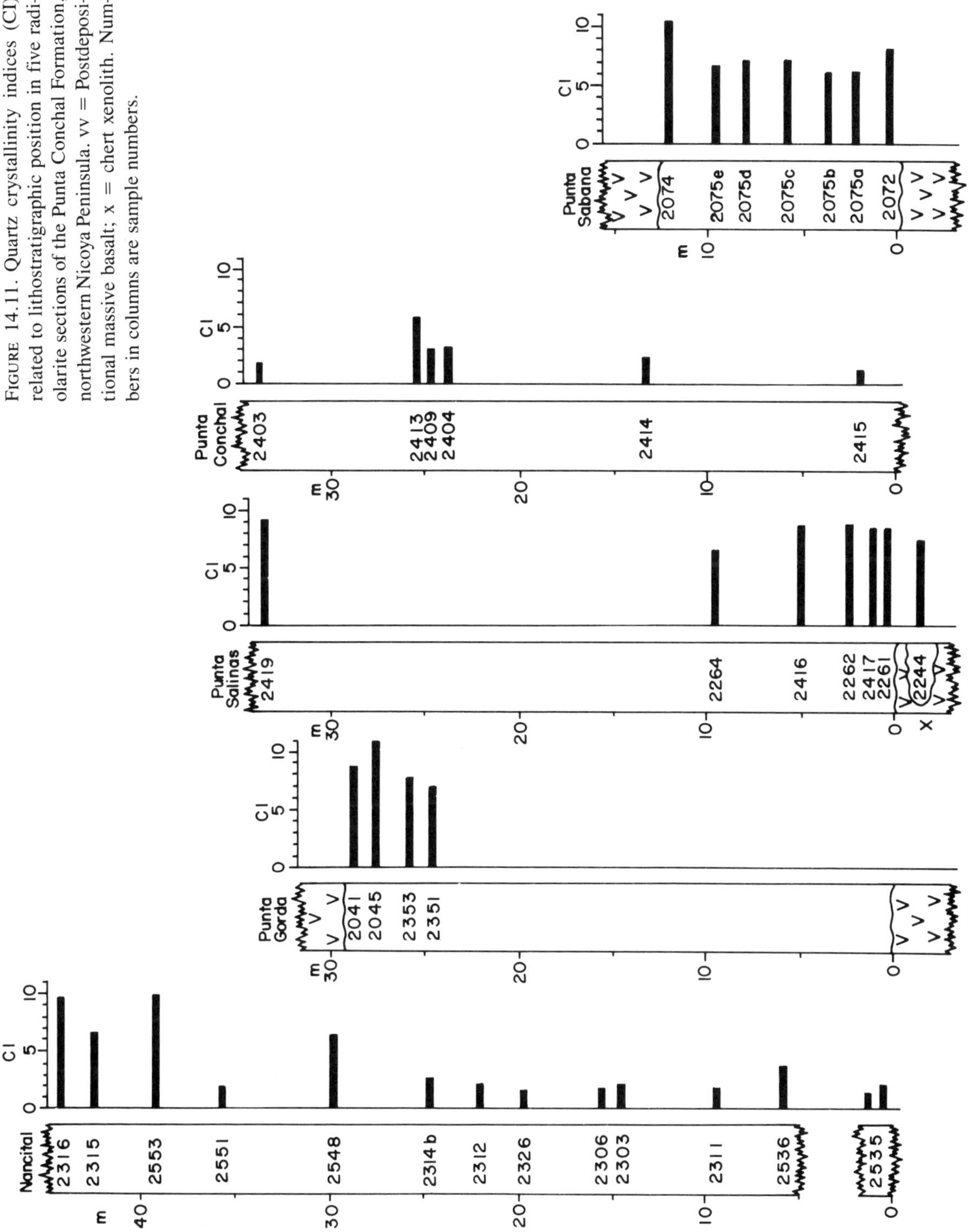

FIGURE 14.11. Quartz crystallinity indices (CI) related to lithostratigraphic position in five radiolarite sections of the Punta Conchal Formation, northwestern Nicoya Peninsula. vv = Postdepositional massive basalt; x = chert xenolith. Numbers in columns are sample numbers.

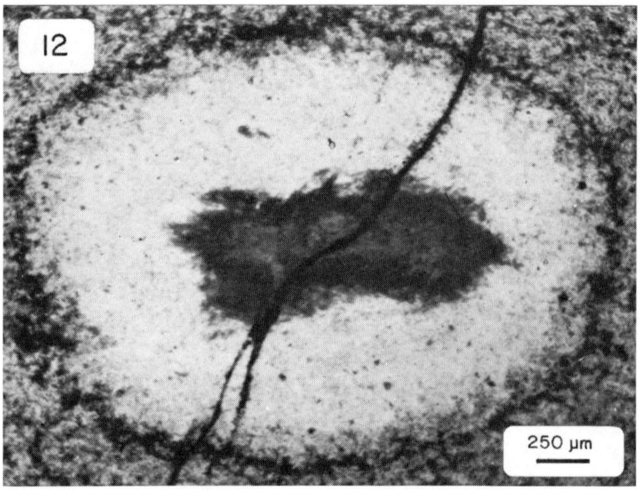

FIGURE 14.12. Chlorite aggregate with chlorite-free halo from chert close to volcanic sill. Sample 2048, Punta Gorda; plane light.

Winkler, 1979). The presence of smectite and ISML in cherts of the Nicoya Complex shows that this boundary was generally not exceeded. Diagenetic heulandite or clinoptilolite cement occurs in places; neither survives the onset of metamorphism (Winkler, 1979).

Thermal metamorphism of Nicoya Complex cherts was clearly caused by intruding basalts of the Upper Nicoya Complex. In many outcrops, dikes, sills, or flows penetrate, overlie, or undercut radiolarite sequences (e.g., Punta Salinas, Punta Sabana, Punta Gorda). Crosscutting, vertical feeder dikes are rare but occur (e.g., at Punta Nombre de Jesús). The basalts show millimeter- to centimeter-wide chilled margins at the contacts with the sedimentary rocks; cherts show centimeter- to tens of meters-wide zoned contact aureoles, whose width and metamorphic grades depend on the thickness of the igneous bodies.

Basaltic magmas reach temperatures of about 1,200°C, and the country rock is subjected to 60% of that 1,200°C (about 725°C) at the contact (Jaeger, 1957). As the levels of intrusion were generally close to the sea floor or the basalt formed as surface flows, the initial temperature of the country rock was not elevated above surface temperature. Consequently, chert close to intrusions or flows should reflect temperature conditions of high-grade metamorphism at very low pressure, if the igneous bodies were voluminous enough to provide long-term heating.

Metamorphism of Radiolarian Chert

The cherts consist of radiolarian tests, diagenetic silica phases, hematite pigment, manganese micronodules, and fine-grained siliciclastic grains including clay minerals and, in places, plagioclase and pyroxene. Because of this composition, parageneses indicating specific metamorphic zones were generally not well developed. Radiolarites are subdivided into three groups based on their behavior during metamorphism: (1) relatively pure chert with variable hematite contents; (2) chert with hematite and clay minerals, dominantly smectite; and (3) chert with hematite, clay minerals, and volcanic detritus. Groups 1, 2, and, to a lesser extent, 3 built up the thick radiolarite sequences; group 3 rocks are typical of small lenses.

Sedimentary rocks of group 1 are mineralogically stable. Quartz is stable up to the highest metamorphic grade encountered, and thermal effects are restricted to grain growth and changes of crystallinity. Hematite is also affected by grain growth with increasing temperature; close to contacts with igneous bodies, it is progressively reduced to magnetite. Group 1 cherts neither indicate the onset of metamorphism nor define its grade.

Metamorphic processes in group 2 are similar to those in group 1. Quartz recrystallizes, and hematite is partly or completely transformed to magnetite. Crystals generally remain smaller than in group 1 owing to clay minerals along the grain boundaries. In contact zones, smectite is completely transformed to chlorite. This process begins during late-stage diagenesis (Dunoyer de Segonzac, 1970). The boundary between diagenesis and metamorphism cannot be defined by group 2 rocks. Chlorite forms platey aggregates or poikiloblastic crystals surrounded by regular halos of chlorite-free "bleached" microquartzite when metamorphic conditions are sufficiently long and intense. Quartz grain sizes are considerably smaller outside the halos, and tiny chlorite flakes, replacing clay minerals, are finely disseminated at the quartz grain boundaries (Fig. 14.12).

The assemblage quartz + chlorite + illite (phengite) is stable up to medium-grade metamorphism (Winkler, 1979), which is probably also true for the quartz + chlorite (in places ± hematite ± magnetite) assemblage in chert. With an appropriate starting composition, staurolite or cordierite would form at the onset of medium-grade metamorphism (Winkler, 1979), which are not observed in our samples. Calculations of the molecular ratios of Al_2O_3, K_2O, Na_2O, FeO, Fe_2O_3, MgO, and CaO (see Winkler, 1976) for two samples of metamorphosed chert of groups 2 and 3 show that their compositions would not have allowed the formation of staurolite or cordierite. Other minerals such as muscovite, which might act as reacting partners of chlorite at the onset of medium-grade metamorphism, are not present in group 2 rocks. Thus, we infer that in a quartz-chlorite host rock, chlorite is stable up to medium- or even high-grade metamorphism. Therefore, it is found close to igneous contacts. The group 2 mineral assemblage does not provide information on maximum temperatures of metamorphism.

Group 3 chert contains plagioclase and minor hornblende and pyroxene; calcite is rarely present in minor amounts. Volcanic components are generally rare in the radiolarian chert of the Punta Conchal Formation. In small chert lenses in basalt, the contents are much higher. These clastic minerals allow additional metamorphic phases to form locally.

Garnet of the grandite group (presumably grossular-andradite mixed crystals) typically occurs in minor amounts, disseminated as greenish idiomorphic, zoned, fine-grained crystals with abnormal birefringence. Garnet probably formed from calcite and the anorthite component of plagioclase. The observed paragenesis quartz + grossular/andradite + plagioclase ± chlorite is stable at low-grade metamorphic conditions, if the fluid phase was poor in CO_2 (Winkler, 1976). If the fluid phase was richer in CO_2 (e.g., $X_{CO_2} = 0.5$; Winkler, 1976), grossular does not appear until higher temperatures. Turner (1968) indicated the stability of the assemblage quartz + grossular under very low pressure during shallow contact metamorphism, as in the Nicoya Complex, to be up to high-grade metamorphic temperatures. The paragenesis is not characteristic of a specific zone but indicates that at least low-grade conditions (320 to 500°C) or even medium- or high-grade conditions were reached.

Diopside apparently formed at a somewhat higher temperature than grossular. We found it in rocks closer to the contacts of igneous rocks. It is colorless to yellowish green or green, and probably formed from chlorite and calcium from plagioclase or calcite. Poikiloblastic chlorite aggregates and crystals were completely or partly transformed to poikiloblastic elongate diopside with tiny quartz inclusions.

The formation temperature of diopside has a broad range depending on the composition of the protolith. Conditions of formation for rocks studied here may have been comparable to those of metamorphosed mafic rocks, as the volcaniclastic admixtures in the cherts show mafic composition.

Turner (1968) and Winkler (1976) stated that diopside mainly forms in mafic as well as siliceous-calcareous dolomite rocks under upper-medium- to high-grade metamorphism. Diopside occurs in the Nicoya Complex cherts with or without grossular/andradite and/or chlorite. Diopside + grossular/andradite is stable up to high-grade metamorphism (Turner, 1968; Winkler, 1976). Additional metamorphic sphene is a typical accessory. Sphene, occurring as aggregates or hypidiomorphic to idiomorphic crystals (e.g., Punta Salinas, Punta Sabana, Punta Gorda, Playa Arbolito xenoliths, and in most of the studied lenses), allows no further characterization of metamorphic grade. Retrograde chlorite, formed from diopside or garnet, is present in places.

The assemblage quartz + diopside + plagioclase ± Ca-garnet ± sphene ± chlorite ± magnetite/hematite probably represents temperatures characteristic of medium- to perhaps high-grade metamorphism at low pressure. This assemblage is typical in the intensely recrystallized Punta Gorda chert and at Punta Nombre de Jesús.

Metamorphism of Siliceous-Calcareous Volcaniclastic Rocks

At Punta Nombre de Jesús, basalt is overlain by 10-to-12-m-thick metamorphic and fine-grained tuffaceous rocks with siliceous and calcareous admixtures. The rocks grade upward into some 3 to 4 m of radiolarian chert. A thick basalt sill overlies the sequence, and some thin sills intruded the rocks. To the northwest, the sequence is cut by a large basalt feeder dike, which incorporates chert xenoliths. The

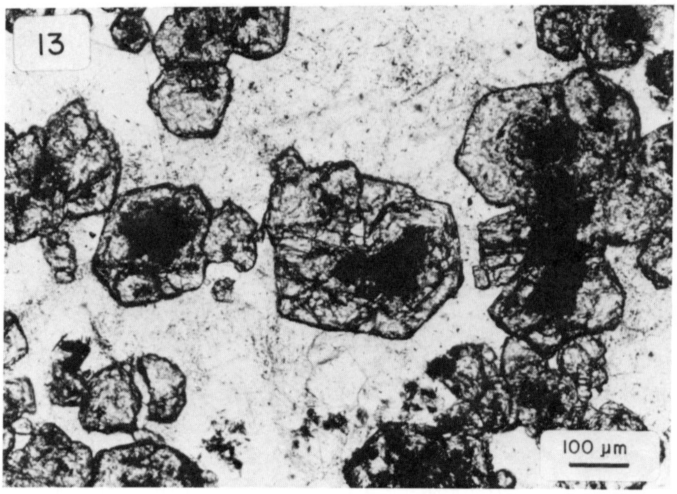

FIGURE 14.13. Idiomorphic, zoned, greenish garnet (probably andradite-grossular mixed crystals) with chloritized centers. Garnet is embedded in calcite. The section is from a calcareous-siliceous, tuffaceous rock. Sample 2427, Punta Nombre de Jesús; plane light.

thickness of the neighboring basalt bodies caused a relatively long period of heating of the sedimentary rocks so that high-grade metamorphic minerals with crystals up to several millimeters formed.

The thin-bedded sequence shows notable differences in the mineralogic composition of beds and laminae due to variable amounts of primary silica (some radiolarians are still visible), carbonate, volcanogenic detritus, volcanic glass shards, and Fe and Mn minerals. Colors created during metamorphism reflect the different mineral parageneses. Piemontite and hematite cause red and violet colors; diopside and garnet, greenish ones. Coarse-grained metamorphic mineral parageneses, indicating the highest local temperatures, are found close to the basalt dike—for example, diopside + scapolite + grossular/andradite ± plagioclase ± sphene ± calcite ± magnetite/hematite.

Diopside forms elongate crystals (up to 2 mm long), in places in sheaflike arrangements. Scapolite shows isometric, commonly idiomorphic poikiloblasts of up to 1 mm with inclusions of garnet, diopside, sphene, magnetite, and hematite. Greenish garnet is present as disseminated, idiomorphic zoned grains (Fig. 14.13), which commonly are not completely isotropic. Sphene forms irregular aggregates and rare idiomorphic crystals. Secondary calcite fills crosscutting fractures and replaces other minerals. Magnetite and hematite occur as isometric crystals (4 to 40 μm) or as laths (up to 1 mm long) concentrated in laminae (relict sedimentary layering). Chlorite has several origins: retrograde, hydrothermal (from late-stage fluids emanating from postdepositional basalts), and weathering. There is no quartz because of the presumably low primary silica contents and the SiO_2-consuming metamorphic reactions.

The scapolite paragenesis is found closest to the magmatic contacts, indicating the greatest heating at this locality, but scapolite cannot define the specific metamorphic temperature reached there. Winkler (1976) included the paragenesis diopside + scapolite + calcite in calcareous rocks in high-grade metamorphism; Turner (1968) reported scapolite to be associated, as in our rocks, with grossular, diopside, and plagioclase in medium-grade metamorphism of calcareous material.

Farther from the dike contact, scapolite disappears, and violet-red piemontite appears as hypidiomorphic to idiomorphic, partly poikiloblastic crystals with diameters of up to 150 μm. Piemontite especially crystallized in Mn-rich laminae, which are probably equivalent to layers with manganese micronodules found in the less metamorphic parts of the sequence. Piemontite is found in parageneses with or without diopside (plagioclase + piemontite + diopside + quartz ± calcite ± sphene or plagioclase + piemontite + grossular/andradite + calcite + quartz), representing low- to medium-grade metamorphic conditions. In places, rare epidote is present (partly coexisting with piemontite)—for example, filling radiolarians (see Gursky, 1984, Fig. 158)—where it formed from a clay mineral precursor with addition of calcium. Epidote occurs only rarely in the Nicoya Complex cherts and basalts, indicating together with other evidence (see Gursky, 1986) a low-pressure metamorphic environment (Miyashiro et al., 1971).

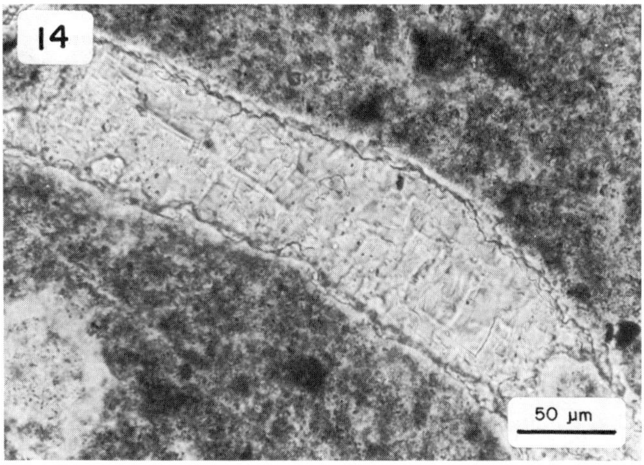

FIGURE 14.14. Barite-filled fractures with quartz rims in radiolarian chert. Sample 2292, 1 km NE of Sardinal; plane light.

Mineralizing Fluids

All the sedimentary rocks of the Nicoya Complex show postdepositional mineralogic changes. Fluids of various origins penetrated the rocks along partly very early-formed joints, fractures, and faults and precipitated predominantly quartz, generally as chalcedony, opal-CT, barite, and zeolites. Less abundant are apophyllite, smectite, chlorite, hematite/magnetite, pyrolusite, hollandite, and, as the last phase, calcite. Minor metasomatic phenomena may occur. Most important were early stage, predominantly diagenetic, low-temperature fluids, locally followed by late-stage, aqueous phases (hydrothermal solutions) emanating from postdepositional magma bodies and migrating into the wall rocks. In places, these wall rocks were still in an early stage of diagenesis when affected by the magma-generated solutions; thus, the two types of fluids may have interacted. Postmagmatic hydrothermal influences may also have occurred in places.

Under diagenetic conditions, quartz was predominantly formed in open fractures and in radiolarians, commonly as thin layers, and in places alternating with opal-CT layers. All of the opal-CT subsequently recrystallized to chalcedony. Only minute inclusions of opaque minerals, which crystallized on surfaces of opal-CT layers and reflect e.g. the lepispheric texture, reveal the former presence of opal-CT. Quartz may be accompanied by iron oxides and either by barite (Fig. 14.14), especially in chert with an insignificant volcaniclastic component (e.g., at Nancital, Punta Conchal, Sardinal), or zeolites. The most common zeolite is stilbite; stellerite and sodium stilbite varieties may occur. In rocks relatively rich in volcaniclastic debris, clinoptilolite-heulandite mixtures are typical.

Approaching igneous contacts, the mineral assemblages filling fractures and in places radiolarians developed a temperature-dependent zonation (Fig. 14.15). Emanations from the magma body may have been an additional influence. The quartz + barite zone (1) reflects the lowest temperatures and is generally the result of diagenetic conditions. Closer to the igneous contacts, i.e., with increasing temperature, a narrow stilbite + barite zone occurs (2), with barite being subordinate; quartz is no longer predominant and ultimately disappears. The next narrow zone (3) shows abundant stilbite and subordinate barite and laumontite. In the next zone (4), barite disappears, and laumontite becomes more abundant than stilbite (typical at Punta Salinas (Fig, 14.16 and Punta Sabana). However, this ideal zonal distribution of minerals can be imperfect or more complicated; e.g., barite may be absent or

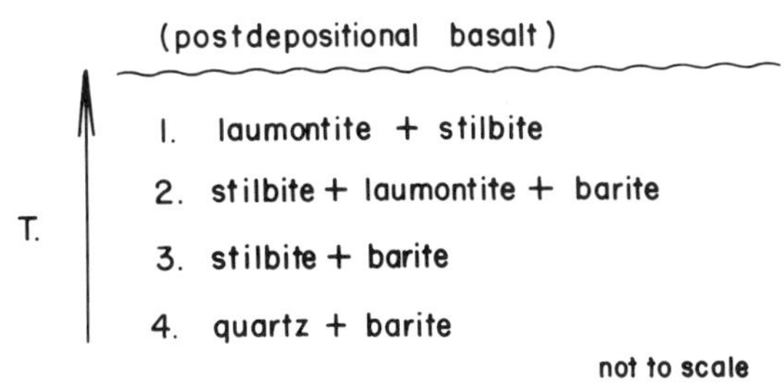

FIGURE 14.15. Temperature-dependent mineral zonation of secondary mineralization in metamorphic contact aureoles in radiolarian chert of the Punta Conchal Formation . Zones 2 and 3 are very narrow and are absent in places.

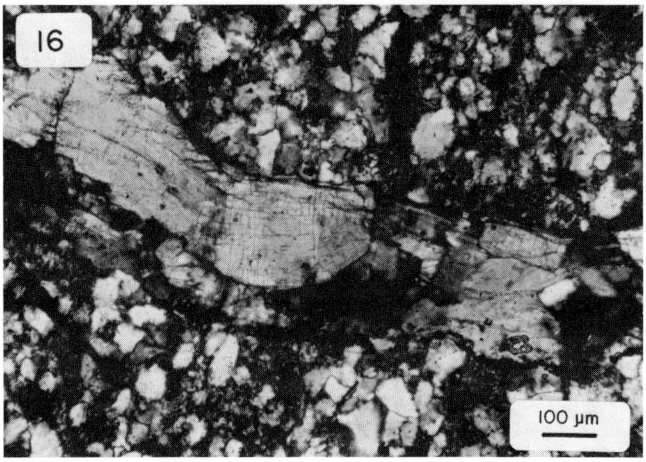

FIGURE 14.16. Fracture filled with stilbite and laumontite (crossed cleavage) in contact-metamorphic radiolarite. Sample 2214, Punta Salinas; crossed nicols.

additional hydrothermal phases (e.g., apophyllite) may be present. In places these veins cut through contact-metamorphic minerals (e.g., garnet) of the sedimentary rock. Vein formation and filling therefore seem to postdate the thermal metamorphism.

In comparison to the thicker radiolarite sequences, small chert bodies such as xenoliths and lenses, which are abundant in the Upper Nicoya Complex, show different reactions to mineralizing fluids. Their assemblages of secondary minerals are strongly dominated by those minerals that also precipitated from residual aqueous fluids within the postdepositional basalts, especially when, during early diagenesis, the sediments/sedimentary rocks were influenced by hot lava. The minerals filling vesicles and cavities in basalt are heulandite, clinoptilolite, heulandite-clinoptilolite mixtures, stilbite, calcite, and locally apophyllite, as well as smectite, chlorite, hematite and goethite; they are also typically in the small chert bodies along with the diagenetic minerals quartz and opal-CT. Opal-CT and chalcedony commonly formed first (e.g., at Playa Arbolito), filling radiolarians and fractures under diagenetic conditions. Next, opal-CT was partly or completely replaced by heulandite, clinoptilolite, or heulandite-clinoptilolite mixtures, forming broad bands in the rock. If the rock was rich in volcaniclastic debris, these zeolites may have been partly diagenetic. These minerals were followed by stilbite and then calcite, as in the basalt. Apophyllite is locally abundant; the other minerals are accessories. If the chert was influenced by the magma in a very early stage of diagenesis, when only incipient opal-CT had formed, this series of minerals then developed as infillings of fractures and radiolarians.

Manganese deposits may have had a different mineralization history. For example, at El Francés, the first hydrothermal mineral was hollandite, which resulted from Ba metasomatism (P. Halbach, oral communication; Technical University of Clausthal, 1982), followed by quartz and some barite, and then pyrolusite + hematite. Quartz crystallinity may be low.

In summary, the radiolarites had variable diagenetic maturities, from incipient to advanced diagenesis, when they were first affected by intrusion or overflow of basaltic magma. The basalts presumably belong to the Oceanic Series of the Upper Nicoya Complex. Metamorphic contact aureoles formed in places and were followed by the injection of fluids of variable but generally relatively low temperature. From these fluids minerals precipitated, which preserved diagenetic textures. Later mineralizations, possibly related to volcanism associated with the younger, Primitive Island–Arc Series of the Upper Nicoya Complex, are not clearly distinguishable from the older ones.

Discussion and Conclusions

Few studies have addressed the thermal alteration of chert. The scarcity of characteristic metamorphic mineral assemblages made quartz-rich sedimentary rocks unattractive for studies. The special geologic relationships in the ophiolite basement of southern Central America, however, with multiple interactions between magma and siliceous rocks, various thermal alteration types, and variable chert mineralogy, provide relatively good conditions for study.

Optical petrography clearly reveals progressive recrystallization depending on intensity and duration of heating, and permits the definition of qualitative recrystallization stages. Because grain growth also occurs under diagenetic conditions, the local geologic relationships and the diagenetic development of the rocks have to be carefully considered. Systematic SEM studies (e.g., Keller et al., 1985) suggest that relatively simple classifications of recrystallization textures may help to identify specific metamorphic grades. The examples presented by Thurston (1972) and Iijima et al. (1978) and the southern Central American cherts demonstrate that progressive recrystallization is typical and deserves more attention.

Changes in quartz crystallinity reflecting progressive lattice ordering are related to rising temperature and precede recrystallization as seen in thin sections. However, two important points need to be considered. The crystallinity index (CI) is not an exact indicator of the degree of recrystallization. Our observations in Costa Rica and Panama suggest that, owing to rock inhomogeneity, grain size, and other factors, CIs represent statistical values rather than uniformly direct responses to specific thermal influences (cf. illite crystallinity). The second point is that CIs can also increase under diagenetic conditions (Elorza et al., 1985) and they probably depend on the presence of other mineral phases (Bustillo and Iglesia, 1979). Nevertheless, there is evidence that quartz crystallinity, grain size, and temperature covary (see Murata and Norman, 1976; Arriortua et al., (1984). Chert near postsedimentary igneous contacts was probably heated to more than 500°C and is characterized by recrystallization stage 4 grading to stage 3 and CIs of generally more than 8. At a distance of several decimeters or meters (depending on magmatic heat transfer) from the contact, stage 3 grades to stage 2, with variable CIs, mainly about 6 to 8. Nonmetamorphic chert (< 200°C) occurring away from postdepositional igneous bodies is classified in stages 1 and 2, with CIs of mostly 1.5 to 3.5 and in places up to 6.

The metamorphic mineral assemblages are generally not diagnostic of p-T conditions but locally permit estimations of metamorphic grade and/or minimum or maximum temperatures. Few siliceous rocks of the Nicoya Complex are rich enough in mafic detritus, calcite, or manganese minerals to produce well-developed metamorphic assemblages. Pressures were always low. The quartz-hematite mineral assemblage in chert is not diagnostic, since the metamorphic hematite-magnetite transformation depends primarily on the lithologic environment (e.g., O_2 fugacity; Winkler, 1979). Quartz-hematite chert with some clay minerals is the most abundant radiolarite of the Nicoya Complex. The smectite and illite-smectite mixed-layer minerals, as well as diagenetic heulandite and clinoptilotite, indicate diagenetic conditions of less than 200°C (Dunoyer de Segonzac, 1970; Winkler, 1979). Illite can be stable under low-grade metamorphic conditions, but its presence in the rocks studied does not allow reliable conclusions. Chlorite represents the range from late-stage diagenesis to medium- or high-grade metamorphism in quartz-chlorite chert and again does not provide a more precise estimation of temperature conditions. Chert with piemontite indicates low- (?) to medium-grade metamorphism; those with garnet and/or diopside, low- to high-grade metamorphism; and scapolite, occurring only immediately adjacent to igneous bodies, represents medium- to high-grade metamorphic temperatures.

Secondary infillings of fractures and relict voids are of various origins reflecting various but generally low-temperature conditions. Away from igneous bodies, secondary minerals in the Nicoya Complex cherts formed at relatively low, nonmetamorphic temperatures (see Hein et al., 1983). Using the information by Jefferis and Voight (1981), we suggest that the temperatures mostly ranged from about 70 to 100°C. Higher temperatures, exceptionally exceeding 150°C, were reached near contacts with postdepositional igneous bodies because of emanating fluids of elevated temperature. From these fluids precipitated minerals, which preserved textures in the cherts of different diagenetic stages from incipient to late.

In conclusion, chert can provide evidence about the grade of thermal alteration if a combination of methods is applied to its study. Keller et al. (1985) gave a good example of the efficacy of using a combination of methods. Development of a chert geothermometer is the next logical step.

Acknowledgments. The research was supported by grants from the Deutsche Forschungs-Gemeinschaft and the Deutscher Akademischer Austauschdienst. We sincerely thank J. Laguna and K. Fecher for assistance in the X-ray and SEM studies. We gratefully acknowledge discussions and field trips with R. Schmidt-Effing and H.G.H. Wildberg, as well as discussions with J. Elorza, P. Halbach, J.R. Hein, and others. The Geological Institutes at Münster and Marburg in the Federal Republic of Germany, San José in Costa Rica, and Linares in Mexico provided logistic and laboratory help. W. Heinrich made helpful comments on the first version of the paper, and J.A. Henry improved its English text. J.R. Hein critically reviewed the second version. J.L. Delgado assisted in the graphics and photographs, and E.Y. Cortés typed the manuscript.

References

Arriortua MA, Elorza J, Amigo JM 91984) Indices de cristalinidad y volumen de la celda unidad de algunos cuarzos diagenéticos presentes en la Cuenca Vasco-Cantábrica. Congreso Español de Geología 2:217–226.

Bustillo MA, Iglesia A La (1979) Sílex: Relación entre índices de cristalinidad y ambientes de sustitución. Boletín Geológico Minero 90(6):588–594.

Cann JR (1979) Metamorphism in the ocean crust. In: Talwani M, Harrison CG, Hayes DE (eds) Maurice Ewing Series, Deep Drilling Results in the Atlantic Ocean: Ocean Crust; American Geophysical Union, Washington, D.C., pp 230–238.

Church TM (1979) Marine barite. In: Burns RG (ed) Marine Minerals. Reviews in Mineralogy 6:175–209.

Dengo G (1962) Estudio geológico de la Región de Guanacaste, Costa Rica. Instituto Geográfico Nacional de Costa Rica, San José, 112 pp.

Dunoyer de Segonzac G (1970) The formation of clay minerals during the diagenesis and low-grade metamorphism: A review. Sedimentology 15:281–346.

Elorza JJ, Arriortua MA, Amigo JM (1985) Indices de cristalinidad en los sílex de carácter turbidítico de Barrika (NE de Bilbao). Boletín Geológico Minero 96(1): 74–81.

Gursky H-J (1984) Die Sedimentgesteine im ophiolithischen Nicoya-Komplex (Ober-Jura bis Alt-Tertiär von Costa Rica): Ihre Verbreitung, Fazies und geologische Geschichte mit besonderer Berücksichtigung der Radiolarite. PhD Thesis, University of Marburg, Marburg, FRG, 394 pp.

Gursky H-J (1985) Composition and origin of the sedimentary rocks in the Nicoya Ophiolite Complex (Jurassic – lower Tertiary, southern Central America) – a survey. Zentralblatt Geologie Paläontologie Teil I:1197–1205.

Gursky H-J, Schmidt-Effing R (1983) Sedimentology of radiolarites within the Nicoya Ophiolite Complex, Costa Rica, Central America. In: Iijima A, Hein JR, Siever R (eds) Developments in Sedimentology 35. Elsevier, Amsterdam, pp 127–142.

Gursky H-J, Schmidt-Effing R, Strebin MM, Wildberg H (1982) The ophiolite sequence in northwestern Costa Rica (Nicoya Complex): Outlines of stratigraphical, geochemical, sedimentological, and tectonical data. Actas 5 Congreso Latinoamericano de Geología, Buenos Aires 3:607–619.

Gursky H-J, Gursky MM, Schmidt-Effing R, Wildberg H (1984) Karten zur Geologie von Nordwest-Costa Rica (Mittelamerika) mit Erläuterungen. Geologica et Palaeontologica 18:173–182.

Gursky MM (1986) Tektonische und thermische Deformationen im ophiolithischen Nicoya-Komplex und seinem sedimentären Auflager (Nicoya-Halbinsel, Costa Rica) and ihre Bedeutung für die geodynamische Entwicklung im südlichen Zentralamerika. PhD Thesis, University of Marburg, Marburg, FRG, 230 pp.

Hein JR, Kuijpers EP, Denyer P, Sliney RE (1983) Petrology and geochemistry of Cretaceous and Paleogene cherts from western Costa Rica. In: Iijima A, Hein JR, Siever R. (eds) Developments in Sedimentology 36. Elsevier, Amsterdam, pp 143–174.

Iijima A, Kakuwa Y, Yamazaki K, Yanagimoto Y (1978) Shallow-sea, organic origin of the Triassic bedded chert in central Japan. Journal of the Faculty of Science, University of Tokyo, Section II, 19(5):369–400.

Jaeger JC (1957) Temperatures outside of a cooling intrusive sheet. American Journal of Science 255:306–318.

Jefferis RG, Voight B (1981) Fracture analysis near the mid-ocean plate boundary, Reykjavik-Hvalfjördur area, Iceland. Tectonophysics 76(3/4):171–236.

Kakuwa Y (1984) Preservation of siliceous skeletons in siliceous rocks. Scientific Papers of the College of Arts and Sciences, University of Tokyo 34(1/2):43–61.

Keller WD, Stone CG, Hoersch AL (1985) Textures of Paleozoic chert and novaculite in the Ouachita Mountains of Arkansas and Oklahoma and their geological significance. Geological Society of America Bulletin 96:1353–1363.

Kuijpers EP (1980) The geologic history of the Nicoya Ophiolite Complex, Costa Rica, and its geotectonic significance. Tectonophysics 68:233–255.

Liou JG, Ernst WG (1979) Oceanic ridge metamorphism of the Eastern Taiwan Ophiolite. Contributions to Mineralogy and Petrology 68:335–348.

Miyashiro A, Shido F, Ewing M (1971) Metamorphism in the mid-Atlantic ridge near 24°E and 30°N. Philosophical Transactions of the Royal Society London A 268: 589–603.

Murata KJ, Norman MB (1976) An index of crystallinity for quartz. American Journal of Science 276:1120–1130.

Spry A (1969) Metamorphic Textures. Pergamon, Oxford, U.K. 350 pp.

Thurston DR (1972) Studies on bedded cherts. Contributions to Mineralogy and Petrology 36:329–334.

Tröger WE (1969) Optische Bestimmung der gesteinsbildenden Minerale Teil 2 Textband. Schweizerbart, Stuttgart, 822 pp.

Tullis J, Yund RA (1982) Grain-growth kinetics of quartz and calcite aggregates. Journal of Geology 90(3):301–318.

Turner FJ (1968) Metamorphic Petrology. McGraw-Hill, New York, 403 pp.

Wildberg HGH (1984) Der Nicoya-Komplex, Costa Rica, Zentralamerika: Magmatismus and Genese eines polygenetischen Ophiolith-Komplexes. Münstersche Forschungen zur Geologie und Paläontologie 62:1–123.

Wildberg H, Gursky H-J, Schmidt-Effing R, Strebin M (1981) Der Ophiolith-Komplex der Halbinsel Nicoya, Costa Rica, Zentralamerika. Zentralblatt Geologie und Paläontolgie Teil I(3/4):195–209.

Winkler HGF (1976) Petrogenesis of Metamorphic Rocks, 3d Ed. Springer, Berlin, 348 pp.

Winkler HGF (1979) Petrogenesis of Metamorphic Rocks, 4th Ed. Springer, Berlin.

Index